Robotics in Civil Engineering

Robotics in Civil Engineering

Editor:

Dr. Tanjina Nur, Ph.D.

ArclerPress

www.arclerpress.com

Robotics in Civil Engineering

Editor: Dr. Tanjina Nur, Ph.D.

708 3rd Avenue, 6th Floor
New York
NY 10017
United States of America
www.arclerpress.com

ISBN: 978-1-68094-376-4

Library of Congress Control Number: 2016950956

About the Editor

Dr. Tanjina Nur, Ph.D.

Tanjina finished her PhD in Civil and Environmental Engineering in 2014 from University of Technology Sydney (UTS). Now she is working as Post-Doctoral Researcher in the Centre for Technology in Water and Wastewater (CTWW) and published about eight International journal papers with 80 citations. Her research interest is wastewater treatment technology using adsorption process.

Contents

List of Contributors

Abdelrahman Osman Elfaki
University of Tabuk, Tabuk 50060, Saudi Arabia
Binladen Research Chair on Quality and Productivity Improvement in the Construction Industry, College of Engineering, University of Hail, Saudi Arabia

Saleh Alatawi
University of Tabuk, Tabuk 50060, Saudi Arabia
Binladen Research Chair on Quality and Productivity Improvement in the Construction Industry, College of Engineering, University of Hail, Saudi Arabia

Eyad Abushandi
University of Tabuk, Tabuk 50060, Saudi Arabia
Binladen Research Chair on Quality and Productivity Improvement in the Construction Industry, College of Engineering, University of Hail, Saudi Arabia

ZhiQiang Chen
Department of Civil and Mechanical Engineering, University of Missouri-Kansas City, 5100 Rockhill Street, Kansas City, MO 64110, USA

Jianfei Chen
Department of Computer Science Electrical Engineering, University of Missouri-Kansas City, 5100 Rockhill Street, Kansas City, MO 64110, USA

J. A. Dourakopoulos
School of Civil Engineering, National Technical University of Athens, Zografou Campus, GR-157 80 Athens, Greece

E. J. Sapountzakis
School of Civil Engineering, National Technical University of Athens, Zografou Campus, GR-157 80 Athens, Greece

Hussam Mansour
Department of Mechanics and Robotics, University of Duisburg-Essen, Lotharstr. 1, 47057 Duisburg, Germany

Anik Keller
Department of Process Engineering and Water Technology, University of Duisburg-Essen, Bismarckstr. 90, 47057 Duisburg, Germany

Rolf Gimbel
Department of Process Engineering and Water Technology, University of Duisburg-Essen, Bismarckstr. 90, 47057 Duisburg, Germany
IWW Water Research Institute, Moritzstr. 26, 45476 Mülheim an der Ruhr, Germany

Wojciech Kowalczyk
Department of Mechanics and Robotics, University of Duisburg-Essen, Lotharstr. 1, 47057 Duisburg, Germany

Enrique Valero
School of Computer Engineering, National University of Distance Education, C/Juan del Rosal, 16, 28040 Madrid, Spain
3D Visual Computing and Robotics Lab, University of Castilla-La Mancha (UCLM), Paseo de la Universidad, 4, 13071 Ciudad Real, Spain

Antonio Adán
3D Visual Computing and Robotics Lab, University of Castilla-La Mancha (UCLM), Paseo de la Universidad, 4, 13071 Ciudad Real, Spain

Carlos Cerrada
School of Computer Engineering, National University of Distance Education, C/Juan del Rosal, 16, 28040 Madrid, Spain

Mario Garzón
Automation and Robotics Center, UPM-CSIC. Calle Jose Gutierrez Abascal, 2 28006 Madrid, Spain

João Valente
Automation and Robotics Center, UPM-CSIC. Calle Jose Gutierrez Abascal, 2 28006 Madrid, Spain

David Zapata
Automation and Robotics Center, UPM-CSIC. Calle Jose Gutierrez Abascal, 2 28006 Madrid, Spain

Antonio Barrientos
Automation and Robotics Center, UPM-CSIC. Calle Jose Gutierrez Abascal, 2 28006 Madrid, Spain

Berardo Naticchia
Department of Architecture Construction and Structures (DACS), Engineering Faculty Polytechnic University of Marche, via Brecce Bianche, 60131 Ancona, Italy

Alberto Giretti
Department of Architecture Construction and Structures (DACS), Engineering Faculty Polytechnic University of Marche, via Brecce Bianche, 60131 Ancona, Italy

Alessandro Carbonari
Department of Architecture Construction and Structures (DACS), Engineering Faculty Polytechnic University of Marche, via Brecce Bianche, 60131 Ancona, Italy

Jianguo Zhou
School of Geodesy and Geomatics, Wuhan University, Wuhan 430079, China

Yaming Xu
School of Geodesy and Geomatics, Wuhan University, Wuhan 430079, China
Key Laboratory of Precise Engineering and Industry Surveying, National Administration of Surveying, Mapping and Geoinformation, Wuhan 430079, China

Tao Zhang
School of Geodesy and Geomatics, Wuhan University, Wuhan 430079, China
Key Laboratory of Precise Engineering and Industry Surveying, National Administration of Surveying, Mapping and Geoinformation, Wuhan 430079, China

Xi Kuai
School of Resource and Environmental Sciences, Wuhan University, Wuhan 430079, China

Lin Li
School of Resource and Environmental Sciences, Wuhan University, Wuhan 430079, China

Heng Luo
School of Resource and Environmental Sciences, Wuhan University, Wuhan 430079, China

Shen Hang
School of Resource and Environmental Sciences, Wuhan University, Wuhan 430079, China

Zhijun Zhang
Tianjin Institute of Surveying and Mapping, Tianjin 300381, China

Yu Liu
School of Resource and Environmental Sciences, Wuhan University, Wuhan 430079, China

Lifeng Wang
Key Laboratory of MEMS of the Ministry of Education, Southeast University, Nanjing 210096, China

Jingjing Ma
Smart Integrated Sensor Engineering Center, Jiangsu R & D Center for Internet of Things, Wuxi 214135, China

Yan Huang
Key Laboratory of MEMS of the Ministry of Education, Southeast University, Nanjing 210096, China

Dan Tang
Key Laboratory of MEMS of the Ministry of Education, Southeast University, Nanjing 210096, China

Qing-An Huang
Key Laboratory of MEMS of the Ministry of Education, Southeast University, Nanjing 210096, China

Lin Li
School of Resource and Environmental Sciences, Wuhan University, Wuhan 430079, China

Hui Xia
School of Resource and Environmental Sciences, Wuhan University, Wuhan 430079, China

Zheng Li
Hubei Institute of Land Surveying and Mapping, Wuhan 430010, China

Zhijun Zhang
Tianjin Institute of Surveying and Mapping, Tianjin 300381, China

Shunsuke Nansai
SUTD-JTC Industrial Infrastructure Innovation Centre, Singapore University of Technology and Design, 8 Somapah Rd, Singapore 487372, Singapore

Rajesh Elara Mohan
Engineering Product Development Pillar, Singapore University of Technology and Design, 8 Somapah Rd, Singapore 487372, Singapore

Hubo Cai
Division of Construction Engineering and Management, School of Civil Engineering, Purdue University, 550 Stadium Mall Drive, West Lafayette, IN 47907, USA

Osama Abudayyeh
Department of Civil and Construction Engineering, Western Michigan University, Kalamazoo, MI 49008, USA

Ikhlas Abdel-Qader

Upul Attanayake
Department of Civil and Construction Engineering, Western Michigan University, Kalamazoo, MI 49008, USA

Joseph Barbera
Department of Civil and Construction Engineering, Western Michigan University, Kalamazoo, MI 49008, USA

Eyad Almaita
Department of Electrical and Computer Engineering, Western Michigan University, Kalamazoo, MI 49009, USA

Masataka Ozaki
Graduate School of Doshisha University, Tatara, Kyotanabe, Kyoto 6100321, Japan

Kei Kakimuma
Graduate School of Doshisha University, Tatara, Kyotanabe, Kyoto 6100321, Japan

Masafumi Hashimoto
Faculty of Science and Engineering, Doshisha University, Tatara, Kyotanabe, Kyoto 6100321, Japan

Kazuhiko Takahashi
Faculty of Science and Engineering, Doshisha University, Tatara, Kyotanabe, Kyoto 6100321, Japan

Jijian Lian
State Key Laboratory of Hydraulic Engineering Simulation and Safety, Tianjin University, 92 Weijin Road, Nankai District, Tianjin 300072, China

Wenjiao Zhang
State Key Laboratory of Hydraulic Engineering Simulation and Safety, Tianjin University, 92 Weijin Road, Nankai District, Tianjin 300072, China

Qizhong Guo
Department of Civil and Environmental Engineering, Rutgers, The State University of New Jersey, Piscataway, NJ 08854, USA

Fang Liu
State Key Laboratory of Hydraulic Engineering Simulation and Safety, Tianjin University, 92 Weijin Road, Nankai District, Tianjin 300072, China

Sung Woo Park
Center for Structural Health Care Technology in Buildings, Yonsei University, Seoul 120-749, Korea

Byung Kwan Oh
Center for Structural Health Care Technology in Buildings, Yonsei University, Seoul 120-749, Korea
Department of Architectural Engineering, Yonsei University, Seoul 120-749, Korea

Hyo Seon Park
Center for Structural Health Care Technology in Buildings, Yonsei University, Seoul 120-749, Korea
Department of Architectural Engineering, Yonsei University, Seoul 120-749, Korea

Hyo Seon Park
Department of Architectural Engineering, Yonsei University, 134 Shinchon-dong, Seoul 110-732, Korea
Center for Structural Health Care Technology in Buildings, Yonsei University, 134 Shinchon-dong, Seoul 110-732, Korea

Hwan Young Lee
Department of Architectural Engineering, Yonsei University, 134 Shinchon-dong, Seoul 110-732, Korea
Center for Structural Health Care Technology in Buildings, Yonsei University, 134 Shinchon-dong, Seoul 110-732, Korea

Se Woon Choi
Center for Structural Health Care Technology in Buildings, Yonsei University, 134 Shinchon-dong, Seoul 110-732, Korea

Yousok Kim
Center for Structural Health Care Technology in Buildings, Yonsei University, 134 Shinchon-dong, Seoul 110-732, Korea

Michaella Chung
Department of Civil and Environmental Engineering, University of California, Berkeley, Davis Hall, Berkeley, CA 94720-5800, USA

Carrick Detweiler
Department of Computer Science and Engineering, University of Nebraska-Lincoln, 220 Schorr Center, Lincoln, NE 68588, USA

Michael Hamilton
Blue Oak Ranch Reserve, Director, University of California, 23100 Alum Rock Falls Road, San Jose, CA 95127, USA

James Higgins
Department of Computer Science and Engineering, University of Nebraska-Lincoln, 220 Schorr Center, Lincoln, NE 68588, USA

John-Paul Ore
Department of Computer Science and Engineering, University of Nebraska-Lincoln, 220 Schorr Center, Lincoln, NE 68588, USA

Sally Thompson
Department of Civil and Environmental Engineering, University of California, Berkeley, Davis Hall, Berkeley, CA 94720-5800, USA

Preface

This book illustrates the robotics and computer hardware and software applications in Civil Engineering. It tries to identify one or multiple problems where the use of automation, robotic technologies, or advanced management techniques can be applied to provide innovative solutions in civil engineering such as construction works, safety measures of construction site, damage inspection, building analysis etc. At beginning, we discuss the using of intelligent techniques in construction project about cost estimation and mobile imaging and computing for intelligent structural damage inspection. Then we present the nonlinear dynamic analysis of plates stiffened by parallel beams with deformable connection and numerical simulation of particle distribution in capillary membrane during backwash. Automatic method for building indoor boundary models from dense point clouds collected by laser scanners and an aerial-ground robotic system for navigation and obstacle mapping in large outdoor areas are also discussed.

In the middle part of the book, we focus on set up of an automated multi-colour system for interior wall painting and a wireless monitoring system for cracks on the surface of reactor containment buildings. Geospatial information categories mapping in a cross-lingual environment: A case study of "Surface Water" categories in Chinese and American topographic maps and passive wireless hermetic environment monitoring system for spray painting workshop are also discussed. We also try to present temporal-spatial evolution analysis of Lake size-distribution in the middle and lower Yangtze River basin using Landsat imagery data and a survey about recent advances and challenges of wall climbing robots.

The last chapters of the book describes bridge deck load testing using sensors and optical survey equipment and laser-based pedestrian tracking in outdoor environments by multiple mobile robots. We also try to find out generation mechanism and prediction model for low frequency noise induced by energy dissipating submerged jets during flood discharge from a high dam and maximum stress estimation model for multi-span Waler beams with deflections at the supports using average strains. At the end, we describe a practical monitoring system for the structural safety of mega-trusses using wireless vibrating wire strain gauges and obtaining the thermal structure of lakes from the air.

Editor
Dr. Tanjina Nur, Ph.D.

INTRODUCTION

The word robot is usually used to refer to a mechanical agent that performs one or more tasks in which it mimics a human or animal agent either through programming or commands. Another word used synonymously is automaton. Virtual robots exist, but are most often called by the abbreviated name bots.

This term comes from the Czech word robota, which means drudgery or servitude. It was coined by Karel Capek, a Czech playwright, in his play R.U.R., which stands for "Rossum's Universal Robots." It was published in 1921, and entered English in 1923. The field of study is referred to as robotics, and people who specialize in it are called roboticists.

Robots can be classified in a number of ways. Creators may use the means of locomotion as their categories, differentiating their creations by whether they are static, on the one hand, or whether they have treads, a propeller, fins, legs, wheels, rotors, or other means of propulsion.

The National Aeronautics and Space Administration (NASA), however, classifies robots in several different ways. First, it classifies them by whether they work on Earth or in space. Second, it classifies them by the industry they work in. Third, it classifies them by the type of jobs they do.

These machines are employed in industries such as manufacturing, medicine, the military, and transportation. They are used widely in assembly operations, in which they complete a range of tasks, including the following:

- arc welding
- diecasting
- fettling machines
- gas welding
- manipulating machine tools
- placement of items into a structure that's being built
- sealant application
- spot welding
- spray painting

Robots are also used for parts inspections, making glass, cleaning, monitoring radiation, sorting, loading and unloading, fastening, forging, and sand blasting. Because they are not human, they can be used in hazardous situations such as firefighting, military warfare, and bomb detection. Surgical robots are under development and robotic hands, for example, are already used in some surgical operations, allowing the human surgeon more control in laparoscopic procedures, those done through a very small incision.

Droid is the name for a type of intelligent robot. The droid comes from a shortening of the word android, which means "an automaton that has features of a human being." Examples of droids include C-3PO and R2-D2 of Star Wars fame. George Lucas, the creator of Star Wars, has trademarked the term.

In the real world, a number of companies are working on android robots that closely resemble humans and are able to interact with real people. Much of the current work is being done in Japan and South Korea. Some South Korean companies hope to make them a household item in the future.

What is Advanced Robotics?

The term "advanced robotics" first can into use in the 1980s. It is used to define any sensor-based robots that attempt to mimic human intelligence. They are used in a variety of fields ranging from manufacturing, nuclear, construction, space and underwater exploration, and health care.

Popular culture is filled with advanced robotics. The robot on the television show Lost in Space and movie robots like Star Wars' C3PO and R2D2 were some of the first seemingly intelligent robots that average people were exposed to. In 1986, movie-going audiences met Johnny 5, the little autonomous robot from the movie Short Circuit. While all of these robots appeared truly autonomous, the robots themselves were largely a combination of puppetry and acting.

By the late 20th century, the science fiction of advanced robotics and the reality of it were beginning to overlap. Advanced robots of the 21st century are considered semi-autonomous. This mean that they are able to perform their tasks with a level of independence not found in automatic machines. Those called general purpose robots are able to perform their required tasks nearly independently. Some can recognize people or objects, talk, monitor environmental quality, pick up supplies and perform other useful tasks.

Robots are able to do these tasks through the use of a sensor. A simple example of this sensor is in room cleaning robots that bump a wall and understand to turn around and try another direction. Lawn mowing robots rely on underground markers to tell them this same information. Some of the most advanced robots are able to actually "see" through the use of infrared or stereo vision.

Some of the most advanced robotics of the 21st century are humanoid robots, meaning they resemble humans in their physical appearance as well as in their actions. They are considered autonomous because they can learn and adapt to changes within their environment. Johnny is more of a reality in the 21st century than moviegoers of the 1980s could have imagined. Robots are being taught everything from how to load a dishwasher to mimicking facial expressions in response to particular types of human interactions.

One of the greatest feats of advanced robotics was seen in the rovers Opportunity and Spirit. Opportunity and Spirit landed on Mars in January of 2004 with the intention of completing an approximately 90-day mission. As of January 2009, five years later, they were still in operation. They landed on the surface with a precision unmatched in previous missions. They operate through communications with the National Aeronautics and Space Administration (NASA) and continue their missions through semi-autonomous interactions with the surface of Mars.

There were estimated to be about 3,540,000 service robots in use in 2006. At that time, there were an additional 950,000 industrial robots. In early 2009, Microsoft founder Bill Gates has predicted that every home will have a robot by 2025. Small robots like room sweepers and surprisingly complicated entertainer robots like Furby have been in homes since the 1990s. Given the advances even since those robots were introduced, the future of advanced robotics certainly seems boundless.

Application of Robotics to Building Construction

The findings of the study can be summarized as follows:

a. Construction is the largest single industry in the United States, both in terms of its output, and the labor force employed in its projects. It is also one of the least advanced, in terms of technological progress and productivity. The largest and the most homogeneous part of the construction output are the buildings which account for 70% of its total value. Any technological innovation which could increase the building construction productivity would therefore have a highly beneficial impact both on the national economy and the welfare of an individual consumer. As such it should be thoroughly examined and encouraged.

b. The robotization is gaining an increasing acceptance in the manufacturing industry. The robot population in the U.S. is growing at an exponential rate and is expected to reach, towards the beginning of the next decade, 100,000 units. This growth is accompanied by a large investment in research and development of the robotic technology. No parallel trend is evident in the construction sector. No robots are employed in construction at present and there is almost no research and development towards this end.

c. The robotization of the construction process is more difficult than that of the manufacturing production, due to dispersion of the construction activities over many sites, distinctive nature of each project, changing work location within the project and the rugged conditions on the building sites. Despite these conditions many building construction activities can be robotized at the present stage of technology with promising economic results given an appropriate design of the robotic equipment and a suitable organization of the construction tasks.

d. The main components of a robotic system which should be examined with reference to application in the building construction industry are the manipulator, the effector, the control unit, the sensors, and the locomotion mechanism. Almost all building activities, except for hauling of large structural members, can be performed with the existing configurations of robot manipulators, which must, however, be made sturdier considering the ruggedness of the building environment. The handling of heavy members can be performed with existing crane like manipulators. The effectors needed for building construction can be classified into two categories: grippers adapted to the nature of the objects to be handled, and specific working tools depending on the task to be performed. The robotic control of building operations has two levels. At the first level it ensures through appropriate preprogramming the performance of a required task (like painting, welding, etc.) at a static location. At a higher level it directs the movement of the robot over the entire work area with an aid of feedback from sensors and appropriate artificial intelligence algorithms. This second level requires, at the present state of technology, an additional development effort for dependable and economically feasible robot's navigation. The control of robot operations can be facilitated by human intervention , either direct, or through a teleoperator. The sensors needed for ensuring required quality of preprogrammed activity at the two control levels belong to the vision, contact and proximity classes. All these sensors require additional development for dependence, sturdiness and adaptation to building conditions for feasible application at the second level of control, and even for many operations at the first level. The mobility of robots over continuous surfaces does not pose any mechanical difficulty and depends entirely on the development of sensors and control units. Human intervention will be required when the robot is to move over discontinuous surfaces like, e.g. from one to another floor in the same building.

e. he building construction activities consist of handling of building components, connecting them to the existing structure and finishing them by application of appropriate mechanical treatment. The most amenable to robotization are the activities which require covering (with liquid or semiliquid substances) or conditioning of large continuous surfaces - vertical or horizontal, e.g. painting, spraying, plastering, trowelling, etc. These activities can be robotized without difficulty at the present stage of technology. Less amenable, but possible to robotize are activities which require moving the effector at different locations in a predetermined pattern - linear or point to point, e.g. welding, bolting, jointing, spreading of fabric rolls, etc. These activities can be robotized in building with a proper adaptation of sensory devices. The least amenable to robotization are activities requiring handling and assembly of components like structural steel members, precast elements, bricks, formwork, sheathing, etc., which require picking, orientation, precise positioning and often temporary supporting of objects.

f. A selective robotization of the building construction works can be attained with minimum development if applied to the first two groups of building activities as enumerated above. A more comprehensive robotization particularly promising under hazardous or harsh environments must, however, involve also assembling tasks. The process can be simplified when using large self-supporting prefabricated components comprising as many building works as possible, and thus minimizing the extent of assembling work on site. The finishing work on site should be limited, as much as possible, too simple, one activity, covering or conditioning tasks.

g. Four types of robots can perform all major building construction tasks. The first type used for handling of large building components may have a configuration of a building crane or a concrete pump boom. It will use grippers for picking and placing of the components, and a control unit with teleported or sensory guidance. The second type can be employed for most of interior finishing, and connecting works. It will use an anthropomorphic arm with a reach of 3-4m and tools according to the type of work to be done. It will operate from static work stations, each one encompassing a room size service area, with a preprogrammed mode of operation. The quality and precision of its work can be monitored by vision, tactile or proximity sensors. The robot may be used either directly by a working crew who will move it from one work station to the next and set it up for operation at each work station, or guided by a teleoperator, or act autonomously with movement between work stations and its setting up for operation at them monitored by sensors and an appropriate artificial intelligence algorithm. The third type employed for finishing of large horizontal surfaces will consist of a self-propelled carriage with the finishing effector mounted underneath. It can also use one of control alternatives, and associated sensory devices as described above. The fourth type will be employed for finishing of vertical exterior walls. It will employ a carriage suspenaed from the roof's parapet with the effecting tool mounted between it and the wall surface.

h. Actual robotization of the building tasks must be preceded by a detailed predevelopment design process which includes analysis of the present construction method, designing several alternatives of robot employment, selecting of conceptual design of system components for each alternative, preparation of appropriate operational procedure for each alternative, and conducting a thorough feasibility study to determine the optimal choice, which will be consequently detailed for development.

i. The development effort will focus on a manipulator with sufficient reach (for the general purpose robot), and a carriage adapted to movement between work stations. Both should be made sturdy enough to operate in rugged building environment but their weight should be reduced as much as possible so as not to impose excessive loads on the structure. The development will include also effectors with their feeding systems for performance of different construction tasks, and sensors for interaction with guiding devices fixed in the structure.

j. The robotization of building construction may be easier within the framework of closed systems using large selections of predesigned components to be prefabricated in plant, and assembled on site. Such systems may employ CAD/CAM programs and procedures for design, managerial planning and robotization - on site and in prefabrication plant, of building projects, for optimal utilization of human, robotic and material resources. An extension of such systems will allow for automated adaptation of existing designs to particular requests of users. It will also allow for automatic adaptation of production resources in the prefabrication factory to the changes in the basic design.

k. A preliminary economic analysis indicates that a value of a construction robot to its user may be, under normal working conditions, approximately $250,000, based on 1983 prices. The value was calculated as the present worth of savings in labor from robot's employment less the associated operation and maintenance costs. It is expected, based on the current cost of manufacturing robots, that this value will exceed the cost of purchasing and installation of a construction robot after its development. This expectation assumes rational utilization of the robots in suitable building projects. Should the robots be used at a rate much lower than the indicated above, or in building projects very ill structured for robotization, they will have difficulty competing economically with the conventional building methods.

l. The value of robots in building construction increases under conditions which impair the performance (and decrease the productivity) of human labor. Such conditions arise when the building is performed under harsh weather (very hot or very cold), in high altitudes, or in extended periods of overtime work. . The economic analysis indicates that the value of robots under such, conditions may be higher to the user by 50% and more than under normal conditions, as discussed before. The value of robots is particularly high when their use can limit human participation in hazardous tasks with a high fatality or injury incidence to workers.

m. Another important benefit expected from robots application is the improved quality of the product due to a higher accuracy and a better control of robot operations. The benefits include savings in materials, in repair work during the construction, in maintenance thereafter, and a higher satisfaction of the user. These benefits are difficult to quantify, however their value is considered by users as important as productivity gains attained with robotization.

n. The economic success of robotization in building will ultimately depend upon the extent of replacement of the human labor. It may also change the contents of many existing tasks. Such changes may be resisted by the workers on site and their unions. The robotization will certainly require changes in the contents of managerial tasks on and off the building site, and the training necessary for their performance. It may also be resisted by the management, especially at the lower levels, in the construction companies. The resistance of labor and management may be avoided or at least mitigated if the implementation will be preceded by a planned effort to secure the cooperation of all parties.

What Does a Robotics Technician Do?

The job of a robotics technician is to design and maintain automated robotic systems. These technicians must be knowledgeable in a variety of areas, such as mechanical repair, electric circuit design, and computer programming. A two year associate's degree is usually required for individuals in this career.

Robotics technicians work in many different locations. Some are employed by factories and automation facilities. Industrial robots in these places usually have a specific job, such as welding or painting. Any time that a robot breaks or malfunctions, a robotics expert is responsible for diagnosing and repairing the problem. Robot techs must be able to work under pressure in order to minimize factory downtime and fix problems quickly.

Malfunctions in robotic systems are not always caused by physical problems, but can also be created by computer errors. A robotics technician must understand the programming code used to control a device. Technicians often must search through many lines of programming code in order to discover a mistake, and then re-program a robot to function properly. The ability to work methodically and patiently is a vital skill for workers in this profession.

Robotics technicians use a variety of diagnostic and repair tools. These include oscilloscopes and multimeters, which are used to measure electrical signals. Technicians must also be familiar with programmable logic controllers and electronics schematics, in order to determine the source of a malfunction.

Training for robot techs begins at the high school level. Individuals who are interested in this career typically take classes in science and mathematics. Many workers also attend a technical college or trade school to obtain an associate's degree in robotics technology. Over a period of two years, robotics students typically learn about microprocessors, hydraulics, electronics, and other technologies used to create robots.

The education of a robotics technician does not always stop at the college level. Many workers receive specialized training from an employer. Companies that use robots often train workers in the detailed specifics of a particular robotic device or model. Technicians gain valuable experience by working hands-on with robots in actual factory locations.

Some technicians become supervisors or trainers, and provide education about robotic systems to other workers. Robotics trainers work for technical companies, trade schools, and colleges, and demonstrate the proper way to use and repair robots. Technicians with high levels of experience can be hired by robot manufacturers, and provide advice on ways to improve new robotic systems.

CHAPTER 1

USING INTELLIGENT TECHNIQUES IN CONSTRUCTION PROJECT COST ESTIMATION

Abdelrahman Osman Elfaki, Saleh Alatawi and Eyad Abushandi

INTRODUCTION

Information technology (IT) plays a crucial role in dealing with challenges in construction projects. Thomas et al. [1] have illustrated the importance of using IT to improve the performance of construction projects. The construction industry faces numerous complicated challenges that go beyond IT. These complicated challenges motivate the use of intelligent techniques to handle those challenges. For instance, intelligent techniques may be used to handle challenges such as (1) selecting the best-qualified prime contractor, (2) predicting project performance at different phases, or (3) estimating risk for cost overruns (running beyond a proper plan may lead to greater risks for many contractors). Recently, the civil engineering community has begun to consider Artificial Intelligence (AI) techniques as an optimal art for handling the above 3 fuzzy and ambiguous challenges [2]. The use of AI in the civil engineering sector has been introduced by Parmee [3], who proposes for AI to tackle problem areas characterised by uncertainty and poor definition.

Cost estimation is the most important preliminary process in any construction project. In the construction industry, cost estimation is the process of predicting the costs required to perform the work within the scope of the project [4]. Accurate cost estimation is crucial to ensure the successful completion of a construction project. Estimating construction cost is an example of a knowledge-intensive engineering task [5]; that is, it depends on the expertise of the human professional. In fact, engineers require several years to develop the necessary expertise to conduct the cost estimation process. The main problem here is that the engineers' expertise is often not documented or authenticated. Hence, this expertise is prone to subjectivity (i.e., defined to an extent by one's personal opinion). According to Shane et al. [6], accuracy and comprehensiveness in cost estimation are delicate issues and can be easily affected by many different parameters; in addition, each parameter must be properly addressed in order to maintain an acceptable level of accuracy during the process. Therefore, estimating construction cost to a fair degree of accuracy is mostly impossible to achieve manually.

On the other hand, inaccurate cost estimation leads to many problems, such as change order, construction delay [7], or even business bankruptcy in the worst scenarios. These two factors (i) the impossibility of conducting cost estimation manually and (ii) the effects of incorrect cost estimation thus encourage researchers and construction companies to investigate intelligent solutions to handle the problem of cost estimation.

This paper investigates and summarises the current use of intelligent solutions in the construction industry. In order to leverage the importance of intelligent solutions in project cost estimation, a list of state-of-the-art methods has been analysed, including machine-learning (ML), rule-based systems (RBS), evolutionary systems (ES), agent-based system (ABS), and hybrid systems (HS).

This paper has been organised as follows: we discuss our research methodology in Section 2. In Section 3, we explore and define the construction cost estimation factors that are used

Citation: Abdelrahman Osman Elfaki, Saleh Alatawi, and Eyad Abushandi, "Using Intelligent Techniques in Construction Project Cost Estimation," Advances in Civil Engineering, vol. 2014, Article ID 107926, 11 pages, 2014. doi:10.1155/2014/107926.

in this survey paper. In Section 4, intelligent techniques that are used in construction cost estimation are classified into five groups, and the main strengths and weaknesses of each group are defined. Each proposal is then analysed. Conclusions and future directions are presented in Section 5.

RESEARCH METHODOLOGY

The importance of cost estimation in the construction industry has been discussed in the previous section. However, there is no doubt that intelligent solutions may solve the dilemma of cost overruns, considering all affecting factors. In fact, there are a huge number of intelligent techniques available to deal with problems in construction management [8]. This motivates the researchers to carry out and analyse intelligent techniques with regard to tackling the construction cost estimation problem. This paper surveys the intelligent solutions employed over the last decade and identifies the directions for future development. This will help to provide more precise and in-depth analysis for the most recent proposals. The analytical process will highlight the research gap in this area. Furthermore, it will open a door for defining the available opportunities for future research.

This research has been divided into three parts, as shown in Figure 1. Firstly, we create a literature review database on the intelligent techniques that have been used in cost estimations over the last decade. In this step, specific journals have been selected based on their specialisation both in construction management and in information technology. These journals are Journal of Computing in Civil Engineering (http://ascelibrary.org/journal/jccee5), Journal of Construction Engineering and Management (http://ascelibrary.org/journal/jcemd4), Itcon (http://www.itcon.org/), Journal of Civil Engineering and Management (http://www.tandfonline.com/toc/tcem20/current#.VFlO8Wdh71U), and Automation in Construction (http://www.journals.elsevier.com/automation-in-construction/). Consequently, the collected papers have been classified based on their applied techniques. Secondly, we present an analysis and discussion of each intelligent technique to clarify its strengths and weaknesses. The strengths and weaknesses of specific intelligent techniques will be inherited by the cost estimation method based on that technique. Additionally, cost affecting factors have been established in order to carry out a specific benchmarking process.

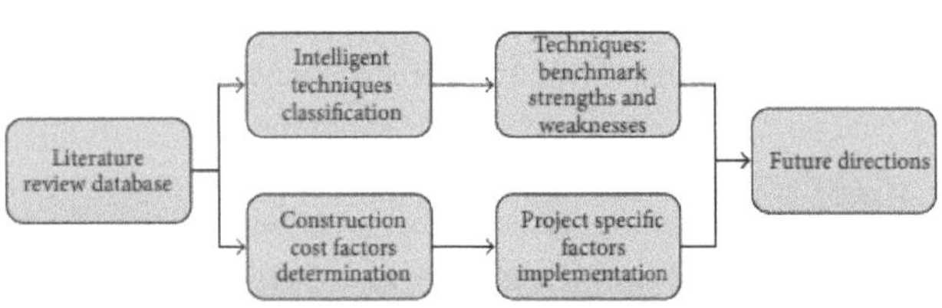

Figure 1: Flow chart illustrating the methodology.

Later, an intensive comparison of the surveyed construction projects' cost estimation methods, based on a proposed benchmark, has been conducted. To analyse each proposal, the research has focused on four points:(i)the intelligent technique in use;(ii)how the proposal's data is collected;(iii)validation of the proposed idea;(iv)the coverage of cost estimation factors.

We will now explain the steps from Figure 1 in detail. The first step is the creation of a literature database from the four journals mentioned earlier. We have used the words "construction cost" as a primary keyword; we have then selected only the proposals that involve the use of intelligent techniques. The second and third steps are parallel. In the second step, the intelligent techniques that have been found in selected proposals are classified based on the scientific concept of each one. The aim of this classification is to define the main features of each class. Step four describes the proposed benchmark, explained in detail in Table 2. This benchmark has been developed on the basis of construction cost factors selected in step five. Step six shows the last step and the main objective of this survey, which is the defining of future directions in the research area of construction cost estimation.

CONSTRUCTION COST FACTORS

According to Shane et al. [6], Oberlender and Trost [9], and Ahiaga-Dagbui and Smith [10], any construction cost estimation should be developed based on specific parameters such as type of project, material costs, likely design and scope changes, ground conditions, duration, size of project, type of client, and tendering method. Therefore, in this paper we have introduced these factors as a benchmark to compare between the cost estimation proposals.

There are various different factors that affect cost estimation in construction projects. These factors can be clustered into two distinct groups [11]: (i) estimator-specific factors and (ii) design and project-specific factors.

Estimator Specific Factors

The cost estimator can be one of the three parties: contractor, consultant, or owner. Based on the estimator's background and experience, cognitive biases or errors in cost estimates may occur accordingly [11]. In many cases, the cost estimator makes decisions based on the likely gains, or losses, of a venture and not necessarily based on the real outcome of the decision [12]. Moreover, the individual estimator may customise pricing based upon best local practices [13], which differ from country to country. For this reason, this paper will focus on design and project-specific factors.

Design and Project-Specific Factors

These factors include project size, type of project, ground conditions, type of client, material costs, likely design and scope changes, duration, tendering method [6, 9, 10], and contract type. In the following paragraphs, these factors are discussed in detail to explore their meanings and functions regarding cost estimation.

Project Size

There is a strong correlation between project size in square feet or metre and the number of labours. However, as the number of labours increases, the cost estimation of some items may have some biases and become more plausible (e.g., production rate estimation or tasks scheduling). There are many empirical studies on how project size can influence cost estimation (e.g., [14, 15]).

Type of Project

Undertaking particular types of projects requires a suitable choice of technology and equipment used, as well as suitable work methods. However, this can limit the choice of materials and size of crew to be employed; consequently, this will affect the project budget.

Project types can be classified under several different categories. In general, there are six major types of construction projects: (1) building construction, (2) special-purpose construction, (3) heavy construction, (4) highway construction, (5) infrastructure construction, and (6) industrial construction.

Ground Conditions

Before tendering, ground condition should be one of the first concerns in any construction project. Without knowing the ground condition, the contractor should still presume to estimate the cost; however, if the assumption is not proper, this will lead to additional costs for bad ground condition.

Type of Client

As each construction project has its own client ideas, roles, and objectives, the characteristics of the contract and bidding behaviour are mainly affected by client type. There are seven different types of clients as classified by Drew et al. [16].(1) Government.(2)Housing Authority.(3)Other public sector clients.(4)Large developers.(5)Large industrial, commercial, and retailing organisations.(6)Medium and small industrial, commercial, and retailing organisations.(7)Other private sector clients.

Material Costs

The material selection-time, type of materials, and their availability in the local market all demonstrate a statistically significant impact on the cost estimation of construction projects. Materials can represent up to 70% of the project construction cost [17]; hence, any methods used to estimate the material cost accurately will reduce wastage and improve the major project's cost and time benefits. In addition, the quantity of material required must be accurately measured from the drawing and is not dependent on the crew performance or work method adopted [13]. However, this factor can vary dramatically and is highly dependent on the performance and work method used by the labours.

Likely Design and Scope Changes

Depending on their level of experience, the client retains more influence over the design and once on site during construction. Certain types of projects require the client to appoint a design firm (Figure 2) to design and inspect the project phases, in order to achieve the standards expected by the client.

On the other hand, the right scope definition phase is highly important in the pre-project planning process. Poor scope definition is recognised by industry practitioners as one of the leading causes of project failure, as a high level of pre-project planning effort can result in around a 20% saving on total costs [18].

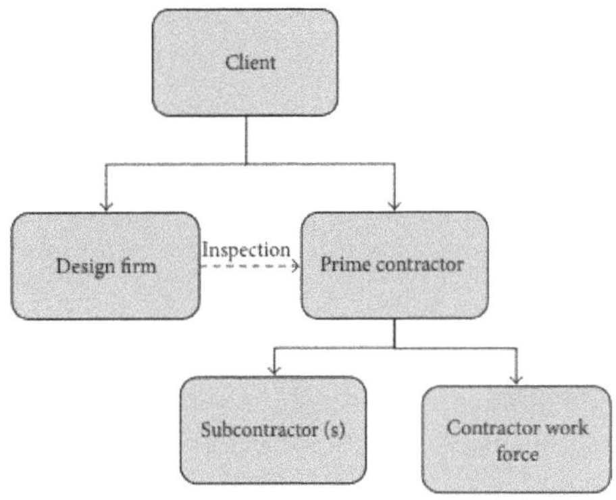

Figure 2: The client appoints a firm to design and inspect the project to meet certain specified (usually, oriented) requirements [50].

Duration

Research has indicated that there is a strong relationship between project cost and construction duration for different construction markets (e.g., [19, 20]). A relationship between completed construction cost and the time taken to complete a construction project was first mathematically established by Bromilow et al. [20]:

$$T = KC^B, \quad (1)$$

where T is the duration of construction period, C is the final project cost, K is a constant value indicating the general level of duration performance, and B is a constant value describing how the duration performance is affected by the size of the construction project measured by its cost.

Figure 3 presents a duration-cost plane frame for small and medium infrastructure projects and identifies three main regions for project scheduling the boundaries of which are defined by general indexed project duration [21].

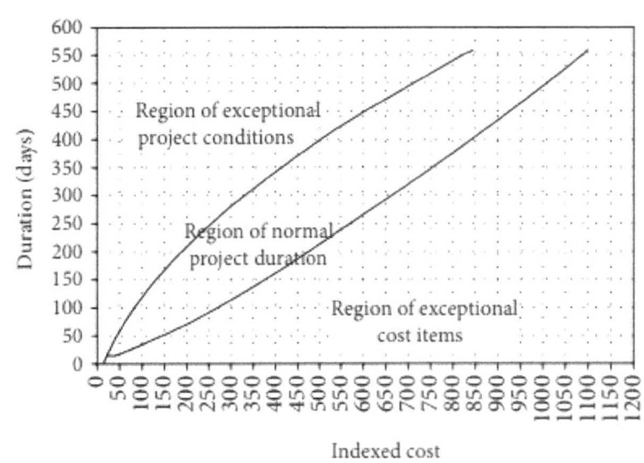

Figure 3: Modelled duration-cost envelope for policy decision support in small and medium projects [21].

Tendering Method

There are five tendering methods, including the following. (i)Open Tendering. Contractors are invited to tender for a contract through local advertisements.(ii)Selective Tendering. Contractors are invited to tender on their proven record in relation to the type and size of contract and their own reliability.

(iii)Negotiated Tendering. Cost reimbursement contract is a variation of this, which can be used when completion time is more important than cost.(iv)Two-Stage Tendering. It is used to bring in a contractor at the design stage, which is useful to advise the architect of any problems with the design of the building. Unit rates would be negotiated on the basis of the original tender.(v)Serial Tendering. Tenders are invited from a selected list on the basis of a typical (notional) bill of quantities. The chosen contractor normally submits the lowest price and undertakes to enter into a series of contracts to carry out the work using the rates in the notional bill of quantities.

The selection of one of the above methods is basically intended to minimise any additional client risk. To achieve this goal, the client must balance four aspects:(i)client needs;(ii) project cost;(iii)completion time;(iv)qualification of the tender to perform the job.

INTELLIGENT CONSTRUCTION PROJECT COST ESTIMATION METHODS

In this section, analysis of the surveyed intelligent construction cost estimation methods was conducted. These methods have been categorised into five groups, based on the intelligent technique that is used in each group: machine-learning (ML), rule-based systems (RBS), evolutionary systems (ES), agent-based system (ABS), and hybrid systems (HS).

At the first step, each group is explored to highlight their strengths and weaknesses. Subsequently, the methods are analysed in depth in terms of coverage of construction cost estimation techniques. In each proposal, four key questions have been highlighted for analysis. These questions are (1) which intelligent technique is used; (2) how the input datasets are collected; (3) how the proposed method is validated; and (4) which construction cost estimation factors are covered.

In the following subsections, firstly, the intelligent techniques employed are discussed, is the findings of which are considered as an answer to the first question. Secondly, each proposal is analysed individually, which answers question 2. The content of Table 1 illustrates the answer of question 3, while the content of Table 2illustrates the answer of question 4.

Table 1: Comparison of proposals based on technique and validation

Proposal	Technique	Validation
Wilmot and Mei [24]	ML: ANN	Have not been mentioned
An et al. [26]	ML: SVM	Comparison with methods for assessing conceptual cost estimates
Petroutsatou et al. [23]	ML: ANN	By comparison with other models in literature
Jafarzadeh et al. [25]	ML: ANN	Have not been mentioned
Hola and Schabowicz [27]	ML: ANN	Have not been mentioned
Son et al. [28]	ML: SVM	Comparison with other techniques such as ANN and a decision tree (DT)
Cheng and Hoang [29]	ML: SVM	Have not been mentioned
Ji et al. [30]	KBS: case-based reasoning	Using case study
Choi et al. [31]	KBS: case-based reasoning	By comparison with previous conceptual cost estimation studies
K. J. Kim and K. Kim [32]	KBS: case-based reasoning	Have not been mentioned
Yildiz et al. [33]	KBS	By doing interview with experts
Lee et al. [34]	KBS: ontology	Comparison with other techniques
Karakas et al. [41]	ABS: MAS	Interview with expert
Rojas and Mukherjee [42]	ABS: multiagent	Have not been mentioned
Kim [36]	KBS: case-based reasoning and analytical hierarchy process	Case study
de Albuquerque et al. [38]	ES: genetic algorithm	Have not been mentioned
Rogalska et al. [37]	ES: hybrid genetic evolutionary algorithm	By comparing the result with case studies from literature
Ghoddousi et al. [35]	ES: genetic algorithm	By comparing the result with case studies from literature
Afshar et al. [39]	ES: ant colony	By comparing the results with case studies in construction optimisation
Zhang and Ng [40]	ES: ant colony	By comparing the results with an academic benchmark
Kim et al. [43]	HS: statistics, CBR, and database	By comparing the result with case study from literature
Cheng et al. [44]	HS: SVM and DE	By comparing the result with other methods
Kim et al. [45]	HS: ANN and GA	Have not been mentioned
Yu and Skibniewski [46]	HS: ANN and fuzzy system	By using a case study of residential building construction projects in China
Williams and Gong [47]	HS: text mining, numerical data, and ensemble classifiers	Have not been mentioned
Cheng et al. [48]	HS: ANN, GA, and fuzzy system	Have not been mentioned
Zhang and Xing [49]	HS: fuzzy and particle swarm optimisation	Have not been mentioned

Table 2: Comparison of the proposals based on design and project-specific factors

Work	Project size	Project type	Ground conditions	Type of client	Likely design and scope changes	Contract type	Material costs	Duration	Tendering method
Wilmot and Mei [24]	Y	Y	Y	Y	Y	Y	Y	Y	N
An et al. [26]	Y	Y	Y	Y	N	N	N	Y	N
Petroutsatou et al. [23]	Y	Y	Y	Y	Y	Y	Y	N	N
Jafarzadeh et al. [25]	Y	Y	Y	Y	Y	N	N	N	N
Hola and Schabowicz [27]	Y	Y	Y	N	N	N	Y	Y	Y
Son et al. [28]	Y	Y	N	Y	Y	N	N	Y	N
Cheng and Hoang [29]	Y	Y	Y	N	N	N	N	Y	N
Ji et al. [30]	Y	Y	Y	N	Y	N	Y	N	N
Choi et al. [31]	Y	Y	Y	Y	N	N	N	N	N
K. J. Kim and K. Kim [32]	Y	Y	Y	Y	N	N	N	N	N
Yildiz et al. [33]	Y	Y	N	N	Y	Y	N	Y	N
Lee et al. [34]	Y	Y	Y	N	Y	N	Y	N	N
Karakas et al. [41]	N	N	Y	Y	Y	N	Y	N	N
Rojas and Mukherjee [42]	Y	Y	Y	N	N	N	N	N	N
Kim [36]	Y	Y	Y	N	N	N	N	N	Y
de Albuquerque et al. [38]	Y	Y	Y	N	N	N	Y	N	N
Rogalska et al. [37]	Y	Y	Y	N	N	N	N	N	N
Ghoddousi et al. [35]	Y	Y	Y	Y	N	N	N	N	N
Afshar et al. [39]	Y	Y	Y	N	N	Y	Y	Y	Y
Zhang and Ng [40]	Y	Y	Y	N	N	N	N	N	N
Kim et al. [43]	Y	Y	Y	N	N	N	Y	Y	N
Cheng et al. [44]	Y	Y	Y	N	N	N	Y	Y	N
Kim et al. [45]	Y	Y	Y	N	N	N	Y	Y	N
Yu and Skibniewski [46]	Y	Y	Y	Y	N	N	Y	Y	N
Williams and Gong [47]	Y	Y	Y	N	N	Y	Y	Y	Y
Cheng et al. [48]	Y	Y	Y	Y	N	N	Y	Y	N
Zhang and Xing [49]	Y	Y	Y	Y	N	Y	Y	Y	Y

Machine Learning (ML) Systems

ML systems have been defined as a construction of a system that can learn from data. In general, the main strengths of ML are (i) the ability to deal with uncertainty, (ii) the ability to work with incomplete data, and (iii) the ability to judge new cases based on acquired experiences from similar cases. On the other hand, the main weakness of ML is the lack of technical justification; that is, the causes beyond the decision are not available. This type of decision is called a black box decision. However, in the construction management, the artificial neural network (ANN) and the support vector machine (SVM) are the most common ML techniques. In the next paragraph, we analyse the construction cost estimation proposals that are based on ML.

One of the earliest papers to introduce the benefits and the implementation of ANN in the civil engineering community is published by Flood and Kartam [22]. This research has opened the door for many proposals that suggest ML as the preferred method to tackle various challenges in the construction industry. Petroutsatou et al. [23] introduced the ANN as a technique for early cost estimation of road tunnel construction. The data collection strategy of this research was based on structured questionnaires from different tunnel construction sites. The main drawback of this research was the ignoring of some of the construction cost factors (for more details, see Table 2). Wilmot and Mei [24] introduced an ANN model for highway construction costs. This research used the following factors as a base for cost estimation: price of labour, price of material, price of equipment, pay item quantity, contract duration, contract location, quarter in which the contract was let, annual bid volume, bid volume variance, number of plan changes, and changes in standards or specifications. The main contribution of this work was that it covered all required factors. Nevertheless, the validation of the proposed method and the data collection process used for training and testing the results were not fully presented. Jafarzadeh et al. [25] proposed the ANN method for predicting seismic retrofit construction costs. This study selected data from 158 earthquake-prone schools. The validation of this method is not clear. An et al. [26] proposed SVM for assessing conceptual cost estimates. Although this proposal is introduced as an assessment tool, still it might be considered as a cost estimation method. The method was developed on the basis of data from 62 completed building construction projects in Korea. Furthermore, Hola and Schabowicz [27] developed an ANN model for determining earthworks' execution times and costs. Basically, this model was developed on the basis of a database created from several studies that were carried out during large-scale earthwork operations on the construction site of one of the largest chemical plants in Central Europe. However, the validation of the presented results is not mentioned.

Son et al. [28] developed a hybrid prediction model that combines principal component analysis (PCA) with a support vector regression (SVR) predictive model for cost performance of commercial building projects. They used 64 related variables to define the pre-project planning stage. They developed their dataset based on information from 84 building projects in South Korea that had been completed within three years of the date at which the study was carried out. Questionnaires and interviews were used as a strategy for data collection. Cheng and Hoang [29] developed cost estimation at completion technique using least squares support vector machine. The data sets that are used in Cheng and Hoang [29] were collected from 13 reinforced concrete building projects executed between 2000 and 2007 by one construction company headquartered in Taiwan.

Knowledge-Based Systems (KBS)

This category includes any technique that uses logical rules for deducing the required conclusions. The main strengths of KBS are (i) the ability to justify any result and (ii) uncomplicated methods (i.e., it is relatively easy to develop KBS). On the other hand, the limitations of KBS are (i) the difficulty of self-learning and (ii) time consumption during the rule acquisition process. Expert system and case-based reasoning are the common techniques used in KBS. The accuracy of case-based reasoning is highly dependent on the number of selected cases. Recently, KBS has been combined with other techniques to handle the limitation of the self-learning process. However, this mixture will be discussed in more detail in the section of this paper that deals with hybrid systems.

Ji et al. [30] proposed case-based reasoning to prepare strategic and conceptual estimations for construction budgeting. The data for this project were collected from 129 military barrack projects. Choi et al. [31] proposed a cost prediction model for public road planning. The research data had been collected from a total of 207 real public road projects. Choi et al. [31] used rough-set theory to control the data collection and a genetic algorithm to optimise the rough-set model. Their work was classified as KBS since the authors implemented the case-based reasoning component in their cost estimation. K. J. Kim and K. Kim [32] developed a cost estimation model using CBR. This research overcomes the uncertainty in choosing the correct case by using a genetic algorithm. For this research, data were collected from 65 projects that constructed 585 bridges over a 5-year period. K. J. Kim and K. Kim [32] focused on construction of national bridges. However, it was not mentioned how the results were validated.

Yildiz et al. [33] developed a knowledge-based risk mapping tool to estimate costs for international construction projects. The required data and cost estimation parameters were collected from related literature. The validation process was performed in the form of expert interviews to get feedback on the developed tool. Lee et al. [34] proposed an ontological inference process for building cost estimation, by automating the process of searching for the most appropriate work items. Ghoddousi et al. [35] proposed a solution for determining total cost, time, and resources for construction projects; this was developed on the basis of a nondominated sorting genetic algorithm.

Kim [36] developed a cost estimation model based on case-based reasoning and analytical hierarchy process (AHP). In this project, data have been selected from literature and only 13 studies have been analyzed. Kim [36] developed his model based on data from high-way construction projects. The validation has been conducted based on case study that contains data from 48 construction projects.

Evolutionary Systems (ES)

ES is a group of intelligent systems concerned with continuous optimisation with heuristics. As the results of ES are generated based on specific heuristics, they are very difficult to generalise, which is considered to be the main limitation of ES. The ability to solve complicated and uncertain problems is the main motivation for researchers to use ES. Evolutionary systems are used mainly as optimisation tools where there are many solutions; however, the ES algorithm assists in obtaining the correct solution.

Rogalska et al. [37] proposed a method based on genetic algorithm to deal with the problem of construction project scheduling. de Albuquerque et al. [38] developed a tool for estimating the cost of concrete structures. This tool is developed based on genetic algorithm. The cost has been estimated in all construction phases, such as manufacture, transport, and erection. Afshar et al. [39] developed a multicolony ant algorithm to solve the time/cost multiobjective optimisation problem. This method estimated both direct and indirect costs. Zhang and Ng [40] developed a Decision Support System (DSS) for cost estimation based on ant colony system. Zhang and Ng [40] used synthetic data to develop their DSS and they do validate their system by comparing it with a standard academic project. However, validation is done. Still Validation with real projects provide more accurate results.

Agent-Based System (ABS)

ABS has been considered as one of the main tracks in Artificial Intelligence, simulating the actions and interactions of autonomous agents with a view of assessing their effects on the system as a whole. In ABS, the generalisation of extracted results is the main challenge.

Karakas et al. [41] developed a multiagent system (MAS) that simulates the negotiation process between contractor and client regarding risk allocation and sharing of cost overruns in construction projects. This MAS was tested by interviewing eight professionals from the construction industry. In addition, Rojas and Mukherjee [42] developed a general multiagent simulation framework that can be used as an effective training environment. This framework could be used to estimate direct and indirect costs for construction projects.

Hybrid Systems (HS)

HS is defined as a collection of techniques used together to solve a specific problem. Usually, researchers use HS to overcome the techniques' individual limitations. Implementation of HS could represent a challenge, due to the unavailability of computational tools that could support its implementation. Furthermore, Kim et al. [43] proposed a hybrid conceptual cost estimating model for large mixed-use building projects. In this proposal, statistical analysis, CBR, and database methodologies were used together as a hybrid methodology. More recently, Cheng et al. [44] proposed a hybrid intelligence system for estimating construction cost. This hybrid system was developed based on support vector machine (SVM) and differential evolution (DE). In this proposal, data were collected across a number of public projects in Taiwan. Kim et al. [45] proposed hybrid models of ANN and GA for cost estimation of residential buildings, in order to predict preliminary cost estimates. In Kim et al.'s proposal, data were collected from residential buildings constructed in the years between 1997 and 2000 in Seoul, Korea. Yu and Skibniewski [46] proposed integrating a neurofuzzy system with conceptual cost estimation to discover cost-related knowledge from residential construction projects. The data used in this proposal was based on historical data from previous construction projects collected by the Ministry of Construction of PRC in the years between 1996 and 2002. Most recently, Williams and Gong [47] proposed text mining, numerical data and ensemble classifiers for estimating construction costs. Data used in this proposal were collected from 121 competitively bid highway projects. These data were collected from California Department of Transportation websites. Cheng et al. [48] proposed web-based conceptual cost estimates for construction projects, using an Evolutionary

Fuzzy Neural Inference Model. Data were collected from 28 construction projects spanning the years from 1997 to 2001 in Taiwan. In this regard, Zhang and Xing [49] proposed a hybrid model for estimating construction costs, based on fuzzy and swarm optimisation. The data were collected from national bridge construction projects.

Table 1 shows the comparison of surveyed proposals, based on two issues. The first issue is the intelligent technique used in a proposal. The second issue is the type of validation that is used to prove the applicability of the proposal. Table 2 shows the comparison of surveyed proposals, based on design and project-specific factors used to estimate construction cost in each proposal. The letter "Y" means that this factor has been considered in this proposal, while the letter "N" means that this factor has not been considered. It is very obvious that there is no proposal that satisfies all the design and project-specific factors. On the other hand, in Table 1, there are some proposals that are provided without clear and scientific validation.

CONCLUSION AND FUTURE DIRECTIONS

In this paper, a survey and analysis were performed on different proposals in order to tackle the problem of developing construction cost estimation based on intelligent techniques. A scientific methodology has been designed to implement this survey. The method of the presented paper was based on two parts. The first part was concerned with a literature survey to examine the current state of intelligent solutions in the construction industry. Regarding this matter, we have chosen exclusively the journals that specialise in both information technology and construction management, within a time frame of ten years. In the research context, a ten-year period is sufficient to surround the directions of research in a specific area.

The second part was concerned with analysis of the proposals collected in the first part. Four key questions were selected to analyse each proposal. These questions are as follows.(i)What is the intelligent technique used?(ii)How is the proposal's data collected?(iii)How is the proposed idea validated?(iv)What are the construction cost estimation factors used?

A justification of the four questions has been provided as follows.(1)Defining the Intelligent Technique Used. This question is used to highlight the general strength and limitations of each proposal, which are reflected by the technique employed.(2)Defining Data Collection Method. This question is used to ensure the degree of accuracy. The degree of accuracy mainly depends on the collected data.(3)Defining the Validation of the Proposed Idea. This question is used to ensure the applicability of the proposed idea.(4)Defining the Commonly Used Cost Estimation Factors. This question is used to ensure the completeness of the proposal.

As mentioned in Section 3, there are two types of construction estimation factors: estimator-specific factors and design and project-specific factors. The first type, estimator-specific factors, depends on estimator expertise and skills and on lack of standardisation. The second type, design and project-specific features, is well defined and established in the civil engineering community. Due to the standardisation and stability of design and project-specific factors, this research paper considered only those factors mentioned in the designed methodology when applying the benchmark.

In conclusion, this paper provides two contributions to this area of knowledge: (1) an analysis of construction cost estimation proposals and (2) a standard survey methodology that can be used in any future surveys that deal with construction cost estimation.

According to the results of this research paper, the research gaps that have been deduced from this survey are as follows. (1)There is a crucial necessity for a cost estimation method that covers all estimation factors from both types; that is, there is a need for one method that involves all "estimator specific" and "design and project-specific" factors. In Table 1, it is obvious that no proposal has a full row of "Y."(2)There is a real need for a standard validation method which can be used to determine the accuracy level of a cost estimation proposal.(3)There are many proposals that suffer from a lack of scientific justification for the results, that is, lack of describing how technically the results have been achieved.

Finally, future research directions are suggested for cost estimation in order to overcome the gaps that have been discussed. These directions are as follows.(1)Providing cost estimation proposals that encourage the acquisition of human expertise: however, this releases the construction cost estimation from human dependability. Computerized expert systems are the better mechanism that might be used to replace human expertise. On another hand, knowledge management models and systems will assist in establishing computerized management systems that are free from the constraints of humanitarian. The main goal of knowledge management systems should be to capture and deal with estimator-specific factors. The first future direction is to encourage researchers and industry experts to adopt the direction of knowledge management systems in construction projects.(2)Providing cost estimation proposals that are developed based on all "design and project-specific" factors: in Section 3, eight "design and project-specific" factors have been mentioned. The second future direction is to encourage researchers and industry experts to develop one integrated construction cost estimation system that works to achieve the all eight "design and project-specific" factors which have been mentioned.(3)Providing a scientific justification for the cost estimation proposals based on real-world data: this will provide an explanation of how the estimates work and gives a justification on estimator's biases. Add the scientific justification for any proposal to increase the level of confidence in it. In addition, providing scientific justification assists in tracing the details of the cost estimation process which increase the level of transparency. Finally, providing scientific justification helps increase the maintainability.(4)Providing a standard benchmark for determining the accuracy level of the construction cost estimation proposals: standard benchmarking leads to establishing a rule of thumb when other means of cost estimation are unavailable. This might be achieved by establishing a database containing information from previous projects. In addition, any future cost estimation models should consider this database "known value" to provide a useful benchmark for how accurately those models can estimate the cost. Using standard benchmark could help in classification, clustering, and ranking of cost estimation proposals.

The limitations of this research paper can be summarised in two points: (a) data was collected from specific journals only; (b) the survey was limited to a ten-year period.

While this paper acknowledges these limitations, it is nevertheless able to provide valid answers on the current state of this area of research and to propose future directions.

ACKNOWLEDGMENTS

The present research work has been undertaken within the "Binladin Research Chair on Quality and Productivity Improvement in the Construction Industry" at the University of Hail and funded by the Saudi Binladin Constructions Group.

REFERENCES

1. S. R. Thomas, S.-H. Lee, J. D. Spencer, R. L. Tucker, and R. E. Chapman, "Impacts of design/information technology on project outcomes," Journal of Construction Engineering and Management, vol. 130, no. 4, pp. 586–597, 2004.
2. H. G. Melhem, "Technical council for computing and information technology," Journal of Computing in Civil Engineering, vol. 22, no. 6, pp. 335–337, 2008.
3. I. C. Parmee, "Computational intelligence and civil engineering-perceived problems and possible solutions," in Towards a Vision for Information Technology in Civil Engineering, I. Flood, Ed., ASCE, Nashville, Tenn, USA, 2003.
4. L. Holm, J. E. Schaufelberger, D. Griffin, and T. Cole, Construction Cost Estimating: Process and Practices, Pearson Education, Upper Saddle River, NJ, USA, 2005.
5. S. Staub-French, M. Fischer, J. Kunz, and B. Paulson, "A generic feature-driven activity-based cost estimation process," Advanced Engineering Informatics, vol. 17, no. 1, pp. 23–39, 2003. · ·
6. J. S. Shane, K. R. Molenaar, S. Anderson, and C. Schexnayder, "Construction project cost escalation factors," Journal of Management in Engineering, vol. 25, no. 4, pp. 221–229, 2009.
7. A. Albogamy, D. Scott, N. Dawood, and G. Bekr, "Addressing crucial risk factors in the middle east construction industries: a comparative study of Saudi Arabia and Jordan," Sustainable Building Conference Coventry University, West Midlands, UK, 2013.
8. T. Bulbul, C. J. Anumba, and J. Messner, "A system of systems approach to intelligent construction systems," in Proceedings of the ASCE International Workshop on Computing in Civil Engineering, pp. 22–32, Austin, Tex, USA, June 2009.
9. G. D. Oberlender and S. M. Trost, "Predicting accuracy of early cost estimates based on estimate quality," Journal of Construction Engineering and Management, vol. 127, no. 3, pp. 173–182, 2001.
10. D. D. Ahiaga-Dagbui and S. D. Smith, "Neural networks for modelling the final target cost of water projects," in Proceedings of the 28th Annual ARCOM Conference, S. D. Smith, Ed., pp. 307–316, Association of Researchers in Construction Management, Edinburgh, UK, September 2012.
11. B. Akinci and M. Fischer, "Factors affecting contractors' risk of cost overburden," Journal of Management in Engineering, vol. 14, no. 1, pp. 67–76, 1998.
12. A.-D. D. Dominic and S. D. Smith, "Rethinking construction cost overruns: cognition, learning and estimation," Journal of Financial Management of Property and Construction, vol. 19, no. 1, pp. 38–54, 2014.
13. N. Sinclair, P. Artin, and S. Mulford, "Construction cost data workbook," in Proceedings of the Conference on the International Comparison Program, World Bank, Washington, DC, USA, 2002.
14. A. Doyle and W. Hughes, "The influence of project complexity on estimating accuracy," in Proceedings of the 16th Annual ARCOM Conference, pp. 623–634, Glasgow Caledonian University, 2000.
15. I. Mahamid, "Early cost estimating for road construction projects using multiple regression techniques," Australasian Journal of Construction Economics and Building, vol. 11, no. 4, pp. 87–101, 2011.
16. D. Drew, M. Skitmore, and H. P. Lo, "The effect of client and type and size of construction work on a contractor's bidding strategy," Building and Environment, vol. 36, no. 3, pp. 393–406, 2001.
17. S. Donyavi and R. Flanagan, "The impact of effective material management on construction site performance for small and medium sized construction enterprises," in Proceedings of the 25th Annual Conference of the Association of Researchers in Construction Management (ARCOM '09), A. R. J. Dainty, Ed., pp. 11–20, Association of Researchers in Construction Management, Nottingham, UK, September 2009.
18. C. Cho and G. Edward, "Building project scope definition using project definition rating index," Journal of Architectural Engineering, vol. 7, no. 4, pp. 115–125, 2001. · ·
19. A. P. Kaka and A. D. F. Price, "Relationship between value and duration of construction projects," Construction Management and Economics, vol. 9, no. 4, pp. 383–400, 1991.
20. F. J. Bromilow, M. F. Hinds, and N. F. Moody, The Time and Cost Performance of Building Contracts 1976–1986, Australian Institute of Quantity Surveyors, Sydney, Australia, 1988.
21. F. Edum-Fotwe, "Developing benchmarks for project schedule risk estimation," in System-Based Vision for Strategic and Creative Design, F. Bontempi, Ed., Swets & Zeitlinger, Lisse, The Netherlands, 2003.
22. I. Flood and N. Kartam, "Neural networks in civil engineering. I: principles and understanding," Journal of Computing in Civil Engineering, vol. 8, no. 2, pp. 131–148, 1994.
23. K. Petroutsatou, E. Georgopoulos, S. Lambropoulos,

and J. P. Pantouvakis, "Early cost estimating of road tunnel construction using neural networks," Journal of Construction Engineering and Management, vol. 138, no. 6, pp. 679–687, 2012.

24. C. G. Wilmot and B. Mei, "Neural network modeling of highway construction costs," Journal of Construction Engineering and Management, vol. 131, no. 7, pp. 765–771, 2005.

25. R. Jafarzadeh, J. M. Ingham, S. Wilkinson, V. González, and A. A. Aghakouchak, "Application of artificial neural network methodology for predicting seismic retrofit construction costs," Journal of Construction Engineering and Management, vol. 140, no. 2, Article ID 04013044, 2014.

26. S.-H. An, U.-Y. Park, K.-I. Kang, M.-Y. Cho, and H.-H. Cho, "Application of support vector machines in assessing conceptual cost estimates," Journal of Computing in Civil Engineering, vol. 21, no. 4, pp. 259–264, 2007.

27. B. Hola and K. Schabowicz, "Estimation of earthworks execution time cost by means of artificial neural networks," Automation in Construction, vol. 19, no. 5, pp. 570–579, 2010.

28. H. Son, C. Kim, and C. Kim, "Hybrid principal component analysis and support vector machine model for predicting the cost performance of commercial building projects using pre-project planning variables," Automation in Construction, vol. 27, pp. 60–66, 2012.

29. M.-Y. Cheng and N.-D. Hoang, "Interval estimation of construction cost at completion using least squares support vector machine," Journal of Civil Engineering and Management, vol. 20, no. 2, pp. 223–236, 2014.

30. S.-H. Ji, M. Park, and H.-S. Lee, "Case adaptation method of case-based reasoning for construction cost estimation in Korea," Journal of Construction Engineering and Management, vol. 138, no. 1, pp. 43–52, 2012.

31. S. Choi, D. Y. Kim, S. H. Han, and Y. H. Kwak, "Conceptual cost-prediction model for public road planning via rough set theory and case-based reasoning," Journal of Construction Engineering and Management, vol. 140, no. 1, Article ID 04013026, 2014. · ·

32. K. J. Kim and K. Kim, "Preliminary cost estimation model using case-based reasoning and genetic algorithms," Journal of Computing in Civil Engineering, vol. 24, no. 6, pp. 499–505, 2010.

33. A. E. Yildiz, I. Dikmen, M. T. Birgonul, K. Ercoskun, and S. Alten, "A knowledge-based risk mapping tool for cost estimation of international construction projects," Automation in Construction, vol. 43, pp. 144–155, 2014.

34. S.-K. Lee, K.-R. Kim, and J.-H. Yu, "BIM and ontology-based approach for building cost estimation," Automation in Construction, vol. 41, pp. 96–105, 2014.

35. P. Ghoddousi, E. Eshtehardian, S. Jooybanpour, and A. Javanmardi, "Multi-mode resource-constrained discrete time-cost-resource optimization in project scheduling using non-dominated sorting genetic algorithm," Automation in Construction, vol. 30, pp. 216–227, 2013.

36. S. Kim, "Hybrid forecasting system based on case-based reasoning and analytic hierarchy process for cost estimation," Journal of Civil Engineering and Management, vol. 19, no. 1, pp. 86–96, 2013.

37. M. Rogalska, W. Bozejko, and Z. Hejducki, "Time/cost optimization using hybrid evolutionary algorithm in construction project scheduling," Automation in Construction, vol. 18, no. 1, pp. 24–31, 2008.

38. A. T. de Albuquerque, M. K. El Debs, and A. M. C. Melo, "A cost optimization-based design of precast concrete floors using genetic algorithms," Automation in Construction, vol. 22, pp. 348–356, 2012.

39. A. Afshar, A. K. Ziaraty, A. Kaveh, and F. Sharifi, "Nondominated archiving multicolony ant algorithm in time—cost trade-off optimization," Journal of Construction Engineering and Management, vol. 135, no. 7, pp. 668–674, 2009.

40. Y. Zhang and S. T. Ng, "An ant colony system based decision support system for construction time-cost optimization," Journal of Civil Engineering and Management, vol. 18, no. 4, pp. 580–589, 2012.

41. K. Karakas, I. Dikmen, and M. T. Birgonul, "Multiagent system to simulate risk-allocation and cost-sharing processes in construction projects," Journal of Computing in Civil Engineering, vol. 27, no. 3, pp. 307–319, 2013.

42. E. M. Rojas and A. Mukherjee, "Multi-agent framework for general-purpose situational simulations in the construction management domain," Journal of Computing in Civil Engineering, vol. 20, no. 3, pp. 165–176, 2006.

43. H.-J. Kim, Y.-C. Seo, and C.-T. Hyun, "A hybrid conceptual cost estimating model for large building projects," Automation in Construction, vol. 25, pp. 72–81, 2012.

44. M.-Y. Cheng, N.-D. Hoang, and Y.-W. Wu, "Hybrid intelligence approach based on LS-SVM and Differential Evolution for construction cost index estimation: a Taiwan case study," Automation in Construction, vol. 35, pp. 306–313, 2013.

45. G. H. Kim, D. S. Seo, and K. I. Kang, "Hybrid models of neural networks and genetic algorithms for predicting preliminary cost estimates," Journal of Computing in Civil Engineering, vol. 19, no. 2, pp. 208–211, 2005.

46. W.-D. Yu and M. J. Skibniewski, "Integrating neurofuzzy system with conceptual cost estimation to discover cost-related knowledge from residential construction projects," Journal of Computing in Civil Engineering, vol. 24, no. 1, pp. 35–44, 2010.

47. T. P. Williams and J. Gong, "Predicting construction cost overruns using text mining, numerical data and ensemble classifiers," Automation in Construction, vol. 43, pp. 23–29, 2014.

48. M.-Y. Cheng, H.-C. Tsai, and W.-S. Hsieh, "Web-

based conceptual cost estimates for construction projects using Evolutionary Fuzzy Neural Inference Model," Automation in Construction, vol. 18, no. 2, pp. 164–172, 2009.

49. H. Zhang and F. Xing, "Fuzzy-multi-objective particle swarm optimization for time-cost-quality tradeoff in construction," Automation in Construction, vol. 19, no. 8, pp. 1067–1075, 2010. · ·

50. S. Nunnally, Construction Methods and Management, Prentice Hall, Englewood Cliffs, NJ, USA, 2007.

CHAPTER 2 MOBILE IMAGING AND COMPUTING FOR INTELLIGENT STRUCTURAL DAMAGE INSPECTION

ZhiQiang Chen and Jianfei Chen

INTRODUCTION

Background and Rationale

Sensing-based inspection technologies are recognized as critical components for solutions to quantitative and risk-informed condition assessment for buildings, bridges, and other civil infrastructure systems [1, 2]. Among numerous sensing technologies, visual inspection is commonly used in engineering practice for the purpose of archiving damage scenes and patterns. For example, visual inspection is considered the predominant approach to condition assessment for the majority of bridge inventories in the United States [3]. In these practices, visual inspection is often companied by the use of digital camera for the purpose of digital archival through photographing (viz., optical imaging). The fundamental basis of optical imaging is that through measuring photonic energy emanating from distant objects that are degraded or damaged, disturbed spatial or spectral patterns on the surface of the objects are recorded in a digital format (i.e., two-dimensional images) [4]. In practice, a large percentage of damage patterns, which are visible on the surfaces of structural or geotechnical members (e.g., cracking, spalling, deformation, or collapse-induced debris), can be captured using a commercial digital camera.

In recent years, many research endeavors attempt to empower visual inspection and optical imaging by quantitative image analysis; hence, automatic or semiautomatic damage detection may be achieved [5–9]. Among these efforts, digital images are captured using digital cameras in the field first; then image processing or pattern recognition methods are employed to detect or classify visible damage, respectively. Research findings in this direction have ultimately driven the development of vehicle-mounted imaging and real-time processing for highway or road surface damage inspection [10, 11].

For conducting image-based condition assessment within a complex or/and hard-access built environment, however, limited technology solutions exist. Figure 1 shows two typical field operation scenes of bridge inspection, wherein the region of inspection interest is considerably hard to access and subject to potential operation risk. Robotic imaging and inspection have been attempted (e.g., climbing robots) [12–14]; particularly, the use of unmanned aerial vehicles (UAVs) for civil infrastructure inspection is on the horizon [15–17]. However, all of them are experimental tools and are heavily customized to a certain built environment. In the current practice, human inspection still dominates the practice for general inspection in a complex or/and hard-access built environment.

In light of the dominance of human inspection, one major limitation to date is the lack of real-time imaging, computing, and analytics capability that can be interactively controlled and utilized by the inspection engineers in the field. The state-of-the-practice visual inspection activities for those complex or hard-access areas usually involve a team of engineers who visually inspect structures and use their discretion to capture images in the field. Towards quantitative assessment after the fieldwork, it is expected that these field images can be digitally processed and damage can be extracted using a suite

Citation: ZhiQiang Chen and Jianfei Chen, "Mobile Imaging and Computing for Intelligent Structural Damage Inspection," Advances in Civil Engineering, vol. 2014, Article ID 483729, 14 pages, 2014. doi:10.1155/2014/483729.

of imaging processing packages. However, one should realize that first most civil engineers are not equipped with sufficient image processing and analysis knowledge and they mostly expect a rapid analysis procedure. Second, once the images are taken away from the field situation, much of the contextual knowledge about the images, such as the damage type of the object and the complexity of the damage scene, is prone to be lost subject to the engineer's memory. On the other hand, such contextual information, if utilized, could greatly alleviate the difficulty in processing of structural damage images that usually come with various complex scenes.

(a) (b)

Figure 1: Bridge inspection in practice (photos taken in 2007 courtesy of Virginia Department of Transportation, VDOT; creative commons rights are reserved to VDOT): (a) steel bridge joint inspection with complex geometric configurations and (b) hard-access inspection of the side of the concrete bridge deck.

The aforementioned limitation can be readily resolved by employing the modern mobile technologies. Mobile technologies are closely associated with the advancement of mobile devices, which can be traced down to the movement of small-form-factor computers in the late last 90s (e.g., various brands of Palms or pocket PCs) [18]. Today's mobile devices tightly integrate high-resolution cameras, multicore computing units, flexible communication and data networks, interactive human-device interfaces, geographical position service, and other auxiliary sensors. Therefore, these devices are commonly regarded as "smart" and are represented by smartphones and smartphone-alike tablet devices (the latter usually lack cellular network support). These smart mobile devices have become truly ubiquitous. There are over one billion mobile devices around the world at the end of 2012 [19], and many of them have embedded cameras. Compared to traditional cameras, the advantage of mobile devices may include easy to carry, easy to share, and easy to modify due to the small form-factor hardware design, the communication capability, and the computing capability of mobile devices. As such, today's mobile devices as everyday imaging utility tend to be much more frequently used than traditional cameras.

We therefore envisage that with the ubiquitous mobile devices (e.g., smartphones) and their mobile imaging and computing (MIC) capability, a new norm for structural damage inspection and condition assessment featuring real-time image collection, computing, and damage analytics is on the horizon. Such a framework, if implemented, may significantly alleviate the cumbersome labor involved in practical visual-based civil infrastructure inspection activities for these complex or hard-access structures (as shown in Figure 1).

Contribution

The key platform considered in this paper is smart devices, including smartphones and mobile tablets, which have appeared as a commonplace device and venue for running smart applications used by the general public. Although a few industrial applications have appeared using smartphones for imaging of structural damage, a comprehensive investigation is not found in the literature. The major contribution of this paper is development of a general mobile imaging and computing framework for civil infrastructure damage inspection that tightly integrated the prominent features of smart devices: in situ imaging, computing, and interactive human-device interfaces for both data logging and visual analytics.

Organization

In the following, first, the related mobile imaging and computing technologies are reviewed. Then software environments and mobile imaging and computing libraries that are crucial for engineering development are introduced. These two sections aim to provide a timely and critical review of the state-of-the-art mobile imaging and computing technologies to readers and developers in the civil engineering communities. Subsequently, the framework for intelligent mobile structural inspection is proposed and described. Then this paper focuses on performance evaluation of the proposed MIC framework using a developed prototype application. A summary of conclusions of this paper is presented in the end.

MIC: DEVELOPMENT ENVIRONMENT AND ENGINEERING APPLICATIONS

Overview of Mobile Development

The advances of modern mobile imaging and computing have been tightly relied on two core technological advances that occurred during the last 20 years: mobile imaging sensors and mobile computing units. Enabled by the high-resolution imaging and high-performance computing power that are built into smart mobile devices, it is possible to use these mobile devices as alternative to professional equipment like camera and workstation. Prosperous research endeavors have also been witnessed in an emerging field, namely, mobile visual computing [20], which is closely related to the novel endeavor in this paper. To develop mobile visual computing applications, the first primary task is to select mobile operating systems (OS), which provide the vital connections between imaging, computing, user interface, and integration of auxiliary sensor data. Two of the mainstream mobile operating systems are Android and iOS [21]. For most smart apps to date, therefore, usually at least two versions are released—one for Android platforms and the other for iOS platforms. In this paper, the Android system is selected as the primary development mobile OS.

Android is open source and Google releases the code under the Apache License. It includes a set of C/C++ libraries used by various components of the Android system. It also includes a set of core libraries that provides most of the functionality using the Java programming language. Every Android application runs in its own process, with its own instance [22]. For application developers, Google Inc. has provided Android software development kit (SDK). The SDK includes a comprehensive set of development tools, including a debugger, software libraries, a handset emulator, documentation, sample code, and tutorials.

OpenCV and Mobile Visual Computing

Most applications on Android platform are written in Java programming language, which can satisfy most mobile and interactive applications. For serious visual computing tasks, however, efficient image and vision libraries are entailed. Modern imaging and vision libraries have been developed for solving various visual computing problems. These libraries include VXL, VLFeat, RAVL, OpenSURF, CImg, ImageJ, and BoofCV. These libraries mainly aims at solving specific problems like feature extraction, classification, or simple image I/O operation. MATLAB-based computer vision toolbox is another tool that can be used for imaging computation, visualization, and programming.

Among these vision libraries, OpenCV is a popular one [23, 24], which provides both computer vision and machine learning programming functions mainly aimed at real-time image processing and understanding. Compared to other libraries mentioned earlier, OpenCV provides the most comprehensive set of optimized algorithms. It has C++, C, Python, and Java interfaces and supports multiple platforms like Windows, Linux, Mac OS, and now Android systems. Figure 2 shows the OpenCV architecture design. Compared to other vision libraries, OpenCV is optimized for mobile platform in real-time processing.

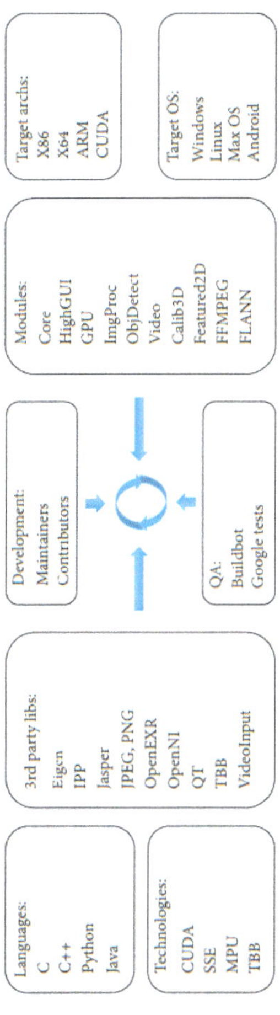

Figure 2: OpenCV architecture and library components.

Among the components in the architecture, HighGUI allows users to interact with the operating system, the file system, and the imaging hardware such as cameras and touch screens. It allows a simple way to query a camera and retrieve the latest image from the camera. The file system focuses on loading and saving images. HighGUI uses an abstract method to load and save images and videos without concerning the particular device. Hence, reading and saving images will only be one line of code in an Android-enabled device. Besides HighGUI, OpenCV library provides over 2500 optimized algorithms that operate on images to accomplish specific tasks, which include basic methods for image enhancement, modification, and transformation and advanced methods for image matching, segmentation, motion tracking, and camera calibration.

Engineering Applications

A plethora of mobile imaging and computing applications has been developed in many areas, such as robotics, automation, medical informatics, and engineering [14, 25–33]. One interesting application is recognition of 2-dimensional quick response (QR) codes, a ubiquitous coding and marking utility to date. In these efforts, mobile devices capture the images of QR patterns and identify in real time [32, 33]. Another noteworthy work by Hull et al. argues that mobile imaging and computing are the most desirable architecture for a series of mobile image recognition problems compared to a mobile-server computing architecture. They further reported a mobile imaging and computing experiment for automatic recognition of articles, photos, and advertisements in newspapers using a mid-range smartphone, in which the average run time for an image is 1.2 secs and the recognition error rate is less than 1% [34]. This paper indicates with concrete evidence for the great potential of mobile imaging and computing when today's smart devices are used.

In civil engineering, mobile computing has been recognized as an important utility for realizing efficient information sharing, design, and construction collaboration. Nagayama et al. built a video-based application with simultaneous acceleration recordings for road condition assessment [35]. By using accelerometer, GPS, and magnetometer sensors on smartphones, a nonintrusive method was introduced by Bhoraskar et al. to detect city traffic and road condition [36]. Combining accelerometer, gyroscope, magnetometer, GPS, and camera on mobile devices, researchers also developed a system to classify and recognize driving styles [37]. Behzadan et al. exploited the location service of mobile devices for realizing context-aware delivery in construction sites [38]. A wearable system for onsite building information modeling and sharing was recently developed through integrating mobile phones and displaying devices for visual analytics [39].

Among these applications, the notation of MIC applications based on the state-of-the-art mobile technologies has four core features: (i) portable generation of digitalized images via mobile imaging sensors, (ii) contextual integration with human-device interface (e.g., touch screens) and other auxiliary sensor data, (iii) real-time or near-real-time computing and analysis using mobile imaging and contextual data, and (iv) deliverable products through mobile communication networks. These features, either all or part of them, are manifested in the previously reviewed mobile applications. It is noted that, however, MIC applications in civil engineering reviewed before are in general lacked behind in terms of applying the state-of-the-art technologies. Among these applications, image-centric in situ computing in the mobile ends (e.g., image analysis-based recognition or information extraction) has not been emphasized due to the limited capability of earlier-generation mobile devices or their specific purposes of applications.

INTELLIGENT MIC METHODOLOGY FOR STRUCTURAL INSPECTION

The research problem in this paper is to conduct intelligent structural damage inspection in a complex or hard-access built environment through exploiting the power of mobile imaging and computing technologies. In the following, an intelligent mobile imaging and computing framework for structural inspection is proposed.

A novel concept is proposed to build this framework, termed engineer-imaging-computing (EIC) unit, which consists of an engineer with his or her equipment in the work zone. As illustrated in Figure 3, each EIC unit has two basic elements, which are the engineer and mobile devices. It assumes that an EIC unit is equipped with at least one Android-based mobile device. Besides using mobile devices as the basic imaging equipment, digital cameras can be used. A basic requirement for regular digital cameras is that they can be wirelessly tethered to the mobile computing device.

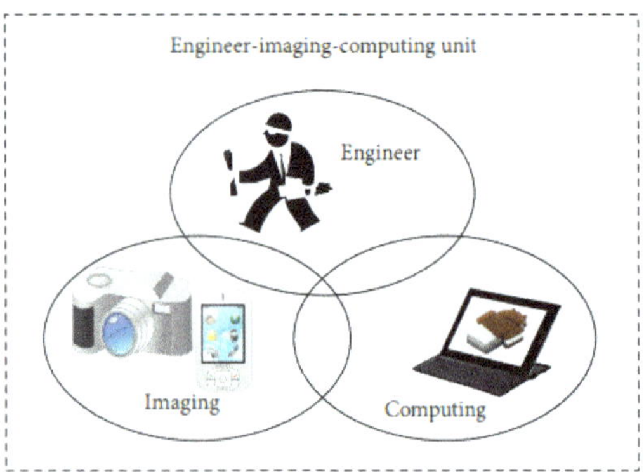

Figure 3: The structure of an engineer, imaging, and computing unit in the field.

Build upon an EIC unit, Figure 4 illustrates the intelligent mobile imaging and computing framework, which emphasizes its operation and data flow. The term of "intelligent" has two meanings relevant to the framework: (1) the potential human-centered interaction between the human and the devices that is based on the human's engineering knowledge and (2) the interactive and context-enabled imaging and computing within an EIC unit.

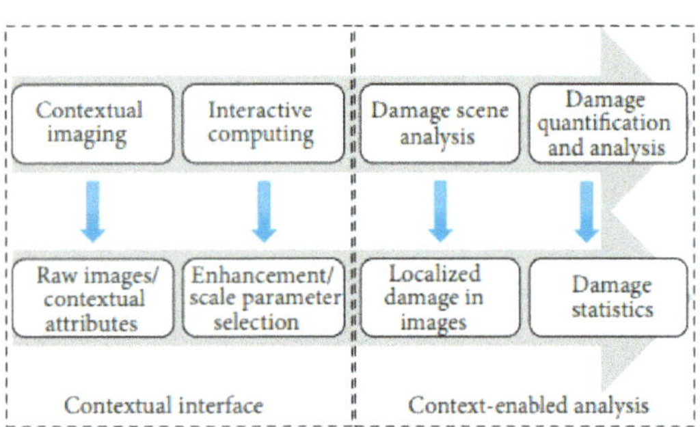

Figure 4: MIC framework for structural inspection: the operation flow (above) and the correspondingly generated dataflow (bottom).

Correspondingly, the operation and dataflow are divided into two components and four functions: (1) contextual interface including the functions of contextual imaging and interactive computing and (2) context-enabled analysis, including the functions of damage scene analysis, quantification, and analytics. Details in fulfilling these two primary function components are presented as follows.

Contextual Interface

The imaging activity in the field includes image and video capturing. This is considered the basic sensing activity in the proposed framework. Contextual interface implies that besides the basic imaging function, an information augmentation process by recording or computing key contextual information is associated with the imaging activity. The contextual information herein has three categories: (1) engineering knowledge that is related to damage scenes, (2) embedded sensor-generated data accompanying the imaging activity, and (3) interactively obtained image processing parameters key to subsequent advanced analysis. This contextual information along with raw or processed imagery data forms complex imagery metadata.

In Table 1, engineering knowledge related and embedded sensor-generated contextual attributes and their values are listed. In the current design, this contextual information in terms of attributes values is implemented and integrated with the imaging interface. For embedded sensor-generated information in Table 1, no user input is needed.

Table 1: Contextual information examples: engineering damage related empirical and embedded sensor-generated information

Category	Contextual attributes	Values (examples)
Engineering damage related information	Structure Type	Building, bridges, and pavement
	Structural element	Column, pier, beam/girder, wall, foundation,
	Structural material	Concrete, wood, steel, masonry, asphalt, soil, and sand
	Damage pattern	Line type: crack, void, and delamination Texture type: pothole, spalling, corrosion, and debris
	Scene complexity	Humidity, vegetation, oil-spill, and dirt-contaminated
Embedded sensor-generated information	Geolocation	Latitudes and longitudes
	Ambient light	Ambient light value
	Acceleration	Motion and gravity acceleration

For engineering knowledge-related information, it is obvious that user input is a straightforward approach to obtaining their values. It is noted that artificial intelligence-enabled automatic recognition is a possible solution that may automatically recognize, for example, the material or damage pattern types given the imagery information. However, we recognize that a fully automatic solution is a difficult task if it is charged to function at any imaging condition for any scene of complexity, considering the objective of recognizing structural element, material types, and damage types at the same time. For simple concrete surface and crack damage recognition, this may be feasible. One may refer to the first author's earlier work, where a feature extraction and neural network-based approach was successfully implemented [40]. On the other hand, if this accurate contextual engineering information is provided for complex damage scenes, it will dramatically improve the robustness and efficiency of the subsequent damage analysis and quantification function.

Another important category of contextual information is image analysis parameters. The interactive computing function implies that human-based gesture and input (e.g., typing or voice-based input) can be used to provide key contextual parameters for advanced image analysis parameters. In the proposed framework, these interactive parameters include important values that are important to the preprocessing steps of the following automatic image-based damage analysis: (1) bounding box parameters for initial damage boundaries and (2) image processing parameters for line- or region-damage type detection. As will be implemented in the next section, this combined approach, integrating engineering knowledge-based context logging and interactive bounding box/processing parameter selection, solves the generally hard visual understanding problem (e.g., recognition and localization in natural images [41]).

Context-Enable Damage Analysis

Given the captured imagery data and the associated contextual values, context-enabled damage analysis is implemented. First, several segmentation and edge or contour detection methods are implemented in the framework for damage scene analysis, which usually results in location of certain damage (e.g., crack or spalling). Again as indicated earlier, no damage recognition is needed herein since it is provided by the contextual input at the imaging interface (Table 1). Therefore, based on this damage type information, either a regional damage detection (e.g., for spalling, pothole, etc.) method or linear damage types (e.g., for cracks) will be initiated automatically to perform the damage localization task.

When location of damaged patterns is determined, which may be in terms of lines or contours in the image domain, geometric parameters summarizing the damage can be easily calculated. However, one should be aware that the imaging spatial domain is only a 2D linear projection of the actual 3D physical object subject to nonlinear imaging distortion. For lens-induced distortion, a camera calibration procedure should be applied before initial imaging [42]. However, we note a special situation that if a planar damage pattern (which may lie in a geometrically complex 3-dimensional object) is parallel to the imaging plane of lens, then the extracted damage patterns in images are approximately only subject to a constant geometric scaling (after distortion correction).

PROTOTYPE IMPLEMENTATION AND DEMONSTRATION FOR CRACK ANALYSIS

Prototype Implementation

Based on the proposed MIC methodology, a prototype application for intelligent structural inspection is developed. Android-based smart devices and systems are chosen for the development environment. The interface of the front-end mobile application is implemented using Android software development kit (SDK) (version 4.1), and the underlying imaging and computing functions are realized through the OpenCV library (version 2.4). Figure 5 shows the human-device interface. One can see that the interface is designed to realize three basic functionalities for realizing: (1) real-time imaging, (2) image analysis parameter selection, and (3) engineering knowledge-based attributes logging. Besides, geospatial location service is used to save the latitude and longitude coordinates when an image is captured, and acceleration values and ambient lighting values are recorded in the background. This interface design collectively realizes the proposed contextual interfaceillustrated in the framework (Figure 5).

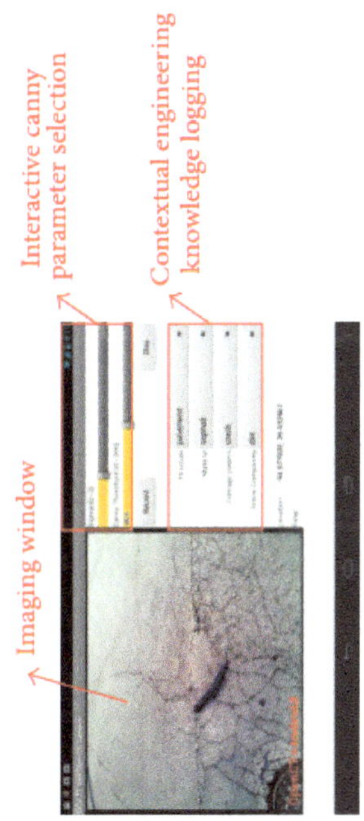

Figure 5: Mobile interface for contextual imaging of structural damage.

Implementation for Mobile Imaging and Computing for Crack Analysis

Crack detection is one of the most common tasks in structural damage inspection activities. The tradition usually follows a routine, in which, first, images are captured using digital cameras, and then digital processing by means of a robust edge detector is performed after the field inspection. Utilizing the proposed intelligent MIC framework and the developed interface (Figure 5), a full implementation of the MIC framework for structural crack analysis is realized in this paper. The working protocol using the developed prototype application is shown in Figure 6.

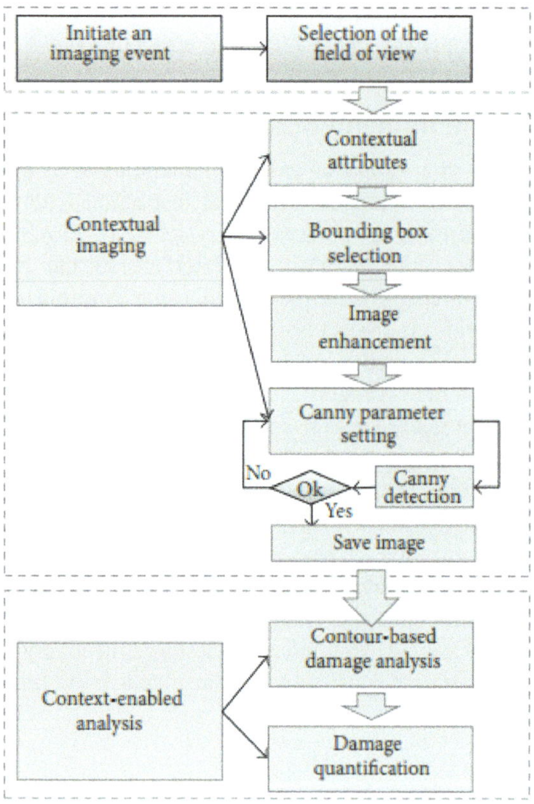

Figure 6: Mobile imaging and computing protocol for a typical crack scene analysis.

In this working protocol, an engineer starts an imaging event by empirically selecting a proper field of view. In the following, technical details regarding how to implement these steps are given below.

(1) Contextual Engineering Attributes Logging. Currently this is realized through touchscreen gesture-based dropdown menu, which is implemented using Android' SDK. Nonetheless, we envision that future development can integrate speech-based logging based on natural language processing libraries (e.g., [43]).

(2) Bounding Box Selection for Damage Localization. Damage patterns in captured images are usually located in one portion of the images. The bounding box envelopes the damage scene providing an approximate manual segmentation of the damage pattern of interest. Therefore, quality damage detection implies that localization of the overall damage pattern is an important initial step. Such localization usually calls for computationally demanding segmentation algorithms. Advanced contour algorithms can be found in the literature (e.g., deterministic level-set or statistical texture-modeling methods) [44, 45]; however, these methods usually entail considerable parameter tuning or training data preparation and high computational demands. Within the imaging window, this is interactively realized first using Android's SDK to obtain the coordinates of the touch gesture within the real-time imaging window; then the OpenCV functions "boundingRect" and "drawContour" are used to create a bounding box, respectively.

(3) Automatic Image Enhancement. In our experiment, this image enhancement step is an important preliminary processing that improves the latter edge detection and contour analysis. The image enhancement is executed automatically for each input true-color image. The process has three steps. First, the image is converted into a gray-level intensity image then a distortion correction is applied using the existing camera correction matrix if it is available. Second, histogram equalization is used to enhance image contrast, which is a widely used low-level technique for image enhancement. Histogram equalization is a nonparametric nonlinear filtering, which stretches out the pixel intensity range of an image by mapping the original histogram distribution to a more uniform distribution of intensity values [46]. The OpenCV function "equalizeHist" is used.

A standard mean-shift filter is applied to further smooth out the image noises and in the meantime to preserve the discontinuity (e.g., edges, lines, and other linear transitions in the image domain). Mean-shift filtering is a widely used common technique for edge-preserving smoothing and image segmentation. A standard mean-shift filter needs one spatial and one range scale parameters, denoted by σ_s and σ_r, respectively. In this paper, constants are used to set the two scale parameters, $(\sigma_s, \sigma_r) = (3, 6)$, which are proved to be a proper setting for preserving edge/line features while moderately smoothing out noise texture for our image database [47]. Correspondingly, the OpenCV function "meanShiftFiltering" is used.

With the above setting, the combined image enhancement is implemented without user intervention, which runs in the background automatically for each image prior to advanced image analysis.

(4) Interactive Canny Edge Parameter Selection. Common surface damage patterns (e.g., line-like cracks or regional damage with a clear boundary) are generally easy to detect using edge detectors. However, edge detection usually needs an appropriate scale factor for defining the detector's filtering kernel size (σ). For a particular detector, for example, the Canny detector, a hysteresis threshold (α) is further needed [48] (note that the Canny function in OpenCV uses two hysteresis threshold parameters, the upper and the lower one; herein, α denotes that the upper threshold is used, and the lower threshold is obtained as $\alpha/2$). Automatic selection of these parameters in general is a challenging problem. Many advanced or automatic scale parameter selection procedures exist [49, 50]. However, these methods only yield statistically "objective" edge detection results, which are still subjected to visual approval and inferior to expert-eye visual result. In light of these, interactive and manual parameter selection for obtaining the smoothing scale parameter (σ) and the hysteresis thresholding parameter (α) for the Canny detection is implemented in the prototype using the Android SDK functions and the selected parameters are fed to the OpenCV function "Canny" [48].

In the implementation, default values for both σ and α are used, which are 2.0 and 40%, respectively, as shown in Figure 5. Our field-based experience is that in most cases, the two values suffice the need; therefore, the users may not need to change them for clearly defined cracks in the images. In case of necessity, the user can first choose to modify the smoothing factor (σ) to a visually suitable detection result. In most cases, the default hysteresis threshold value (α) does not need to modify significantly. This setting largely reduces the interactive time cost.

(5) Contour Analysis and Damage Quantification. After saving the enhanced image and key contextual information from the previous steps, the subsequent image scene analysis and damage quantification are automatic. Contour analysis can be realized using advanced geometric modeling (e.g., [51, 52]). In the current implementation, a simple contour analysis algorithm is adopted, which computes contours based on any binary imagery [53]. Hence, the OpenCV function "findContours" in the OpenCV library is used. In our implementation, the binary output from a Canny detector is used to generate binary inputs. Therefore, the same scale and threshold parameters interactively selected previously are used in the contour analysis. It is noted that when the Canny detection returns 1-pixel wide line artifacts, the contour analysis outputs lines as well. We particularly indicate that in reality if the crack width is small, user can take photos at a closer distance to avoid this situation.

One merit of using this simple contour analysis is that many structural contour-based calculation functions are abundant, which are ready to output important geometric parameters for detected contours, such as area and perimeter. These geometric parameters provide important statistics for summarizing the damage quantitatively.

It is noted that if the camera is positioned parallel to the surface, there will be only a scale factor that accounts for the geometric correction between the image domain and the actual damage scene. Considering other complex situations, we will suggest to use a check board (which should be used for image calibration first) in the imaging field. This check board will be used to extract the projective matrix for the imagery domain, which leads to correct transformation of the images and, hence, accurate damage extraction (such as crack width, [6]). This camera calibration and image correction study is beyond the scope of this paper.

Concrete Crack Imagery Database

To evaluate the performance of this MIC methodology and the application prototype, a large database of concrete crack images is established. The images are gathered from two ways. First, we searched the Internet with certain key words and filters and obtained 140 concrete surface and crack images (which are free to be used by Google copyright terms [54]). Second, the developed prototype application was in two Android-based smart devices to capture road surface crack images, which resulted in an imagery database with 110 images. The two resulting imagery databases are also found in our research website as an open-source project for community-based development [55]. Figures 7(a) and 7(b) show part of the image samples, one in term of an image gallery and the other in terms of searchable images within a Google map, respectively. It is noted that both databases possess images with complex scenes.

(a) (b)

Figure 7: Concrete crack imagery database: (a) a database based Internet searching and (b) a database from field-based imaging using the developed application.

Experimental Results

Effectiveness of Context-Enabled Visual Analysis

To demonstrate the effectiveness of context-enabled visual analysis due to the contextual logging and interactive selection of key analysis parameters, Figure 8 illustrates an example wherein an image with a complex scene is selected. In Figure 8(a), one can see that the damage of interest is a crack in the foreground that is discolored and contaminated by humid. Without selecting a bounding box and visually selected Canny parameters, the contour analysis will produce a result as shown in Figure 8(b). When a bounding box is preselected, contour analysis will be focused on the region of interest wherein an elongated crack is found; however, many noisy contours are found in the result (Figure 8(c)). In Figure 8(c), a relatively optimal parameter setting resulting from the interactive parameter selection using a touched-based smartphone is applied. One can see that visually satisfactory crack detection result is obtained within the bounding box. As such, simple postprocessing by summarizing the geometric shapes of contours will produce much more accurate crack statistics (Figure 8(d)).

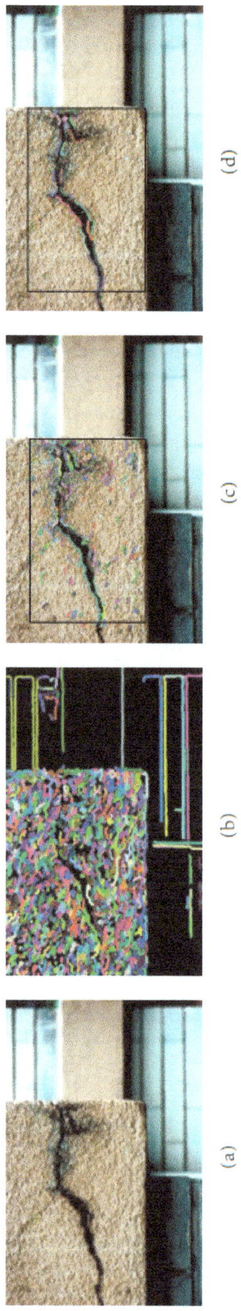

Figure 8: Demonstration of a mobile imaging and computing-based crack damage analysis: (a) original true-color image, (b) contour analysis without a bounding box or optimal Canny parameters($\sigma = 1.0$; $\alpha = 10$), (c) analysis with a bounding box without optimal Canny parameters, and (d) analysis with a bounding box and optimally selected parameters ($\sigma = 2.0$; $\alpha = 40$).

Context-Enabled Visual Analysis and Damage Quantification

Table 2 reports the results of 10 cases of context-enabled visual analysis and damage quantification results (the first five images are from the Internet search-based database and the latter five are based on the field experiment, and all images are resampled to have a width of 256 with a variable height). Besides the ultimate contour analysis output, we report one intermediate result—Canny edge detection result with optimal interactively selected parameters and a selected preliminary bounding box. Based on the subsequent contour analysis, simple damage quantification is conducted. It is noted that without considering a scale/dimension conversion, the reported crack lengths and areas (Len, Area) in the able are in terms of pixels and squared pixels, respectively.

Table 2: Context-enabled image analysis and quantification for concrete crack damage

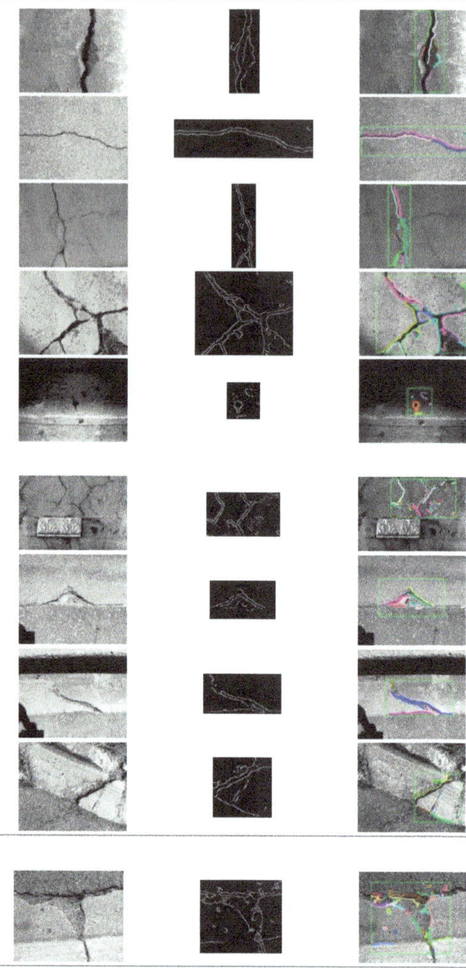

Number	Raw image	Edge detection	Contour analysis	(Len, area)
1				(564, 84)
2				(559, 51)
3				(538, 105)
4				(1330, 379)
5				(129, 30)
6				(724, 211)
7				(369, 95)
8				(435, 89)
9				(483, 132)
10				(1342, 328)

The images shown in Table 2 manifest scene complexity of different levels. For some images, for example, Images numer 2 or number 4, bounding box may be not necessary due to the less noisy image background and the clearly developed crack feature, or because the captured damage fills the image. For many other images, for example, Images number 6, 8, and 10, a bounding box is necessary to define an initial boundary of the damage. Contour analysis results provide damage mapping in terms of either closed contours or single edges. As mentioned earlier, the contour analysis results actually return coordinates of detected contours, which enable fast computation of geometric parameters of the damage. The last column of Table 2 reports the summed lengths and areas of detected damage for all the images. These statistics, when translated into real units (through a simple geometric calibration process), provide key decision-making aids during field damage assessment.

Computational Performance Evaluation

One factor that is critical to success of mobile imaging and computing for improved and intelligent structural damage inspection is its computational performance. To evaluate the performance of mobile imaging and computing, we differentiate two major operations in the proposed MIC framework as well as in the developed prototype. One is interactive parameter selection or engineering knowledge logging. The other is automatic image analysis running in the background of a smart device. For the former, there is no need for evaluating its computational efficiency, since it is part of user-engaged manual operation amid field-based decision-making. Therefore, the performance evaluation is focused on the image analysis components. Therefore, the most obvious and relevant task is that whether mobile computing can encourage or discourage real-time decision-making (such as onsite condition assessment) in field inspection activities. The major parameter is time consumption for a particular computing task. Then the primary criterion is to justify that if the time cost is less than 1.0 sec—if any analysis is greater than 1.0 sec then users may feel delay of operation which discourages real-time decision-making.

In the following, we conduct a comparative study by comparing the computational performance of the image analysis methods when implemented using a smart device and a workstation. In this study, two representative computing platforms are used. The smart mobile device is the Google's Nexus 7 tablet, which features an Nvidia 1.6 GHz quad-core CPU, 1 G RAM, and an Android 4.3 operating system. The workstation is a Linux computer featuring Intel Core 2 Duo CPU E8400 with 3 GHz frequency, 4 G RAM, and Ubuntu Linux 12. The two visual analysis methods are implemented in the prototype as presented earlier.

(1)Image enhancement includes the conjunct use of histogram equalization and mean-shift filtering. However, we note that the method of histogram equalization (the function "equalizeHist" in OpenCV) is a fairly low-cost algorithm, which processes images regardless of regular image sizes with time negligible to users (for images with sizes up to 576 × 720, the time consumption is about 0.3×10^{-3} to 0.9×10^{-3} sec, or 1 to 9 millisec using the workstation, and about 1.5 to 5.0 millisec using the smart tablet). On the other hand, the time-cost and efficiency of the mean-shift method highly depend on image sizes. Considering the previous range of image sizes, the time cost of mean-shift filtering may be up to the order of one second. Therefore, in this paper, only the mean-shift filtering (using "meanShiftFiltering" in OpenCV with $\sigma_s = 3$ and $\sigma_r = 6$) is evaluated for both computational platforms.

(2) Contour analysis uses an embedded Canny detector for binary edge generation. Due to the large number of images, optimal Canny parameters are not sought from the gesture-based interactive selection using the interface shown in Figure 5. Instead a standard Canny parameter set (σ, α) is used for the contour analysis in both platforms ($\sigma = 2$, $\alpha = 40\%$). In addition, bounding box is not used; rather, the contour analysis is conducted over the total image domain. For contour analysis, two performance measures are used: one is the time cost and the other is the number of contours. The rationale for outputting the number of contours is that it is a better indicator of the underlying scene complexity in the image.

The performance evaluation is tested on the imagery database based on the Internet search. The rationale is that, first, image size is obviously the primary factor that affects the computational time; second, images taken in the field vary in sizes when different imaging devices are used or user may choose different resizing ratios to decrease the image sizes. The Internet-searched images with different contents vary in sizes from 200 × 300 (about 0.05 Megapixel) to 1024 × 1024 (or 1 Megapixel). Therefore, this imagery database serves as an appropriate testing database. Seventy images are chosen and tested on the two platforms for both the image analysis procedures above. In the meantime, since a mean-shift filter uses an iterative procedure to locate local maxima of density functions, it is possible that when image sizes are fixed, image contents may affect the computational time. Therefore, the need of performing a variability evaluation of time cost using the same sized images with different scene complexities is warranted.

Figure 9 presents the scatter plots of the running time from the image enhancement procedure (i.e., using the mean-shift filtering method) against the image size of input images, resulting from the smart tablet and the workstation, respectively. First of all, one can see that the image sizes in terms of mega pixels statistically predict well the time consumption. Comparing between the tablet and the workstation, the regression analysis returns that the efficiency of using the smart tablet is about 1376 millisec/Megapixel, whereas it is 625 millisec/Megapixel for the workstation. The tablet is about 45% of the efficiency of the workstation, when the mean-shift filtering function is used. This is not a surprising result considering the lower-end computing hardware configuration of the mobile device. In the meantime, when small to moderate-sized images are used, for example, 480 × 320 (or 0.15 mega pixels), the time consumption from Figure 9 is about 200 millisec if the smart device is used.

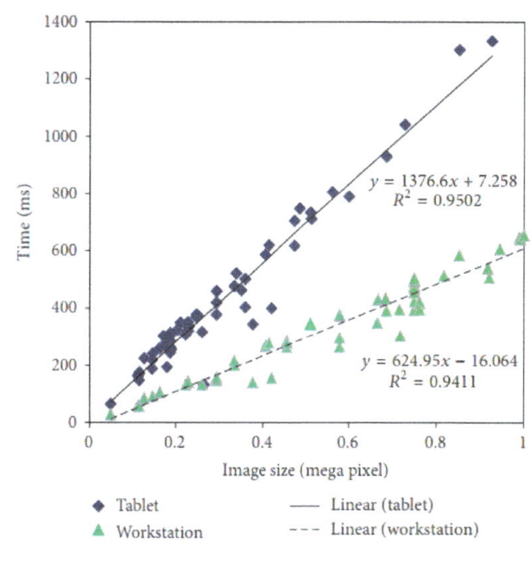

Figure 9: Computational time for image enhancement with different image size.

To further evaluate the potential variability of time consumption on image contents, Figure 10 illustrates the time consumption results of using two different sizes of images representing small to moderate sizes. When images with a size of 480 × 320 are used, the image enhancement time ranges from 200 to 290 milliseconds, with an average value of 250 milliseconds and a standard deviation of 14.4 milliseconds. For images of 720 × 576 (0.4 mega pixels), the running time is 600 to 780 milliseconds with an average value of 710 milliseconds and a standard deviation of 36.7 milliseconds. This experiment shows the overall consistency of predicting time consumption using a fixed image size. Therefore, it is ready to reason that if an image has a small to moderate size, the time cost is statistically less than 800 millisec (or 0.8 sec) using the smart tablet. This statistical study indicates that performing the proposed low-level image enhancement does not prevent the overall interactive imaging and computing as well as real-time decision-making.

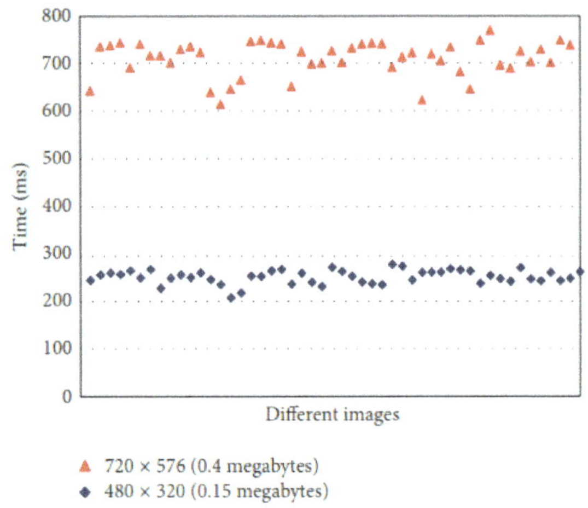

Figure 10: Computational time variability for fixed-size images for image enhancement in the tested smart tablet.

For the damage quantification analysis using the contour detection method, Figure 11 illustrates the running time against the image size. However, in this case, the image size is not a good predicator of the potential running time reflected by the considerable scatterings in the plot. Nonetheless, it seems that when the image size is less than 0.4 Megapixels, the contour analysis costs less than 300 millisec when the tablet is used. This is a promising result in light of realizing real-time imaging and computing. In addition, one can observe with a weak statistical significance that the workstation is about 10 times faster than the tablet (based on the first-order polynomial coefficients of the 2nd order regression results).

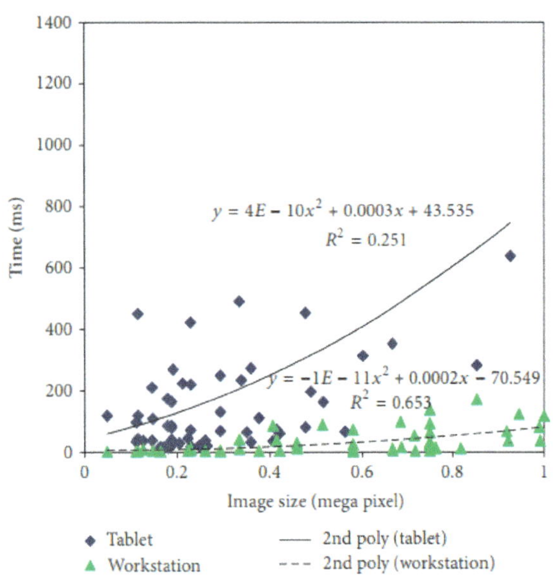

Figure 11: Computational time for contour analysis with different image size.

It is found that the number of contours in the images, which is a direct measure of the scene complexity in the image domain, is a much better predictor of the time consumption. In Figure 12, the scatter plots of time consumption and number of contours are plotted. Using second-order polynomial fitting, the underlying plots yield much improved statistical significance (R^2>0.98 in both cases). The plots indicate that it is the number of contours not the image size that can accurately predict the time consumption. First, the difference between the smart tablet and the workstation is obvious. For the workstation, the time cost relatively slowly grows against the number of contour (or equivalently the scene complexity), whereas in the case of the tablet, the time cost increases rapidly as the number of the contours increases. For example, regardless of the image size, if the number of contours reduces from 300 to 100, the time cost changes from 600 millisec to about 60 millisec using the tablet. This difference is due to the underlying different hardware and software configurations. When the number of contours is less than 100, one can reason that the proposed contour analysis will not impose significant computational hurdle from realizing real-time imaging, computing, and decision computing. This observation implies that edge/line-preserving smoothing should be seriously considered to remove noisy contours prior to any advanced image analysis for damage quantification when a smart mobile device is used.

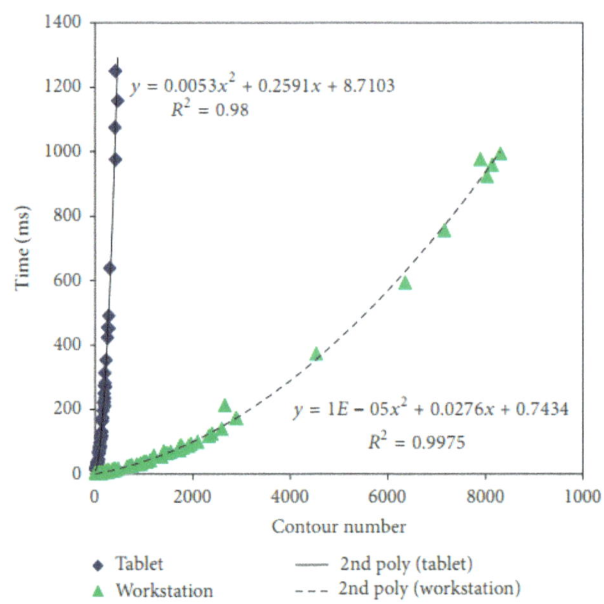

Figure 12: Computational time for contour analysis with different scene complexity.

CONCLUSIONS

The main purpose of this paper is to explore the potential of modern mobile imaging and computing for its use in structural damage inspection in the routine practice of civil infrastructure management. In this paper, we address the critical need that arises from performing intelligent and efficient inspection within a complex or hard-access build environment, wherein visual inspection dominates the practice.

Through developing a mobile imaging and computing prototype, the proposed mobile damage inspection is verified and evaluated in terms of both practical feasibility and computational performance for enabling real-time field-based decision-making. Particularly, the following technical conclusions are obtained using the state-of-the-art smart mobile device (e.g., the commercial Google Nexus 7 tablet).(1)Interactively logged and selected contextual parameters, including types of damage, preliminary image analysis parameters, and bounding boxes, are essential towards yielding quality damage detection and quantification. If these parameters are otherwise not available, significant analysis challenges will be encountered for processing structural damage patterns as complex natural scenes of the built environment.(2)In most applications, it is suggested that image sizes should be at a moderate level, for example, 576 × 720 (or 0.4 Megabytes). Such level of image sizes is considered appropriate for processing common damage patterns (e.g., cracks) using smart devices.(3)Typical image enhancement should be conducted and can be implemented in real-time in mobile devices while accompanying the imaging and computing activity during field inspection.(4)Complexity of image contents can be reduced through two operations: (i) through the selection of bounding boxes and (ii) through image smoothing such as the mean-shift filtering in this paper. The two operations can significantly reduce the number of contours thus enabling real-time damage analysis.With these technical conclusions, we finally conclude that mobile imaging and computing via today's smart devices possess tremendous potential for realizing intelligent and efficient structural damage inspection, and such smart applications are specially justified when the inspection areas are complex or hard to access.

Last, it should be pointed that for quantitative damage extraction using mobile devices is subject to more image-based development. In practice, the common concrete damage statistics include length and area, which are readily to be extracted using the methods in this paper. For crack, width is an important index as well. However, for crack width, it is a complex image analysis problem involving extraction of the central line of a crack. One approach has been proposed in the first author's previous work [6], which could be implemented for mobile smart devices. However, these potential quantitative extraction and methods are subject to field validation, which is not in the scope of this paper.

REFERENCES

1. F. N. Catbas and A. E. Aktan, "Condition and damage assessment: issues and some promising indices," Journal of Structural Engineering, vol. 128, no. 8, pp. 1026–1036, 2002.
2. B. R. Ellingwood, "Risk-informed condition assessment of civil infrastructure: state of practice and research issues," Structure and Infrastructure Engineering, vol. 1, no. 1, pp. 7–18, 2005.
3. M. Moore, B. Phares, B. Graybeal, D. Rolander, and G. Washer, Reliability of Visual Inspection for Highway Bridges, Volume I, Federal Highway Administration, 2001.
4. M. J. Olsen, Z. Chen, T. Hutchinson, and F. Kuester, "Optical techniques for multiscale damage assessment," Geomatics, Natural Hazards and Risk, vol. 4, no. 1, pp. 49–70, 2013.
5. C. Koch and I. Brilakis, "Pothole detection in asphalt pavement images," Advanced Engineering Informatics, vol. 25, no. 3, pp. 507–515, 2011.
6. Z. Chen and T. C. Hutchinson, "Image-based framework for concrete surface crack monitoring and quantification," Advances in Civil Engineering, vol. 2010, Article ID 215295, 18 pages, 2010.
7. S. R. Cruz-Ramírez, Y. Mae, T. Arai, T. Takubo, and K. Ohara, "Vision-based hierarchical recognition for dismantling robot applied to interior renewal of buildings," Computer-Aided Civil and Infrastructure Engineering, vol. 26, no. 5, pp. 336–355, 2011.
8. T. Nishikawa, J. Yoshida, T. Sugiyama, and Y. Fujino, "Concrete crack detection by multiple sequential image filtering," Computer-Aided Civil and Infrastructure Engineering, vol. 27, no. 1, pp. 29–47, 2012.
9. M. R. Jahanshahi and S. F. Masri, "Adaptive vision-based crack detection using 3D scene reconstruction for condition assessment of structures," Automation in Construction, vol. 22, pp. 567–576, 2012.
10. H. Cheng and C. Glazier, Automated Real-Time Pavement Crack Deflection/Classification System, IDEA Program, Transportation Research Board, 2002.
11. S. Ghanta, R. Birken, and J. Dy, "Automatic road surface defect detection from grayscale images," in Proceedings of the SPIE 8347, Nondestructive Characterization for Composite Materials, Aerospace Engineering, Civil Infrastructure, and Homeland Security, vol. 83471E, 2012.
12. C. Balaguer, A. Giménez, J. M. Pastor, V. M. Padrón, and M. Abderrahim, "Climbing autonomous robot for inspection applications in 3D complex environments," Robotica, vol. 18, no. 3, pp. 287–297, 2000.
13. J.-K. Oh, G. Jang, S. Oh et al., "Bridge inspection robot system with machine vision," Automation in Construction, vol. 18, no. 7, pp. 929–941, 2009.
14. S.-N. Yu, J.-H. Jang, and C.-S. Han, "Auto inspection system using a mobile robot for detecting concrete cracks in a tunnel," Automation in Construction, vol. 16, no. 3, pp. 255–261, 2007.
15. C. Eschmann, C.-M. Kuo, C.-H. Kuo, and C. Boller, "Unmanned aircraft systems for remote building inspection and monitoring," in Proceedings of the 6th European Workshop on Structural Health Monitoring (EWSHM '12), pp. 1179–1186, Dresden, Germany, July 2012.
16. S. Rathinam, Z. W. Kim, and R. Sengupta, "Vision-based monitoring of locally linear structures using an unmanned aerial vehicle," Journal of Infrastructure Systems, vol. 14, no. 1, pp. 52–63, 2008.
17. N. Metni and T. Hamel, "A UAV for bridge inspection: visual servoing control law with orientation limits," Automation in Construction, vol. 17, no. 1, pp. 3–10, 2007.
18. G. H. Forman and J. Zahorjan, "The challenges of mobile computing," Computer, vol. 27, no. 4, pp. 38–47, 1994.
19. Mobithinking, "Global mobile statistics 2013 Part A: mobile subscribers; handset market share; mobile

operators," 2013, http://mobithinking.com/mobile-marketing-tools/latest-mobile-stats/a.

20. K. Pulli, W.-C. Chen, N. Gelfand et al., "Mobile visual computing," in Proceedings of the International Symposium on Ubiquitous Virtual Reality (ISUVR '09), pp. 3–6, July 2009.

21. I. Gartner, "Gartner says worldwide sales of mobile phones," 2013, http://www.gartner.com/newsroom/id/2017015.

22. D. Ehringer, "The dalvik virtual machine architecture," Tech. Rep., 2010.

23. G. Bradski and A. Kaehler, Learning OpenCV: Computer Vision with the OpenCV Library, O'Reilly Media, 2008.

24. OpenCV, "Open Source Computer Vision," 2013, http://opencv.org/.

25. W.-H. Cho and T.-C. Kim, "Image enhancement technique using color and edge features for mobile systems," in Digital Photography VII, Proceedings of SPIE, San Francisco, Calif, USA, January 2011.

26. T. Yeh, K. Grauman, K. Tollmar, and T. Darrell, "A picture is worth a thousand keywords: image-based object search on a mobile platform," in Proceedings of the Extended Abstracts on Human Factors in Computing Systems (CHI '05), pp. 2025–2028.

27. R. Andrade, A. V. Wangenheim, and M. K. Bortoluzzi, "Wireless and PDA: a novel strategy to access DICOM-compliant medical data on mobile devices," International Journal of Medical Informatics, vol. 71, no. 2-3, pp. 157–163, 2003.

28. Y. Kondo, "Medical image transfer for emergency care utilizing internet and mobile phone," Nippon Hoshasen Gijutsu Gakkai Zasshi, vol. 58, no. 10, pp. 1393–1401, 2002.

29. Z. Tu and R. Li, "Automatic recognition of civil infrastructure objects in mobile mapping imagery using a markov random field model," in Proceedings of the 19th Congress of ISPRS, pp. 16–23, 2000.

30. C. Tao, R. Li, and M. A. Chapman, "Automatic reconstruction of road centerlines from mobile mapping image sequences," Photogrammetric Engineering and Remote Sensing, vol. 64, no. 7, pp. 709–716, 1998.

31. D. Wagner, G. Reitmayr, A. Mulloni, T. Drummond, and D. Schmalstieg, "Real-time detection and tracking for augmented reality on mobile phones," IEEE Transactions on Visualization and Computer Graphics, vol. 16, no. 3, pp. 355–368, 2010.

32. Y. Liu, J. Yang, and M. Liu, "Recognition of QR Code with mobile phones," in Chinese Control and Decision Conference, 2008 (CCDC '08), pp. 203–206, Yantai, China, July 2008.

33. A. Sun, Y. Sun, and C. Liu, "The QR-code reorganization in illegible snapshots taken by mobile phones," in Proceedings of the International Conference on Computational Science and Its Applications (ICCSA '07), pp. 532–538, Kuala Lampur, Malaysia, August 2007.

34. J. J. Hull, X. Liu, B. Erol, J. Graham, and J. Moraleda, "Mobile image recognition: architectures and tradeoffs," in Proceedings of the 11th Workshop on Mobile Computing Systems Applications, pp. 84–88, 2010.

35. T. Nagayama, A. Miyajima, S. Kimura, Y. Shimada, and Y. Fujino, "Road condition evaluation using the vibration response of ordinary vehicles and synchronously recorded movies," in Proceedings of the SPIE Smart Structures and Materials+ Nondestructive Evaluation and Health Monitoring, p. 86923A, March 2013.

36. R. Bhoraskar, N. Vankadhara, B. Raman, and P. Kulkarni, "Wolverine: traffic and road condition estimation using smartphone sensors," in Proceedings of the 4th International Conference on Communication Systems and Networks (COMSNETS '12), pp. 1–6, Bangalore, India, January 2012.

37. D. A. Johnson and M. M. Trivedi, "Driving style recognition using a smartphone as a sensor platform," in Proceedings of the 14th IEEE International Intelligent Transportation Systems Conference (ITSC '11), pp. 1609–1615, Washington, DC, USA, October 2011.

38. A. H. Behzadan, Z. Aziz, C. J. Anumba, and V. R. Kamat, "Ubiquitous location tracking for context-specific information delivery on construction sites," Automation in Construction, vol. 17, no. 6, pp. 737–748, 2008.

39. K. C. Yeh, M. H. Tsai, and S. C. Kang, "The iHelmet: an AR-nhanced wearable display for BIM information," in Mobile and Pervasive Computing in Construction, pp. 149–168, Wiley-Blackwell, 2012.

40. Z. Chen, R. R. Derakhshani, C. Halmen, and J. T. Kevern, "A texture-based method for classifying cracked concrete surfaces from digital images using neural networks," in Proceedings of the International Joint Conference on Neural Network (IJCNN '11), pp. 2632–2637, San Jose, Calif, USA, August 2011.

41. D. A. Forsyth and J. Ponce, Computer Vision: A Modern Approach, Prentice Hall, New York, NY, USA, 2002.

42. Z. Zhang, "A flexible new technique for camera calibration," IEEE Transactions on Pattern Analysis and Machine Intelligence, vol. 22, no. 11, pp. 1330–1334, 2000.

43. C. Sphinx, "Open source toolkit for speech recognition," 2011, http://cmusphinx.sourceforge.net/.

44. A. K. Jain and F. Farrokhnia, "Unsupervised texture segmentation using Gabor filters," Pattern Recognition, vol. 24, no. 12, pp. 1167–1186, 1991.

45. L. A. Vese and T. F. Chan, "A multiphase level set framework for image segmentation using the Mumford and Shah model," International Journal of Computer Vision, vol. 50, no. 3, pp. 271–293, 2002.

46. H. D. Cheng and X. J. Shi, "A simple and effective histogram equalization approach to image enhancement," Digital Signal Processing: A Review Journal, vol. 14, no. 2, pp. 158–170, 2004.

47. D. Comaniciu and P. Meet, "Mean shift analysis and applications," in Proceedings of the 7th IEEE International Conference on Computer Vision (ICCV '99), pp. 1197–1203, September 1999.

48. J. Canny, "A computational approach to edge detection," IEEE Transactions on Pattern Analysis and

Machine Intelligence, vol. 8, no. 6, pp. 679–698, 1986. ·

49. R. Koren and Y. Yitzhaky, "Automatic selection of edge detector parameters based on spatial and statistical measures," Computer Vision and Image Understanding, vol. 102, no. 2, pp. 204–213, 2006.

50. Y. Yitzhaky and E. Peli, "A method for objective edge detection evaluation and detector parameter selection," IEEE Transactions on Pattern Analysis and Machine Intelligence, vol. 25, no. 8, pp. 1027–1033, 2003.

51. S. Kichenassamy, A. Kumar, P. Olver, A. Tannenbaum, and A. Yezzi, "Gradient flows and geometric active contour models," in Proceedings of the 5th International Conference on Computer Vision, pp. 810–815, 1995.

52. V. Caselles, R. Kimmel, and G. Sapiro, "Geodesic active contours," International Journal of Computer Vision, vol. 22, no. 1, pp. 61–79, 1997.

53. S. Suzuki and K. be, "Topological structural analysis of digitized binary images by border following," Computer Vision, Graphics and Image Processing, vol. 30, no. 1, pp. 32–46, 1985. ·

54. Google, Find Content to Reuse, Google, 2013, https://support.google.com/websearch/answer/29508.

55. Z. Chen, "UMKC Concrete Damage Imagery Database," 2013, http://lasir.umkc.edu/cdid/.

CHAPTER 3

NONLINEAR DYNAMIC ANALYSIS OF PLATES STIFFENED BY PARALLEL BEAMS WITH DEFORMABLE CONNECTION

J. A. Dourakopoulos and E. J. Sapountzakis

INTRODUCTION

Stiffened plate panels are structural elements of practical importance in applications such as ship superstructures, bridge decks, and aircraft structures. Stiffening of the plate provides the benefit of added load-carrying capability with a relatively small additional weight penalty, while buckling is prevented especially in case of inplane loading. The unique peculiarities of the aforementioned structures are obtained due to the behavior of the bond between the plate and the beams; however, this bond is the usual reason why these structures are prone to failure. Moreover, since these stiffened plates are frequently subjected to dynamic loading such as air blasts or underwater explosions, a clear understanding of the dynamic response requires development of an efficient dynamic analysis capability.

When the deflections of the structure are small, a wide range of linear analysis tools, such as modal analysis, can be used and some analytical results are possible. As the deflections become larger, the induced geometric nonlinearities result in effects that are not observed in linear systems, making therefore the determination of an analytical solution extremely difficult and in most cases impossible. Moreover, having in mind the importance of weight saving in engineering structures, the study of nonlinear effects in the analysis (large deflection analysis) of stiffened plates becomes essential. These nonlinearities result from retaining the squares of the slopes in the strain-displacement relations (intermediate nonlinear theory), avoiding in this way the inaccuracies arising from a linear or a linearized second-order analysis.

The linear dynamic behavior of stiffened plates has been widely studied employing the Rayleigh-Ritz method [1–4], the transfer matrix method [5], the finite difference method [6–9], the finite element method [10–12], the finite strip method [13], and the boundary element method (BEM) [14–16]. To the authors' knowledge, a limited amount of technical literature is also available on the nonlinear dynamic analysis of stiffened plate systems. According to this, the integral equation method has been applied for the large deflection analysis of clamped laterally loaded skew plates with stiffener parallel to the skew directions [17]. Moreover, the finite element method has been widely used for the free large-amplitude flexural vibration of stiffened [18] and laminated plates [19, 20] and for forced vibrations under instantaneous loading taking into account both geometric and material nonlinearities [21]. The spline finite strip method has also been applied for the examination of nonlinear transient vibration of stiffened plates [22] and the Galerkin method has been employed for the solution of discrete equations derived after the application of Fourier series for the deflection of the plate [23, 24]. Nevertheless, in all of the aforementioned research efforts restrictions are encountered, arising from the ignorance of the tractions in the fictitious plate-beams interfaces, of the nonuniform distribution of the interface transverse shear force, or of the nonuniform torsional response of the beams.

The behavior of composite slab-and-beam structure is affected significantly by the deformability of the shear connection. Therefore, much research effort has been done concerning the linear dynamic analysis of composite beams

Citation: J. A. Dourakopoulos and E. J. Sapountzakis, "Nonlinear Dynamic Analysis of Plates Stiffened by Parallel Beams with Deformable Connection," Advances in Civil Engineering, vol. 2014, Article ID 942763, 22 pages, 2014. doi:10.1155/2014/942763.

with deformable connections. Wu et al examined the free vibration of partial interaction composite beams with axial force and proposed an approximate simple expression to predict the fundamental frequency of the partial-interaction composite members with axial force [26]. Xu and Wu investigated the static, dynamic, and buckling behavior of partial-interaction composite members by taking into account the influences of rotary inertia and shear deformations [27]. Moreover, they also employed the state-space method in order to develop an exact two-dimensional plane stress model of composite beams with interlayer slips [28, 29].

On the other hand, less research effort has been made on the forced vibration of partial-interaction composite beams. Girhammar and Pan [30] investigated the dynamic behavior of partial-interaction composite beams without axial force and presented the orthogonality relation of vibration modes. They also obtained the analytical expressions for the transient response of composite beams under impulsive and step loads. Adam et al. [31] decomposed the dynamic response into two parts, the quasi-static component solved by integrating the influence function and the complementary component solved by the mode superposition. Recently, Girhammar et al. [32] employed Hamilton's principle to derive the governing equations for dynamic problems of composite beams under generalized boundary conditions. At the same time, they studied the forced vibration of composite beams by virtue of mode superposition. However, in the literature, large deflection analysis of partial interaction has received limited attention and focused especially on the static behavior of the structure [33–38].

In this paper, a general solution to the geometrically nonlinear dynamic analysis of plates stiffened by arbitrarily placed parallel beams of arbitrary doubly symmetric cross-section with deformable connections subjected to arbitrary dynamic loading is presented. The solution is based on the structural model proposed by Sapountzakis and Mokos in [39] and Dourakopoulos and Sapountzakis in [40], according to which the stiffening beams are isolated from the plate by sections in the lower outer surface of the plate, making the hypothesis that the plate and the beams can slip in all directions of the connection without separation (i.e., uplift is neglected) and taking into account the arising tractions in all directions at the fictitious interfaces. Due to the slip, relative inplane displacements between beam and plate arise, assuming however that they are continuously in contact and taking into account the arising tractions in all directions at the fictitious interfaces. These tractions are integrated with respect to each half of the interface width resulting in two interface lines, along which the loading of the beams and the additional loading of the plate are defined. The unknown distribution of the aforementioned integrated tractions is established by applying continuity conditions in all directions at the two interface lines taking into account their relationship with the interface slip through the shear connector stiffness. Any distribution of connectors in each direction of the interfaces can be handled. The utilization of two interface lines for each beam enables the nonuniform torsional response of the beams to be taken into account as the angle of twist is indirectly equated with the corresponding plate slope. Six boundary value problems are formulated and solved using the analog equation method (AEM) [41], a BEM-based method. The essential features and novel aspects of the present formulation are summarized as follows.(i)The adopted model permits the evaluation of the inplane shear forces (longitudinal and transverse) at the interfaces in both directions, taking into account the influence of interface slip, the knowledge of which is very important in the design of shear connectors in plate-beam structures.(ii)Utilization of two interface lines permits the nonuniform distribution of the inplane forces along the interface width to be taken into account.(iii)Both free and forced transverse vibrations are considered taking into account geometric nonlinearities.(iv)The stiffened plate is of arbitrary shape and is subjected to arbitrary dynamic loading, while both the number and the placement of the parallel beams are also arbitrary (eccentric beams are included).(v)The cross-section of the stiffening beams is an arbitrary doubly symmetric thin or thick-walled one. The formulation does not stand on the assumption of a thin-walled structure according to which torsional and warping rigidities can be evaluated employing closed-form expressions without significant error as long as the cross-section thickness is small [42]. Therefore, in this paper, the cross-section's torsional and warping rigidities are evaluated "exactly" in a numerical sense employing the BEM (requiring only boundary discretization for the cross-sectional analysis).(vi)The plate and the beams are supported by the most general boundary conditions including elastic support or restraint.(vii)The nonuniform torsion in which the stiffening beams are subjected is taken into account by solving the corresponding problem and by comprehending the arising twisting and warping in the corresponding displacement continuity conditions. The distributed warping moment arising from the nonuniform distribution of longitudinal inplane forces is also taken into account.(viii)Contrary to previous research efforts where the numerical analysis is based on BEM using a lumped mass assumption model after evaluating the flexibility matrix at the mass nodal points, in this work, a distributed mass model is employed.

Based on the numerical solution developed, several examples with great practical interest are presented, demonstrating the effectiveness, wherever possible, the accuracy, and the range of applications of the proposed method.

STATEMENT OF THE PROBLEM

Let us consider a thin plate of homogeneous, isotropic, and linearly elastic material with modulus of elasticity E_p, volume mass density ρ_p, shear modulus G_p, and Poisson's ratio]p having constant thickness h_p and occupying the two-dimensional multiple connected region Ω of the x_y plane bounded by the piecewise Γ_j ($j = 0, 1, 2, \dots, K$) boundary curves. The plate is stiffened by a set of $i = 1, 2, \dots, I$ arbitrarily placed parallel beams of arbitrary doubly symmetric cross-section of area A_b^i and length l^i. The material of the beams is considered to be homogeneous, isotropic, and linearly elastic with modulus of elasticity E_b^i, mass density ρ_b^i, shear modulus G_b^i, and Poisson's ratio ν_b^i. Therefore, material nonlinearities are not considered in the following analysis.

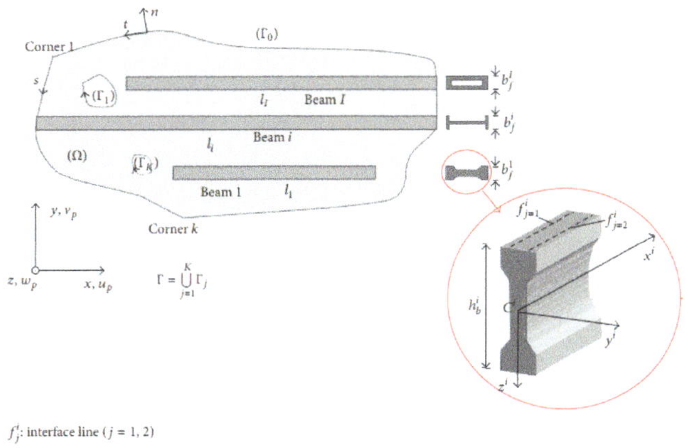

Figure 1: Two-dimensional region Ω occupied by the plate.

For the sake of convenience the x axis is taken parallel to the beams of length $l^{\,i}$ which may have either internal or boundary point supports. The stiffened plate is subjected to the arbitrary lateral dynamic load $g = g(\mathrm{x}, t)$, x : $\{x, y\}$, $t \geq 0$. Owing to this loading, the plate and the beams can slip in all directions of the connection without separation (i.e., uplift is neglected). For the analysis of the aforementioned problem, a global coordinate system Oxy for the analysis of the plate and local coordinate ones $C^i\,x^i\,y^i$ corresponding to the centroid axes of each beam are employed (Figure 1).

The solution of the problem at hand is approached employing the improved model proposed by Sapountzakis and Mokos in [39] and Dourakopoulos and Sapountzakis in [40]. According to this model, the stiffening beams are isolated from the plate by sections in its lower outer surface, taking into account the arising tractions at the fictitious interfaces (Figure 2). Integration of these tractions along each half of the width of the ith beam results in line forces per unit length in all directions in two interface lines, which are denoted by q^i_{xj}, q^i_{yj}, and q^i_{zj} $(j = 1, 2)$, encountering in this way the nonuniform distribution of the interface transverse shear forces q^i_y, which in previous models [43] was ignored. The aforementioned integrated tractions result in the loading of the ith beam and the additional loading of the plate. Their distribution is unknown and can be established by imposing displacement continuity conditions in all directions along the two interface lines, enabling in this way the nonuniform torsional response of the beams to be taken into account. Thus, the arising additional loading (Figure 3) at the middle surface of the plate and the loading along the centroid axis (coinciding with the shear center axis), of each beam, can be summarized as follows.

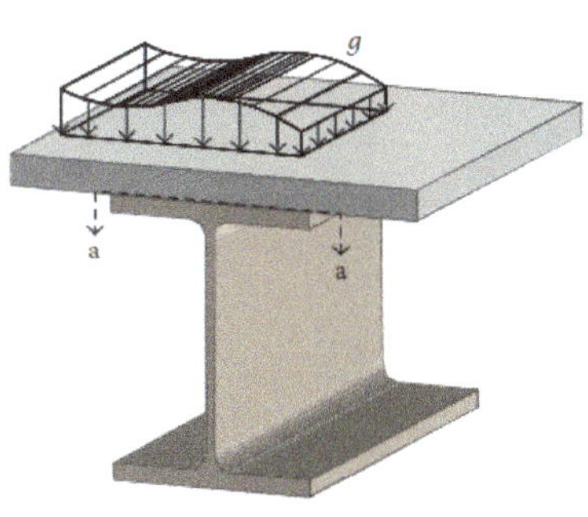

(a)

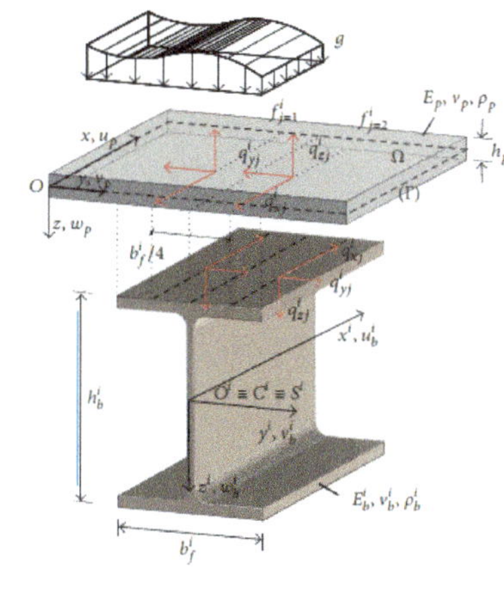

(b)

Figure 2: Thin elastic plate stiffened by beams (a) and isolation of the beams from the plate (b).

(a) In the Plate (at the Traces of the Two Interface Lines j = 1, 2 of the ith Plate-Beam Interface)

(i) An inplane line body force q^i_{xj} at the middle surface of the plate.

(ii) An inplane line body force q^i_{yj} at the middle surface of the plate.

(iii) A lateral line load q^i_{zj}

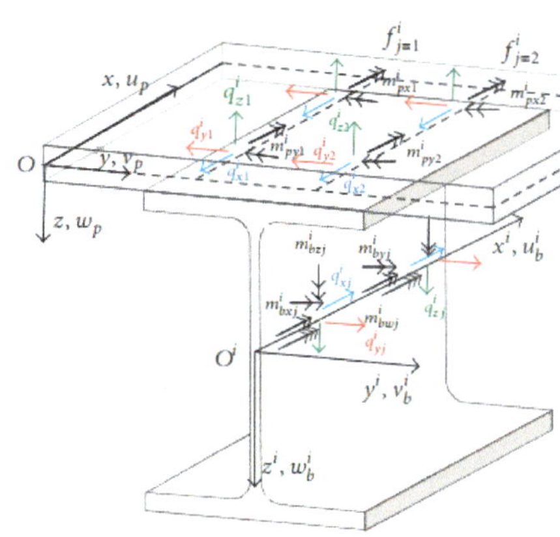

Figure 3: Structural model and directions of the additional loading of the plate and the ith beam.

(iv) A lateral line load $\partial m^i_{pyj}/\partial x$ due to the eccentricity of the component q^i_{xj} from the middle surface of the plate. $m^i_{pyj} = q^i_{xj} h_p/2$ is the bending moment.

(v) A lateral line load $\partial m^i_{pxj}/\partial y$ due to the eccentricity of the component q^i_{yj} from the middle surface of the plate. $m^i_{pxj} = q^i_{yj} h_p/2$ is the bending moment.

(b) In Each (ith) Beam ($C^i\,x^i\,y^i\,z^i$ System of Axes)

(i) An axially distributed line load q^i_{xj} along the beam centroid axis $C^i\,x$.

(ii) A transversely distributed line load q^i_{yj} along the beam centroid axis $C^i x$.

(iii) A perpendicularly distributed line load q^i_{zj} along the beam centroid axis $C^i x$.

(iv) A distributed bending moment $m^i_{byj} = q^i_{xj} e^i_{zj}$ along $C^i y^i$ local beam centroid axis due to the eccentricities q^i_{zj} of the components q^i_{xj} from the beam centroid axis. $e^i_{z1} = e^i_{z2} = -h^i_b/2$ are the eccentricities.

(v) A distributed bending moment $m^i_{bzj} = -q^i_{xj} e^i_{yj}$ along $C^i z^i$ local beam centroid axis due to the eccentricities q^i_{yj} of the components q^i_{xj} from the beam centroid axis. $e^i_{y1} = -b^i_f/4$ and $e^i_{y2} = b^i_f/4$ are the eccentricities.

(vi) A distributed twisting moment $m^i_{bxj} = q^i_{zj} e^i_{yj} - q^i_{yj} e^i_{zj}$ along $C^i x^i$ local beam shear center axis due to the eccentricities q^i_{zj} and q^i_{yj} of the components q^i_{yj} and q^i_{zj} from the beam shear center axis, respectively. $e^i_{z1} = e^i_{z2} = -h^i_b/2$ and $e^i_{y1} = -b^i_f/4$ and $e^i_{y2} = b^i_f/4$ are the eccentricities.

(vii) A distributed warping moment $m^i_{bwj} = -q^i_{xj}(\varphi^{Pi}_S)_{fj}$', along $C^i x^i$ local beam shear center axis, was ignored in previous models [39, 43]. $(\varphi^{Pi}_S)_{fj}$ is the value of the primary warping function φ^{Pi}_S with respect to the shear center of the beam cross-section (coinciding with its centroid) at the point of the jth interface line of the ith plate-beam interface.

On the basis of the above considerations the response of the plate and the beams may be described by the following boundary value problems.

(a) For the Plate. The analysis of the plate is based on the Von Karm´ an plate theory, according to which the deflection ´ of the plate cannot be regarded as small as compared to the plate thickness, while it remains small in comparison with the rest dimensions of the plate. Due to this assumption, geometrical nonlinearities should be taken into account and the displacement field of an arbitrary point of the plate, as implied by the Kirchhoff hypothesis, is given as

$$\bar{u}_p(x,y,z,t) = u_p(x,y,t) - z\frac{\partial w_p(x,y,t)}{\partial x}, \tag{1a}$$

$$\bar{v}_p(x,y,z,t) = v_p(x,y,t) - z\frac{\partial w_p(x,y,t)}{\partial y}, \tag{1b}$$

$$\bar{w}_p(x,y,z,t) = w_p(x,y,t), \tag{1c}$$

where $\bar{u}_p, \bar{v}_p, \bar{w}_p, \mathbf{x} : \{x, y\}, t \geq 0$ are the time dependent inplane and transverse displacement components of an arbitrary point of the plate and $u_p = u_p(x, t)$, $V_p = V_p(x, t)$, and $w_p = w_p(x, t)$, x : $\{x, y\}$, $t \geq 0$ are the corresponding components of a point at its middle surface. Employing the strain-displacement relations of the threedimensional elasticity for moderate large displacements [44, 45], the strain components can be written as

$$\varepsilon_{xx} = \frac{\partial \bar{u}_p}{\partial x} + \frac{1}{2}\left(\frac{\partial \bar{w}_p}{\partial x}\right)^2, \tag{2a}$$

$$\varepsilon_{yy} = \frac{\partial \bar{v}_p}{\partial y} + \frac{1}{2}\left(\frac{\partial \bar{w}_p}{\partial y}\right)^2, \tag{2b}$$

$$\gamma_{xy} = \frac{\partial \bar{u}_p}{\partial y} + \frac{\partial \bar{v}_p}{\partial x} + \frac{\partial \bar{w}_p}{\partial x}\frac{\partial \bar{w}_p}{\partial y}, \tag{2c}$$

$$\varepsilon_{zz} = \gamma_{xz} = \gamma_{yz} = 0. \tag{2d}$$

Substituting (1a)–(1c) and (2a)–(2d) to the stress-strain relations defined by the Hooke's law

$$\begin{Bmatrix} S_{xx} \\ S_{yy} \\ S_{xy} \end{Bmatrix} = \begin{bmatrix} \dfrac{E_p}{(1-\nu_p)^2} & \dfrac{E_p \nu_p}{(1-\nu_p)^2} & 0 \\ \dfrac{E_p \nu_p}{(1-\nu_p)^2} & \dfrac{E_p}{(1-\nu_p)^2} & 0 \\ 0 & 0 & \dfrac{E_p}{2(1+\nu_p)} \end{bmatrix} \begin{Bmatrix} \varepsilon_{xx} \\ \varepsilon_{yy} \\ \gamma_{xy} \end{Bmatrix} \tag{3}$$

the nonvanishing components of the second Piola-Kirchhoff stress tensor are obtained as

$$S_{xx} = \frac{E_p}{(1-\nu_p^2)} \times \left\{\left[\frac{\partial u_p}{\partial x} - z\frac{\partial^2 w_p}{\partial x^2} + \frac{1}{2}\left(\frac{\partial w_p}{\partial x}\right)^2\right] + \nu_p\left[\frac{\partial v_p}{\partial y} - z\frac{\partial^2 w_p}{\partial y^2} + \frac{1}{2}\left(\frac{\partial w_p}{\partial y}\right)^2\right]\right\}, \tag{4a}$$

$$+\nu_p\left[\frac{\partial v_p}{\partial y} - z\frac{\partial^2 w_p}{\partial y^2} + \frac{1}{2}\left(\frac{\partial w_p}{\partial y}\right)^2\right]\right\},$$

$$S_{yy} = \frac{E_p}{(1-\nu_p^2)}\left\{\nu_p\left[\frac{\partial u_p}{\partial x} - z\frac{\partial^2 w_p}{\partial x^2} + \frac{1}{2}\left(\frac{\partial w_p}{\partial x}\right)^2\right] + \left[\frac{\partial v_p}{\partial y} - z\frac{\partial^2 w_p}{\partial y^2} + \frac{1}{2}\left(\frac{\partial w_p}{\partial y}\right)^2\right]\right\}, \tag{4b}$$

$$S_{xy} = \frac{E_p}{2(1+\nu_p)}\left(\frac{\partial u_p}{\partial y} + \frac{\partial v_p}{\partial x} - 2z\frac{\partial^2 w_p}{\partial x \partial y} + \frac{\partial w_p}{\partial x}\frac{\partial w_p}{\partial y}\right). \tag{4c}$$

Subsequently, integrating the stress components over the plate thickness, the stress resultants acting on the plate are written as

$$N_{px} = C\left[\frac{\partial u_p}{\partial x} + \nu_p\frac{\partial v_p}{\partial x} + \frac{1}{2}\left(\frac{\partial w_p}{\partial x}\right)^2 + \frac{1}{2}\nu_p\left(\frac{\partial w_p}{\partial y}\right)^2\right], \tag{5a}$$

$$N_{py} = C\left[\frac{\partial v_p}{\partial x} + \nu_p\frac{\partial u_p}{\partial x} + \frac{1}{2}\left(\frac{\partial w_p}{\partial y}\right)^2 + \frac{1}{2}\nu_p\left(\frac{\partial w_p}{\partial x}\right)^2\right], \tag{5b}$$

$$N_{pxy} = C\frac{1-\nu_p}{2}\left(\frac{\partial u_p}{\partial y} + \frac{\partial v_p}{\partial x} + \frac{\partial w_p}{\partial y}\frac{\partial w_p}{\partial x}\right), \tag{5c}$$

$$M_{px} = -D\left(\frac{\partial^2 w_p}{\partial x^2} + \nu_p\frac{\partial^2 w_p}{\partial y^2}\right), \tag{5d}$$

$$M_{py} = -D\left(\frac{\partial^2 w_p}{\partial y^2} + \nu_p\frac{\partial^2 w_p}{\partial x^2}\right), \tag{5e}$$

$$M_{pxy} = -D\left(1-\nu_p\right)\frac{\partial^2 w_p}{\partial x \partial y}, \tag{5f}$$

where $C = E_p h_p/(1-\nu_p^2)$ and $D = E_p h_p^3/12(1-\nu_p^2)$ are the membrane and bending rigidities of the plate, respectively.

On the basis of Hamilton's principle, the system of partial differential equations of motion of the plate in terms of the stress resultants is obtained as

$$\frac{\partial N_{px}}{\partial x} + \frac{\partial N_{pxy}}{\partial y} - \rho_p h_p \ddot{u}_p = \sum_{i=1}^{I}\left(\sum_{j=1}^{2} q_{xj}^i \delta\left(y-y_j\right)\right), \tag{6a}$$

$$\frac{\partial N_{py}}{\partial y} + \frac{\partial N_{pxy}}{\partial x} - \rho_p h_p \ddot{v}_p = \sum_{i=1}^{I}\left(\sum_{j=1}^{2} q_{yj}^i \delta\left(y-y_j\right)\right), \tag{6b}$$

$$\begin{aligned} &-\frac{\partial^2 M_{px}}{\partial x^2} - 2\frac{\partial^2 M_{pxy}}{\partial x \partial y} - \frac{\partial^2 M_{py}}{\partial y^2} - N_{px}\frac{\partial^2 w_p}{\partial x^2} \\ &\quad - 2N_{pxy}\frac{\partial^2 w_p}{\partial x \partial y} - N_{py}\frac{\partial^2 w_p}{\partial y^2} \\ &\quad + \rho_p h_p \ddot{w}_p - \rho_p h_p \ddot{u}_p \frac{\partial w_p}{\partial x} - \rho_p h_p \ddot{v}_p \frac{\partial w_p}{\partial y} \\ &\quad - \frac{\rho_p h_p^3}{12}\frac{\partial^2 \ddot{w}_p}{\partial x^2} - \frac{\rho_p h_p^3}{12}\frac{\partial^2 \ddot{w}_p}{\partial y^2} \\ &= g - \sum_{i=1}^{I}\left[\sum_{j=1}^{2}\left(q_{zj}^i + \frac{\partial m_{pxj}^i}{\partial y} + \frac{\partial m_{pyj}^i}{\partial x} - q_{xj}^i \frac{\partial w_{pj}^i}{\partial x}\right.\right. \\ &\quad \left.\left. - q_{yj}^i \frac{\partial w_{pj}^i}{\partial y}\right)\delta\left(y-y_j\right)\right], \end{aligned} \tag{6c}$$

where $\delta(y-y_i)$ is the Dirac's delta function in the y direction. Employing relations (5a)–(5f), the governing differential equations (6a)–(6c) in the domain Ω can be expressed in terms of the displacement components as

$$\begin{aligned} &G_p h_p\left[\nabla^2 u_p + \frac{1+\nu_p}{1-\nu_p}\frac{\partial}{\partial x}\left(\frac{\partial u_p}{\partial x} + \frac{\partial v_p}{\partial y}\right)\right. \\ &\quad + \left(\frac{2}{1-\nu_p}\frac{\partial^2 w_p}{\partial x^2} + \frac{\partial^2 w_p}{\partial y^2}\right)\frac{\partial w_p}{\partial x} \\ &\quad \left. + \frac{1+\nu_p}{1-\nu_p}\frac{\partial^2 w_p}{\partial x \partial y}\frac{\partial w_p}{\partial y}\right] - \rho_p h_p \ddot{u}_p \\ &= \sum_{i=1}^{I}\left(\sum_{j=1}^{2} q_{xj}^i \delta\left(y-y_j\right)\right), \end{aligned} \tag{7a}$$

$$\begin{aligned} &G_p h_p\left[\nabla^2 v_p + \frac{1+\nu_p}{1-\nu_p}\frac{\partial}{\partial y}\left(\frac{\partial u_p}{\partial x} + \frac{\partial v_p}{\partial y}\right)\right. \\ &\quad + \left(\frac{2}{1-\nu_p}\frac{\partial^2 w_p}{\partial y^2} + \frac{\partial^2 w_p}{\partial x^2}\right)\frac{\partial w_p}{\partial y} \\ &\quad \left. + \frac{1+\nu_p}{1-\nu_p}\frac{\partial^2 w_p}{\partial x \partial y}\frac{\partial w_p}{\partial x}\right] - \rho_p h_p \ddot{v}_p \\ &= \sum_{i=1}^{I}\left(\sum_{j=1}^{2} q_{yj}^i \delta\left(y-y_j\right)\right), \end{aligned} \tag{7b}$$

$$\begin{aligned} &D\nabla^4 w_p - C \\ &\quad \times \left\{\left[\left(\frac{\partial u_p}{\partial x} + \frac{1}{2}\left(\frac{\partial w_p}{\partial x}\right)^2\right)\right.\right. \\ &\quad \left. + \nu_p\left(\frac{\partial v_p}{\partial y} + \frac{1}{2}\left(\frac{\partial w_p}{\partial y}\right)^2\right)\right] \\ &\quad \times \frac{\partial^2 w_p}{\partial x^2} + \left(1-\nu_p\right) \\ &\quad \cdot \left(\frac{\partial u_p}{\partial y} + \frac{\partial v_p}{\partial x} + \frac{\partial w_p}{\partial x}\frac{\partial w_p}{\partial y}\right)\frac{\partial^2 w_p}{\partial x \partial y} \\ &\quad + \left[\left(\frac{\partial v_p}{\partial y} + \frac{1}{2}\left(\frac{\partial w_p}{\partial y}\right)^2\right)\right. \\ &\quad \left.\left. + \nu_p\left(\frac{\partial u_p}{\partial x} + \frac{1}{2}\left(\frac{\partial w_p}{\partial x}\right)^2\right)\right] \cdot \frac{\partial^2 w_p}{\partial y^2}\right\} \\ &\quad + \rho_p h_p \ddot{w}_p - \rho_p h_p \ddot{u}_p \frac{\partial w_p}{\partial x} - \rho_p h_p \ddot{v}_p \frac{\partial w_p}{\partial y} \\ &\quad - \frac{\rho_p h_p^3}{12}\frac{\partial^2 \ddot{w}_p}{\partial x^2} - \frac{\rho_p h_p^3}{12}\frac{\partial^2 \ddot{w}_p}{\partial y^2} \\ &= g - \sum_{i=1}^{I}\left[\sum_{j=1}^{2}\left(q_{zj}^i + \frac{\partial m_{pxj}^i}{\partial y} + \frac{\partial m_{pyj}^i}{\partial x} - q_{xj}^i \frac{\partial w_{pj}^i}{\partial x}\right.\right. \\ &\quad \left.\left. - q_{yj}^i \frac{\partial w_{pj}^i}{\partial y}\right)\delta\left(y-y_j\right)\right]. \end{aligned} \tag{7c}$$

The governing differential equations (7a)–(7c) are also subjected to the pertinent boundary conditions of the problem at hand

$$a_{p1}u_{pn} + a_{p2}N_{pn} = a_{p3}, \tag{8a}$$

$$\beta_{p1}u_{pt} + \beta_{p2}N_{pt} = \beta_{p3}, \tag{8b}$$

$$\gamma_{p1}w_p + \gamma_{p2}R_{pn} = \gamma_{p3}, \tag{8c}$$

$$\delta_{p1}\frac{\partial w_p}{\partial n} + \delta_{p2}M_{pn} = \delta_{p3}, \tag{8d}$$

$$\varepsilon_{1k}w_p + \varepsilon_{2k}\left\|Tw_p\right\|_k = \varepsilon_{3k}, \quad \varepsilon_{2k} \neq 0 \tag{8e}$$

and to the initial conditions

$$w_p(\mathbf{x}, 0) = w_{p0}(\mathbf{x}), \tag{9a}$$

$$\dot{w}_p(\mathbf{x}, 0) = \overline{w}_{p0}(\mathbf{x}), \tag{9b}$$

where a_{pl}, β_{pl}, γ_{pl}, and δ_{pl} ($l = 1, 2, 3$) are functions specified at the boundary Γ; ε_{lk} ($l = 1, 2, 3$) are functions specified at the k corners of the plate; $w_{p0}(\mathrm{x})$, $\overline{w}_{p0}(\mathbf{x})$, and x : $\{x, y\}$ are the initial deflection and velocity of the points of the middle surface of the plate; u_{pn}, u_{pt} and N_{pn}, N_{pt} are the boundary membrane displacements and forces in the normal and tangential directions to the boundary, respectively; R_{pn} and M_{pn} are the effective reaction along the boundary and the bending moment normal to it, respectively, which by employing intrinsic coordinates (i.e., the distance along the outward normal n to the boundary and the arc length s) are written as

$$R_{pn} = -D\left[\frac{\partial}{\partial n}\nabla^2 w_p - (\nu_p - 1)\frac{\partial}{\partial s}\left(\frac{\partial^2 w_p}{\partial s \partial n} - \kappa\frac{\partial w_p}{\partial s}\right)\right] + N_{pn}\frac{\partial w_p}{\partial n} + N_{pt}\frac{\partial w_p}{\partial s}, \tag{10a}$$

$$M_{pn} = -D\left[\nabla^2 w_p + (\nu_p - 1)\left(\frac{\partial^2 w_p}{\partial s^2} + \kappa\frac{\partial w_p}{\partial n}\right)\right] \tag{10b}$$

in which $\kappa(s)$ is the curvature of the boundary. Finally, $\|Tw_p\|_k$ is the discontinuity jump of the twisting moment Tw_p at the corner k of the plate, while Tw_p along the boundary is given by the following relation:

$$Tw_p = D(\nu_p - 1)\left(\frac{\partial^2 w_p}{\partial s \partial n} - \kappa\frac{\partial w_p}{\partial s}\right). \tag{11}$$

The boundary conditions (8a)–(8d) are the most general boundary conditions for the plate problem including also elastic support, while the corner condition (8e) holds for free or transversely elastically restrained corners k. It is apparent that all types of the conventional boundary conditions can be derived from these equations by specifying appropriately the functions apl, β_{pl}, γ_{pl}, and δ_{pl} ($l = 1, 2, 3$) (e.g., for a clamped edge it is $a_{p1} = \beta_{p1} = \gamma_{p1} = \delta_{p1} = 1$, $a_{p2} = a_{p3} = \beta_{p2} = \beta_{p3} = \gamma_{p2} = \gamma_{p3} = \delta_{p2} = \delta_{p3} = 0$).

(b) For Each (ith) Beam. Each beam undergoes transverse deflection with respect to z^i and y^i axes and axial deformation and nonuniform angle of twist along x^i axis. Based on the Bernoulli theory, the displacement field of an arbitrary point of a cross-section (taking into account moderate large displacements and considering the angle of rotation of twist to have relatively small values) can be derived with respect to those of its centroid as

$$\overline{u}_b^i(x^i, y^i, z^i, t) = u_b^i(x^i, t) - y^i\theta_{bz}^i(x^i, t) + z^i\theta_{by}^i(x^i, t) + \frac{\partial\theta_{bx}^i}{\partial x^i}\varphi_S^{Pi}(y^i, z^i, t), \tag{12a}$$

$$\overline{v}_b^i(x^i, y^i, z^i, t) = v_b^i(x^i, t) - z^i\theta_{bx}^i(x^i, t), \tag{12b}$$

$$\overline{w}_b^i(x^i, y^i, z^i, t) = w_b^i(x^i, t) + y^i\theta_{bx}^i(x^i, t), \tag{12c}$$

$$\theta_{by}^i(x^i, t) = -\frac{\partial w_b^i(x^i, t)}{\partial x^i}, \tag{12d}$$

$$\theta_{bz}^i(x^i, t) = \frac{\partial v_b^i(x^i, t)}{\partial x^i}, \tag{12e}$$

where $\overline{u}_b^i$, $\overline{v}_b^i$, and $\overline{w}_b^i$ are the axial and transverse displacement components with respect to the $C^i x^i y^i z^i$ system of axes; $u_b^i = u_b^i(x^i)$, $v_b^i = v_b^i(x^i)$, and $w_b^i = w_b^i(x^i)$ are the corresponding components of the centroid C^i; $\theta_{by}^i = \theta_{by}^i(x^i)$ and $\theta_{bz}^i = \theta_{bz}^i(x^i)$ are the angles of rotation of the cross-section due to bending, with respect to its centroid; $\partial\theta_{bx}^i/dx^i$ denotes the rate of change of the angle of twist $\theta_{bx}^i(x^i)$ regarded as the torsional curvature and φ_S^{Pi} is the primary warping function with respect to the cross-section's shear center (coinciding with its centroid). Employing again the strain-displacement relations of the three-dimensional elasticity for moderate displacements [44, 45], the strain components are given as

$$\varepsilon_{xx} = \frac{\partial\overline{u}_b^i}{\partial x^i} + \frac{1}{2}\left[\left(\frac{\partial\overline{v}_b^i}{\partial x^i}\right)^2 + \left(\frac{\partial\overline{w}_b^i}{\partial x^i}\right)^2\right], \tag{13a}$$

$$\gamma_{xz} = \frac{\partial\overline{w}_b^i}{\partial x^i} + \frac{\partial\overline{u}_b^i}{\partial z^i} + \left(\frac{\partial\overline{v}_b^i}{\partial x^i}\frac{\partial\overline{v}_b^i}{\partial z^i} + \frac{\partial\overline{w}_b^i}{\partial x^i}\frac{\partial\overline{w}_b^i}{\partial z^i}\right), \tag{13b}$$

$$\gamma_{xy} = \frac{\partial\overline{v}_b^i}{\partial x^i} + \frac{\partial\overline{u}_b^i}{\partial y^i} + \left(\frac{\partial\overline{v}_b^i}{\partial x^i}\frac{\partial\overline{v}_b^i}{\partial y^i} + \frac{\partial\overline{w}_b^i}{\partial x^i}\frac{\partial\overline{w}_b^i}{\partial y^i}\right), \tag{13c}$$

$$\varepsilon_{yy} = \varepsilon_{zz} = \gamma_{yz} = 0. \tag{13d}$$

Employing the Hooke's stress-strain law and integrating the arising stress components over the beam's cross-section after ignoring the nonlinear terms with respect to the angle of twist and its derivatives, the stress resultants of the beam are derived as

$$N_b^i = E_b^i A_b^i \left[\frac{\partial u_b^i}{\partial x^i} + \frac{1}{2}\left(\left(\frac{\partial v_b^i}{\partial x^i} \right)^2 + \left(\frac{\partial w_b^i}{\partial x^i} \right)^2 \right) \right], \tag{14a}$$

$$M_{by}^i = -E_b^i I_y^i \frac{\partial^2 w_b^i}{\partial x^{i2}}, \tag{14b}$$

$$M_{bz}^i = E_b^i I_z^i \frac{\partial^2 v_b^i}{\partial x^{i2}}, \tag{14c}$$

$$M_{bt}^{Pi} = G_b^i I_t^i \frac{\partial \theta_{bx}^i}{\partial x^i}, \tag{14d}$$

$$M_{bw}^i = -E_b^i C_S^i \frac{\partial^2 \theta_{bx}^i}{\partial x^{i2}}, \tag{14e}$$

where M_{bt}^{Pi} is the primary twisting moment [15] resulting from the primary shear stress distribution; M_{bw}^i is the warping moment due to torsional curvature. Furthermore I_y^i and I_z^i are the principal moments of inertia; I_S^i is the polar moment of inertia, while I_t^i and C_S^i are the torsion and warping constants of the ith beam with respect to the cross-section's shear center (coinciding with its centroid), respectively, given as [46]

$$I_t^i = \int_\Omega \left(y^{i2} + z^{i2} + y^i \frac{\partial \varphi_S^{Pi}}{\partial z^i} - z^i \frac{\partial \varphi_S^{Pi}}{\partial y^i} \right) d\Omega, \tag{15a}$$

$$C_S^i = \int_\Omega \left(\varphi_S^{Pi} \right)^2 d\Omega. \tag{15b}$$

On the basis of Hamilton's principle, the differential equations of motion in terms of displacements are obtained as

$$-\frac{\partial N_b^i}{\partial x^i} + \rho_b^i A_b^i \ddot{u}_b^i = \sum_{j=1}^{2} q_{xj}^i, \tag{16a}$$

$$-N_b^i \frac{\partial^2 v_b^i}{\partial x^{i2}} + \frac{\partial^2 M_{bz}^i}{\partial x^{i2}} + \rho_b^i A_b^i \ddot{v}_b^i - \rho_b^i I_z^i \frac{\partial^2 \ddot{v}_b^i}{\partial x^2} - \rho_b^i A_b^i \ddot{u}_b^i \frac{\partial v_b^i}{\partial x^i} = \sum_{j=1}^{2} \left(q_{yj}^i - q_{xj}^i \frac{\partial v_b^i}{\partial x^i} - \frac{\partial m_{bzj}^i}{\partial x^i} \right), \tag{16b}$$

$$-N_b^i \frac{\partial^2 w_b^i}{\partial x^{i2}} - \frac{\partial^2 M_{by}^i}{\partial x^{i2}} + \rho_b^i A_b^i \ddot{w}_b^i - \rho_b^i I_y^i \frac{\partial^2 \ddot{w}_b^i}{\partial x^2} - \rho_b^i A_b^i \ddot{u}_b^i \frac{\partial w_b^i}{\partial x^i} = \sum_{j=1}^{2} \left(q_{zj}^i - q_{xj}^i \frac{\partial w_b^i}{\partial x^i} + \frac{\partial m_{byj}^i}{\partial x^i} \right), \tag{16c}$$

$$-\frac{\partial M_{bt}^i}{\partial x^i} - \frac{\partial^2 M_{bw}^i}{\partial x^{i2}} + \rho_b^i I_S^i \ddot{\theta}_{bx}^i - \rho_b^i C_S^i \frac{\partial^2 \ddot{\theta}_{bx}^i}{\partial x^2} = \sum_{j=1}^{2} \left[m_{bxj}^i + \frac{\partial m_{bwj}^i}{\partial x^i} \right]. \tag{16d}$$

Substituting the expressions of the stress resultants of (14a)–(14e) in (16a)–(16d), the differential equations of motion are obtained as

$$-E_b^i A_b^i \left(\frac{\partial^2 u_b^i}{\partial x^{i2}} + \frac{\partial w_b^i}{\partial x^i}\frac{\partial^2 w_b^i}{\partial x^{i2}} + \frac{\partial v_b^i}{\partial x^i}\frac{\partial^2 v_b^i}{\partial x^{i2}} \right) + \rho_b^i A_b^i \ddot{u}_b^i = \sum_{j=1}^{2} q_{xj}^i, \tag{17a}$$

$$E_b^i I_z^i \frac{\partial^4 v_b^i}{\partial x^{i4}} - N_b^i \frac{\partial^2 v_b^i}{\partial x^{i2}} - \rho_b^i I_z^i \frac{\partial^2 \ddot{v}_b}{\partial x^2} + \rho_b^i A_b^i \ddot{v}_b - \rho_b^i A_b^i \ddot{u}_b^i \frac{\partial v_b^i}{\partial x^i} = \sum_{j=1}^{2} \left(q_{yj}^i - q_{xj}^i \frac{\partial v_b^i}{\partial x^i} - \frac{\partial m_{bzj}^i}{\partial x^i} \right), \tag{17b}$$

$$E_b^i I_y^i \frac{\partial^4 w_b^i}{\partial x^{i4}} - N_b^i \frac{\partial^2 w_b^i}{\partial x^{i2}} - \rho_b^i I_y^i \frac{\partial^2 \ddot{w}_b}{\partial x^2} + \rho_b^i A_b^i \ddot{w}_b - \rho_b^i A_b^i \ddot{u}_b^i \frac{\partial w_b^i}{\partial x^i} = \sum_{j=1}^{2} \left(q_{zj}^i - q_{xj}^i \frac{\partial w_b^i}{\partial x^i} + \frac{\partial m_{byj}^i}{\partial x^i} \right), \tag{17c}$$

$$E_b^i C_S^i \frac{\partial^4 \theta_{bx}^i}{\partial x^{i4}} - G_b^i I_t^i \frac{\partial^2 \theta_{bx}^i}{\partial x^{i2}} + \rho_b^i I_S^i \ddot{\theta}_{bx}^i - \rho_b^i C_S^i \frac{\partial^2 \ddot{\theta}_{bx}^i}{\partial x^2} = \sum_{j=1}^{2} \left[m_{bxj}^i + \frac{\partial m_{bwj}^i}{\partial x^i} \right]. \tag{17d}$$

Moreover, the corresponding boundary conditions of the ith beam at its ends $x^i = 0, l^i$ are given as

$$a_{b1}^i u_b^i + \alpha_{b2}^i N_b^i = \alpha_{b3}^i, \tag{18}$$

$$\beta_{b1}^i v_b^i + \beta_{b2}^i R_{by}^i = \beta_{b3}^i, \tag{19a}$$

$$\bar{\beta}_{b1}^i \theta_{bz}^i + \bar{\beta}_{b2}^i M_{bz}^i = \bar{\beta}_{b3}^i, \tag{19b}$$

$$\gamma_{b1}^i w_b^i + \gamma_{b2}^i R_{bz}^i = \gamma_{b3}^i, \tag{20a}$$

$$\bar{\gamma}_{b1}^i \theta_{by}^i + \bar{\gamma}_{b2}^i M_{by}^i = \bar{\gamma}_{b3}^i, \tag{20b}$$

$$\delta_{b1}^i \theta_{bx}^i + \delta_{b2}^i M_{bt}^i = \delta_{b3}^i, \tag{21a}$$

$$\bar{\delta}_{b1}^i \frac{d\theta_{bx}^i}{dx^i} + \bar{\delta}_{b2}^i M_{bw}^i = \bar{\delta}_{b3}^i \tag{21b}$$

and the initial conditions as

$$w_b^i\left(x^i,0\right) = w_{b0}^i\left(x^i\right), \tag{22a}$$

$$\dot{w}_b^i\left(x^i,0\right) = \overline{w}_{b0}^i\left(x^i\right), \tag{22b}$$

where the angles of rotation of the cross-section due to bending θ_{by}^i, θ_{bz}^i are given from (12d) and (12e), R_{by}^i, R_{bz}^i and M_{bz}^i, M_{by}^i *by* are the reactions and bending moments with respect to y^i, z^i axes, respectively, which after applying the aforementioned simplifications are given as

$$R_{by}^i = N_b^i \frac{\partial v_b^i}{\partial x^i} - E_b^i I_z^i \frac{\partial^3 v_b^i}{\partial x^{i3}}, \tag{23a}$$

$$R_{bz}^i = N_b^i \frac{\partial w_b^i}{\partial x^i} - E_b^i I_y^i \frac{\partial^3 w_b^i}{\partial x^{i3}}, \tag{23b}$$

$$M_{bz}^i = E_b^i I_z^i \frac{\partial^2 v_b^i}{\partial x^{i2}}, \tag{24a}$$

$$M_{by}^i = -E_b^i I_y^i \frac{\partial^2 w_b^i}{\partial x^{i2}}, \tag{24b}$$

and M_{bt}^i, M_{bw}^i are the torsional and warping moments at the boundaries of the beam, respectively, given as

$$M_{bt}^i = G_b^i I_t^i \frac{\partial \theta_{bx}^i}{\partial x^i} - E_b^i C_S^i \frac{\partial^3 \theta_{bx}^i}{\partial x^{i3}}, \tag{25a}$$

$$M_{bw}^i = -E_b^i C_S^i \frac{\partial^2 \theta_{bx}^i}{\partial x^{i2}}. \tag{25b}$$

Finally, $\alpha_{bk}^i, \beta_{bk}^i, \overline{\beta}_{bk}^i, \gamma_{bk}^i, \overline{\gamma}_{bk}^i, \delta_{bk}^i, \overline{\delta}_{bk}^i$ $(k = 1,2,3)$ are functions specified at the ith beam ends ($x^i = 0,\ l^i$). The boundary conditions (18)–(21b) are the most general boundary conditions for the beam problem including also the elastic support. It is apparent that all types of the conventional boundary conditions (clamped, simply supported, free, or guided edge) can be derived from these equations by specifying appropriately the aforementioned coefficients.

Equations (7a)–(17d) constitute a set of seven coupled and nonlinear partial differential equations including thirteen unknowns, namely $u_p, v_p, w_p, u_b^i, v_b^i, w_b^i, \theta_{bx}^i, q_{x1}^i, q_{y1}^i, q_{z1}^i, q_{x2}^i$, q_{y2}^i, and q_{z2}^i. Six additional equations are required, which result from the displacement continuity conditions in the directions of x, y^i, and z^i local axes along the two interface lines of each (ith) plate-beam interface. Taking into account the displacement fields expressed by (1a)–(1c) and (12a)–(12e) the displacement continuity conditions [47] can be expressed as follows.

In the direction of x^i local axis:

$$u_{p1}^i - u_b^i = \frac{h_p}{2}\frac{\partial w_{p1}^i}{\partial x} + \frac{h_b^i}{2}\frac{\partial w_b^i}{\partial x^i} + \frac{b_f^i}{4}\frac{\partial v_b^i}{\partial x^i} + \frac{\partial \theta_{bx}^i}{\partial x^i}\left(\varphi_S^{Pi}\right)_{f1} + \frac{q_{x1}^i}{k_{x1}^i}\left[1 - \frac{1}{2}\left(\frac{\partial w_b^i}{\partial x^i}\right)^2\right] \tag{26a}$$

along interface line 1 $\left(f_{j=1}^i\right)$,

$$u_{p2}^i - u_b^i = \frac{h_p}{2}\frac{\partial w_{p2}^i}{\partial x} + \frac{h_b^i}{2}\frac{\partial w_b^i}{\partial x^i} - \frac{b_f^i}{4}\frac{\partial v_b^i}{\partial x^i} + \frac{\partial \theta_{bx}^i}{\partial x^i}\left(\varphi_S^{Pi}\right)_{f2} + \frac{q_{x2}^i}{k_{x2}^i}\left[1 - \frac{1}{2}\left(\frac{\partial w_b^i}{\partial x^i}\right)^2\right] \tag{26b}$$

along interface line 2 $\left(f_{j=2}^i\right)$.

In the direction of y^i local axis:

$$v_{p1}^i - v_b^i = \frac{h_p}{2}\frac{\partial w_{p1}^i}{\partial y} + \frac{h_b^i}{2}\theta_{bx}^i + \frac{q_{y1}^i}{k_{y1}^i}\left[1 - \frac{1}{2}\left(\theta_{bx}^i\right)^2\right] \tag{27a}$$

along interface line 1 $\left(f_{j=1}^i\right)$,

$$v_{p2}^i - v_b^i = \frac{h_p}{2}\frac{\partial w_{p2}^i}{\partial y} + \frac{h_b^i}{2}\theta_{bx}^i + \frac{q_{y2}^i}{k_{y2}^i}\left[1 - \frac{1}{2}\left(\theta_{bx}^i\right)^2\right] \tag{27b}$$

along interface line 2 $\left(f_{j=2}^i\right)$.

In the direction of z^i local axis:

$$w_{p1}^i - w_b^i = -\frac{b_f^i}{4}\theta_{bx}^i \quad \text{along interface line 1 } \left(f_{j=1}^i\right), \tag{28a}$$

$$w_{p2}^i - w_b^i = \frac{b_f^i}{4}\theta_{bx}^i \quad \text{along interface line 2 } \left(f_{j=2}^i\right), \tag{28b}$$

where $\left(\varphi_S^{Pi}\right)_{fj}$ is the value of the primary warping function with respect to the shear center of the beam cross-section (coinciding with its centroid) at the point of the jth interface line of the ith plate-beam interface f_j^i and k_{xj}^i, k_{yj}^i are the stiffness of the arbitrarily distributed shear connectors along the x^i and y^i directions, respectively. It is noted that k_{xj}^i = $k_{xj}^i(s_{xj}^i)$ and k_{yj}^i = $k_{yj}^i(s_{yj}^i)$ can represent any linear or nonlinear relationship between the inplane interface forces and the interface slip s_j^i in the corresponding direction. In all of the aforementioned equations the values of the primary warping function $\varphi_S^{Pi}(y^i, z^i)$ should be set having the appropriate algebraic sign corresponding to the local beam axes.

INTEGRAL REPRESENTATIONS-NUMERICAL SOLUTION

The solution of the presented dynamic problem requires the integration of the set of (7a)–(7c) and (17a)–(17d) subjected to the prescribed boundary and initial conditions. Moreover, the displacement continuity conditions should also be fulfilled. Due to the nonlinear and coupling character of the equations of motion, an analytical solution is out of question. Therefore, a numerical solution is derived employing the analog equation method [41], a BEM-based method. Contrary to previous research efforts where the numerical analysis is based on BEM using a lumped mass assumption model after evaluating the flexibility matrix at the mass nodal points [15], in this work, a distributed mass model is employed.

In the following sections, the plate and beam problems represented by (7a)–(7c) and (17a)–(17d), respectively, are examined independently and the connection between these problems is achieved employing the displacement continuity conditions.

For the Plate Displacement Components u_p, V_p, w_p.

According to the precedent analysis, the large deflection analysis of the plate becomes equivalent to establishing the inplane displacement components u_p and V_p having continuous partial derivatives up to the second order with respect to x, y, and the deflection wp having continuous partial derivatives up to the fourth order with respect to x, y, and all displacement components having derivatives up to the second order with respect to t. Moreover, these displacement components must satisfy the boundary value problem described by the nonlinear and coupled governing differential equations of equilibrium (equations (7a)–(7c)) inside the domain, the conditions (equations (8a)–(8e)) at the boundary Γ and the initial conditions (equations (9a)-(9b)). Equations (7a)–(7c) and (8a)–(8e) are solved using the analog equation method [41]. More specifically, setting as $u_{p1} = u_p$, $u_{p2} = V_p$, $u_{p3} = w_p$ and applying the Laplacian operator to u_{p1}, u_{p2} and the biharmonic operator to u_{p3} yields

$$\nabla^2 u_{pi} = p_{pi}(x, y, t), \quad (i = 1, 2), \tag{29a}$$

$$\nabla^4 u_{p3} = p_{p3}(x, y, t). \tag{29b}$$

Equations (29a) and (29b) are called analog equations and they indicate that the solution of (7a)–(7c) and (8a)–(8e) can be established by solving (29a) and (29b) under the same boundary conditions (equations (8a)– (8e)), provided that the fictitious load distributions $p_{pi}(x, y)$ $(i = 1, 2, 3)$ are first established. These distributions can be determined using BEM. Following the procedure presented in [47], the unknown boundary quantities $u_{pi}(P,t)$, $u_{pi,x}(P,t)$, $u_{pi,xx}(P,t)$, and $u_{pi,xxx}(P,t)$ ($P \in$ boundary, $i = 1, 2, 3$) can be expressed in terms of p_{pi} $(i = 1, 2, 3)$ after applying the integral representations of the displacement components u_i $(i = 1, 2, 3)$ and their derivatives with respect to x, y to the boundary of the plate.

By discretizing the boundary of the plate into N boundary elements and the domain Ω into M domain cells and employing the constant element assumption (as the numerical implementation becomes very simple and the obtained results are of high accuracy), the application of the integral representations and the corresponding one of the Laplacian $\nabla^2 u_{p3}$ and of the boundary conditions (8a)– (8e) to the N boundary nodal points results in a set of $8\times N$ nonlinear algebraic equations relating the unknown boundary quantities with the fictitious load distributions p_{pi} $(i = 1, 2, 3)$ that can be written as

$$\begin{bmatrix} \mathbf{E}_{p11} & 0 & 0 \\ 0 & \mathbf{E}_{p22} & 0 \\ 0 & 0 & \mathbf{E}_{p33} \end{bmatrix} \begin{Bmatrix} \mathbf{d}_{p1} \\ \mathbf{d}_{p2} \\ \mathbf{d}_{p3} \end{Bmatrix} + \begin{Bmatrix} 0 \\ \mathbf{D}_{p1}^{nl} \\ 0 \\ \mathbf{D}_{p2}^{nl} \\ 0 \\ 0 \\ \mathbf{D}_{p3}^{nl} \end{Bmatrix} = \begin{Bmatrix} 0 \\ \alpha_{p3} \\ 0 \\ \beta_{p3} \\ 0 \\ 0 \\ \gamma_{p3} \\ \delta_{p3} \end{Bmatrix}, \tag{30}$$

Where

$$\mathbf{E}_{p11} = \begin{bmatrix} \widehat{\mathbf{A}}_1 & \mathbf{H}_1 & \widehat{\mathbf{H}}_2 \\ 0 & \mathbf{D}_{p22} & \mathbf{D}_{p23} \end{bmatrix}, \tag{31a}$$

$$\mathbf{E}_{p22} = \begin{bmatrix} \widehat{\mathbf{A}}_1 & \mathbf{H}_1 & \widehat{\mathbf{H}}_2 \\ 0 & \mathbf{D}_{p44} & \mathbf{D}_{p45} \end{bmatrix}, \tag{31b}$$

$$\mathbf{E}_{p33} = \begin{bmatrix} \mathbf{A}_2 & \mathbf{H}_1 & \mathbf{H}_2 & \mathbf{G}_1 & \mathbf{G}_2 \\ \mathbf{A}_1 & 0 & 0 & \mathbf{H}_1 & \mathbf{H}_2 \\ 0 & \mathbf{D}_{p78} & \mathbf{D}_{p79} & 0 & \mathbf{D}_{p711} \\ 0 & \mathbf{D}_{p88} & \mathbf{D}_{p89} & \mathbf{D}_{p810} & 0 \end{bmatrix}. \tag{31c}$$

$\mathbf{D}_{p22}$ to $\mathbf{D}_{p810}$ are $N\times N$ rectangular known matrices including the values of the functions $\alpha_{pj}, \beta_{pj}, \gamma_{pj}, \delta_{pj}$ $(j = 1, 2)$ of (8a)–(8d); $\alpha_{p3}, \beta_{p3}, \gamma_{p3}$, and δ_{p3} are $N\times 1$ known column matrices including the boundary values of the functions $\alpha_{p3}, \beta_{p3}, \gamma_{p3}, \delta_{p3}$, of (8a)-(8d); $\mathbf{H}_i$ $(i = 1, 2)$, $\widehat{\mathbf{H}}_2$, and $\mathbf{G}_i$ $(i = 1, 2)$ are rectangular $N\times N$ known coefficient matrices resulting from the values of kernels at the boundary elements of the plate; $\mathbf{A}_1$, $\widehat{\mathbf{A}}_1$, and $\mathbf{A}_2$ are $N\times M$ rectangular known matrices originating from the integration of kernels on the domain cells of the plate; $\mathbf{D}_{pi}^{nl}$ $(i = 1, 2)$ are $N\times 1$ and $\mathbf{D}_{p3}^{nl}$ is $2N \times 1$ column matrices containing the nonlinear terms included in the expressions of the boundary conditions (equations (8a)–(8d)). It is noted that the derivatives of the unknown boundary quantities with respect to the arc length s, appearing in (8a)–(8d), are approximated employing appropriate central, backward, or forward finite difference schemes. Finally

$$\mathbf{d}_{pi} = \{\mathbf{p}_{pi} \ \ \hat{\mathbf{u}}_{pi} \ \ \hat{\mathbf{u}}_{pi,n}\}^T, \quad (i = 1, 2), \tag{32a}$$

$$\mathbf{d}_{p3} = \{\mathbf{p}_{p3} \ \ \hat{\mathbf{u}}_{p3} \ \ \hat{\mathbf{u}}_{p3,n} \ \ \nabla^2\hat{\mathbf{u}}_{p3} \ \ (\nabla^2\hat{\mathbf{u}}_{p3})_{,n}\}^T \tag{32b}$$

are generalized unknown vectors, where

$$\hat{\mathbf{u}}_{pi} = \left\{ (u_{pi})_1 \ (u_{pi})_2 \ \cdots \ (u_{pi})_N \right\}^T, \quad (i = 1, 2, 3) \tag{33a}$$

$$\hat{\mathbf{u}}_{pi,n} = \left\{ \left(\frac{\partial u_{pi}}{\partial n} \right)_1 \ \left(\frac{\partial u_{pi}}{\partial n} \right)_2 \ \cdots \ \left(\frac{\partial u_{pi}}{\partial n} \right)_N \right\}^T, \quad (i = 1, 2, 3) \tag{33b}$$

$$\nabla^2 \hat{\mathbf{u}}_{p3} = \left\{ (\nabla^2 u_{p3})_1 \ (\nabla^2 u_{p3})_2 \ \cdots \ (\nabla^2 u_{p3})_N \right\}^T, \tag{33c}$$

$$(\nabla^2 \hat{\mathbf{u}}_{p3})_{,n} = \left\{ \left(\frac{\partial \nabla^2 u_{p3}}{\partial n} \right)_1 \ \left(\frac{\partial \nabla^2 u_{p3}}{\partial n} \right)_2 \ \cdots \ \left(\frac{\partial \nabla^2 u_{p3}}{\partial n} \right)_N \right\}^T \tag{33d}$$

are vectors including the unknown boundary values of the respective boundary quantities and $\mathbf{p}_{pi} = \left\{ (p_{pi})_1 \ (p_{pi})_2 \ \cdots \ (p_{pi})_M \right\}^T$, $(i = 1, 2, 3)$, are vectors containing the M unknown nodal values of the fictitious loads at the domain cells of the plate. In the case that the boundary Γ has k free or transversely elastically restrained corners, k additional equations must be satisfied together with (30). These additional equations result from the application of the corner condition (8e) on the k corners following the procedure presented in [48].

Discretization of the integral representations of the displacement components u_i and their derivatives with respect to x, y [49] and application to the M domain nodal points yields

$$\mathbf{u}_{pi} = \mathbf{B}_{pi} \mathbf{d}_{pi}, \quad (i = 1, 2, 3), \tag{34a}$$

$$\mathbf{u}_{pi,x} = \mathbf{B}_{pi,x} \mathbf{d}_{pi}, \quad (i = 1, 2, 3), \tag{34b}$$

$$\mathbf{u}_{pi,y} = \mathbf{B}_{pi,y} \mathbf{d}_{pi}, \quad (i = 1, 2, 3), \tag{34c}$$

$$\mathbf{u}_{pi,xx} = \mathbf{B}_{pi,xx} \mathbf{d}_{pi}, \quad (i = 1, 2, 3), \tag{34d}$$

$$\mathbf{u}_{pi,yy} = \mathbf{B}_{pi,yy} \mathbf{d}_{pi}, \quad (i = 1, 2, 3), \tag{34e}$$

$$\mathbf{u}_{pi,xy} = \mathbf{B}_{pi,xy} \mathbf{d}_{pi}, \quad (i = 1, 2, 3), \tag{34f}$$

where $\mathbf{B}_{pi}, \mathbf{B}_{pi,x}, \mathbf{B}_{pi,y}, \mathbf{B}_{pi,xx}, \mathbf{B}_{pi,yy}, \mathbf{B}_{pi,xy}$ $(i = 1, 2)$ are $M \times (2N + M)$ and $\mathbf{B}_{p3}, \mathbf{B}_{p3,x}, \mathbf{B}_{p3,y}, \mathbf{B}_{p3,xx}, \mathbf{B}_{p3,yy}, \mathbf{B}_{p3,xy}$ are $M \times (4N + M)$ known coefficient matrices. In evaluating the domain integrals of the kernels over the domain cells, singular and hypersingular integrals arise. They are computed by transforming them to line integrals on the boundary of the cell [50, 51]. The final step of the AEM is to apply (7a)–(7c) to the M nodal points inside Ω, yielding the following equations:

$$\begin{aligned} G_p h_p \Bigg[\mathbf{p}_{p1} &+ \frac{1 + \nu_p}{1 - \nu_p} \left(\mathbf{B}_{p1,xx} \mathbf{d}_{p1} + \mathbf{B}_{p2,xy} \mathbf{d}_{p2} \right) \\ &+ \left\{ \frac{2}{1 - \nu_p} \left(\mathbf{B}_{p3,xx} \mathbf{d}_{p3} \right)_{dg} \mathbf{B}_{p3,x} \mathbf{d}_{p3} + \left(\mathbf{B}_{p3,yy} \mathbf{d}_{p3} \right)_{dg} \cdot \mathbf{B}_{p3,x} \mathbf{d}_{p3} \right\} \\ &+ \frac{1 + \nu_p}{1 - \nu_p} \left(\mathbf{B}_{p3,xy} \mathbf{d}_{p3} \right)_{dg} \mathbf{B}_{p3,y} \mathbf{d}_{p3} \Bigg] \\ &- \rho_p h_p \mathbf{B}_{p1} \ddot{\mathbf{d}}_{p1} = \mathbf{Z} \mathbf{q}_x, \end{aligned} \tag{35a}$$

$$\begin{aligned} G_p h_p \Bigg[\mathbf{p}_{p2} &+ \frac{1 + \nu_p}{1 - \nu_p} \left(\mathbf{B}_{p1,xy} \mathbf{d}_{p1} + \mathbf{B}_{p2,yy} \mathbf{d}_{p2} \right) \\ &+ \left\{ \frac{2}{1 - \nu_p} \left(\mathbf{B}_{p3,yy} \mathbf{d}_{p3} \right)_{dg} \mathbf{B}_{p3,y} \mathbf{d}_{p3} + \left(\mathbf{B}_{p3,xx} \mathbf{d}_{p3} \right)_{dg} \cdot \mathbf{B}_{p3,y} \mathbf{d}_{p3} \right\} \\ &+ \frac{1 + \nu_p}{1 - \nu_p} \left(\mathbf{B}_{p3,xy} \mathbf{d}_{p3} \right)_{dg} \mathbf{B}_{p3,x} \mathbf{d}_{p3} \Bigg] \\ &- \rho_p h_p \mathbf{B}_{p2} \ddot{\mathbf{d}}_{p2} = \mathbf{Z} \mathbf{q}_y, \end{aligned} \tag{35b}$$

$$\begin{aligned} D \mathbf{p}_{p3} - C \times \Bigg\{ \Bigg[& \left(\mathbf{B}_{p1,x} \mathbf{d}_{p1} \right)_{dg} \mathbf{B}_{p3,xx} \mathbf{d}_{p3} \\ &+ \frac{1}{2} \left(\mathbf{B}_{p3,x} \mathbf{d}_{p3} \right)_{dg} \left(\mathbf{B}_{p3,x} \mathbf{d}_{p3} \right)_{dg} \mathbf{B}_{p3,xx} \mathbf{d}_{p3} \\ &+ \nu_p \left(\mathbf{B}_{p2,y} \mathbf{d}_{p2} \right)_{dg} \mathbf{B}_{p3,xx} \mathbf{d}_{p3} + \frac{1}{2} \nu_p \left(\mathbf{B}_{p3,y} \mathbf{d}_{p3} \right)_{dg} \\ &\times \left(\mathbf{B}_{p3,y} \mathbf{d}_{p3} \right)_{dg} \mathbf{B}_{p3,xx} \mathbf{d}_{p3} \Bigg] \\ &+ (1 - \nu_p) \end{aligned}$$

$$\times\Big[\left(\mathbf{B}_{p1,y}\mathbf{d}_{p1}\right)_{dg}\mathbf{B}_{p3,xy}\mathbf{d}_{p3} + \left(\mathbf{B}_{p2,x}\mathbf{d}_{p2}\right)_{dg}\mathbf{B}_{p3,xy}\mathbf{d}_{p3} + \left(\mathbf{B}_{p3,x}\mathbf{d}_{p3}\right)_{dg}\cdot\left(\mathbf{B}_{p3,y}\mathbf{d}_{p3}\right)_{dg}\mathbf{B}_{p3,xy}\mathbf{d}_{p3}\Big]$$
$$+\Big[\left(\mathbf{B}_{p2,y}\mathbf{d}_{p2}\right)_{dg}\mathbf{B}_{p3,yy}\mathbf{d}_{p3} + \frac{1}{2}\left(\mathbf{B}_{p3,y}\mathbf{d}_{p3}\right)_{dg}\cdot\left(\mathbf{B}_{p3,y}\mathbf{d}_{p3}\right)_{dg}\mathbf{B}_{p3,yy}\mathbf{d}_{p3} + \nu_p\left(\mathbf{B}_{p1,x}\mathbf{d}_{p1}\right)_{dg}\mathbf{B}_{p3,yy}\mathbf{d}_{p3} + \frac{1}{2}\nu_p\left(\mathbf{B}_{p3,x}\mathbf{d}_{p3}\right)_{dg}\cdot\left(\mathbf{B}_{p3,x}\mathbf{d}_{p3}\right)_{dg}\mathbf{B}_{p3,yy}\mathbf{d}_{p3}\Big]\Big\}$$
$$+\rho_p h_p \mathbf{B}_{p3}\ddot{\mathbf{d}}_{p3} - \rho_p h_p\left(\mathbf{B}_{p1}\ddot{\mathbf{d}}_{p1}\right)_{dg}\mathbf{B}_{p3,x}\mathbf{d}_{p3} - \rho_p h_p\left(\mathbf{B}_{p2}\ddot{\mathbf{d}}_{p2}\right)_{dg}\mathbf{B}_{p3,y}\mathbf{d}_{p3} - \frac{\rho_p h_p^3}{12}\mathbf{B}_{p3,xx}\mathbf{d}_{p3} - \frac{\rho_p h_p^3}{12}\mathbf{B}_{p3,yy}\mathbf{d}_{p3}$$
$$= \mathbf{g} - \mathbf{Z}\mathbf{q}_z - \mathbf{Z}\mathbf{X}_y\mathbf{q}_y - \mathbf{Z}\mathbf{X}_x\mathbf{q}_x + \left(\mathbf{Z}\mathbf{q}_x\right)_{dg}\mathbf{B}_{p3,x}\mathbf{d}_{p3} + \left(\mathbf{Z}\mathbf{q}_y\right)_{dg}\mathbf{B}_{p3,y}\mathbf{d}_{p3}, \tag{35c}$$

where $\mathbf{q}_x = \{\mathbf{q}_{x1}\ \mathbf{q}_{x2}\}^T$, $\{\mathbf{q}_y\} = \{\mathbf{q}_{y1}\ \mathbf{q}_{y2}\}^T$, and $\{\mathbf{q}_z\} = \{\mathbf{q}_{z1}\ \mathbf{q}_{z2}\}^T$ are vectors of dimension $2L$ including the unknown $q^i_{xj}, q^i_{yj}, q^i_{zj}$ $(j = 1, 2)$ interface forces; $2L$ is the total number of the nodal points at each plate-beam interface; Z is a position $M \times 2L$ matrix which converts the vectors q_x, q_y, q_z into corresponding ones with length M; the symbol $(\cdot)_{dg}$ indicates a diagonal $M\times M$ matrix with the elements of the included column matrix. Matrices X_x and X_y of dimension $2L \times 2L$ result after approximating the bending moment derivatives of $m^i_{pyj} = q^i_{xj}h_p/2$, $m^i_{pxj} = q^i_{yj}h_p/2$, respectively, using appropriately central, backward, or forward differences.

For the Beam Displacement Components u^i_b, v^i_b, and w^i_b and for the Angle of Twist θ^i_{bx}.

According to the precedent analysis, the nonlinear dynamic analysis of the ith beam reduces in establishing the displacement components $u^i_b(x^i,t)$ and $v^i_b(x^i,t)$, $w^i_b(x^i,t)$, $\theta^i_{bx}(x^i,t)$ having continuous derivatives up to the second order and up to the fourth order with respect to x, respectively, and also having derivatives up to the second order with respect to t. Moreover, these displacement components must satisfy the coupled governing differential equations (17a)–(17d) inside the beam, the boundary conditions at the beam ends (18)–(21b), and the initial conditions (22a) and (22b).

Let $u^i_{b1} = u^i_b(x^i,t)$, $u^i_{b2} = v^i_b(x^i,t)$, $u^i_{b3} = w^i_b(x^i,t)$, and $u^i_{b4} = \theta^i_{bx}(x^i,t)$ be the sought solutions of the problem represented by (17a)–(17d). Differentiating with respect to x these functions, two and four times, respectively, yields

$$\frac{\partial^2 u^i_{b1}}{\partial x^{i2}} = p^i_{b1}\left(x^i,t\right), \tag{36a}$$

$$\frac{\partial^4 u^i_{bj}}{\partial x^{i4}} = p^i_{bj}\left(x^i,t\right), \quad (j = 2,3,4). \tag{36b}$$

Equations (36a) and (36b) are quasi-static and indicate that the solution of (17a)–(17d) can be established by solving (36a) and (36b) under the same boundary conditions (equations (18)–(21b)) provided that the fictitious load distributions p^i_{bj} (x, t) $(j = 1, 2, 3, 4)$ are first established. These distributions can be determined following the procedure presented in [49] and employing the constant element assumption for the load distributions p^i_{bj} along the L internal beam elements (as the numerical implementation becomes very simple and the obtained results are of high accuracy). Thus, the integral representations of the displacement components u^i_{bj} $(j = 1, 2, 3, 4)$ and their derivatives with respect to xi when applied to the beam ends $(0, l)$, together with the boundary conditions (18)–(21b), are employed to express the unknown boundary quantities $u^i_{bj}(\zeta^i), u^i_{bj,x}(\zeta^i), u^i_{bj,xx}(\zeta^i)$, and $u^i_{bj,xxx}(\zeta^i)$ $(\zeta^i = 0, l^i)$ in terms of p^i_{bj} $(j = 1,2,3,4)$. Thus, the following set of 28 nonlinear algebraic equations for the ith beam is obtained as

$$\begin{bmatrix} \mathbf{E}^i_{b11} & 0 & 0 & 0 \\ 0 & \mathbf{E}^i_{b22} & 0 & 0 \\ 0 & 0 & \mathbf{E}^i_{b33} & 0 \\ 0 & 0 & 0 & \mathbf{E}^i_{b44} \end{bmatrix} \begin{Bmatrix} \mathbf{d}^i_{b1} \\ \mathbf{d}^i_{b2} \\ \mathbf{d}^i_{b3} \\ \mathbf{d}^i_{b4} \end{Bmatrix} + \begin{Bmatrix} 0 \\ \mathbf{D}^{i\,nl}_{b1} \\ 0 \\ 0 \\ \mathbf{D}^{i\,nl}_{b2} \\ 0 \\ 0 \\ \mathbf{D}^{i\,nl}_{b3} \\ 0 \\ 0 \\ 0 \end{Bmatrix} = \begin{Bmatrix} 0 \\ \boldsymbol{\alpha}^i_{b3} \\ 0 \\ 0 \\ \boldsymbol{\beta}^i_{b3} \\ 0 \\ 0 \\ \boldsymbol{\gamma}^i_{b3} \\ 0 \\ 0 \\ \boldsymbol{\delta}^i_{b3} \end{Bmatrix} \tag{37}$$

where $\mathbf{E}^i_{b11}$ and $\mathbf{E}^i_{b22}, \mathbf{E}^i_{b33}, \mathbf{E}^i_{b44}$ are known matrices of dimension $4 \times (N + 4)$ and $8 \times (N + 8)$, respectively; $\mathbf{D}^{i\,nl}_{b1}$ is a 2×1 and $\mathbf{D}^{i\,nl}_{b1}j$ $(j = 2, 3)$ are 4×1 column matrices containing the nonlinear terms included in the expressions of the boundary conditions (equations (18)–(20b)); $\boldsymbol{\alpha}^i_{b3}$ and $\boldsymbol{\beta}^i_{b3}, \boldsymbol{\gamma}^i_{b3}, \boldsymbol{\delta}^i_{b3}$ are 2×1 and 4×1 known column matrices, respectively, including the boundary values of the functions a^i_{b3} and $\beta^i_{b3}, \bar{\beta}^i_{b3}, \gamma^i_{b3}, \bar{\gamma}^i_{b3}, \delta^i_{b3}, \bar{\delta}^i_{b3}$ of (18)–(21b). Finally

$$\mathbf{d}^i_{b1} = \left\{\mathbf{p}^i_{b1} \quad \hat{\mathbf{u}}^i_{b1} \quad \hat{\mathbf{u}}^i_{b1,x}\right\}^T, \tag{38a}$$

$$\mathbf{d}^i_{bj} = \left\{\mathbf{p}^i_{bj} \quad \hat{\mathbf{u}}^i_{bj} \quad \hat{\mathbf{u}}^i_{bj,x} \quad \hat{\mathbf{u}}^i_{bj,xx} \quad \hat{\mathbf{u}}^i_{bj,xxx}\right\}^T, \quad (j = 2, 3, 4) \tag{38b}$$

are generalized unknown vectors, where

$$\hat{\mathbf{u}}^i_{bj} = \left\{u^i_{bj}(0,t) \quad u^i_{bj}\left(l^i,t\right)\right\}^T, \quad (j = 1, 2, 3, 4), \tag{39a}$$

$$\hat{\mathbf{u}}^i_{bj,x} = \left\{\frac{\partial u^i_{bj}(0,t)}{\partial x^i} \quad \frac{\partial u^i_{bj}\left(l^i,t\right)}{\partial x^i}\right\}^T, \quad (j = 1, 2, 3, 4), \tag{39b}$$

$$\hat{\mathbf{u}}^i_{bj,xx} = \left\{\frac{\partial^2 u^i_{bj}(0,t)}{\partial x^{i2}} \quad \frac{\partial^2 u^i_{bj}\left(l^i,t\right)}{\partial x^{i2}}\right\}^T, \quad (j = 2, 3, 4), \tag{39c}$$

$$\hat{\mathbf{u}}^i_{bj,xxx} = \left\{\frac{\partial^3 u^i_{bj}(0,t)}{\partial x^{i3}} \quad \frac{\partial^3 u^i_{bj}\left(l^i,t\right)}{\partial x^{i3}}\right\}^T, \quad (j = 2, 3, 4) \tag{39d}$$

are vectors including the two unknown boundary values of the respective boundary quantities and $p^i_{bj} = \left\{(p^i_{bj})_1 \ (p^i_{bj})_2 \ \cdots \ (p^i_{bj})_L\right\}^T$ $(j = 1, 2, 3, 4)$ are vectors including the L unknown nodal values of the fictitious loads.

Discretization of the integral representations of the unknown quantities u^i_{bj} $(j = 1, 2, 3, 4)$ and those of their derivatives with respect to xi inside the beam $x^i \in (0, l^i)$ and application to the L collocation nodal points yields

$$\mathbf{u}^i_{b1} = \mathbf{B}^i_{b1}\mathbf{d}^i_{b1}, \tag{40a}$$

$$\mathbf{u}^i_{b1,x} = \mathbf{B}^i_{b1,x}\mathbf{d}^i_{b1}, \tag{40b}$$

$$\mathbf{u}^i_{bj} = \mathbf{B}^i_{bj}\mathbf{d}^i_{bj}, \quad (j = 2, 3, 4), \tag{40c}$$

$$\mathbf{u}^i_{bj,x} = \mathbf{B}^i_{bj,x}\mathbf{d}^i_{bj}, \quad (j = 2, 3, 4), \tag{40d}$$

$$\mathbf{u}^i_{bj,xx} = \mathbf{B}^i_{bj,xx}\mathbf{d}^i_{bj}, \quad (j = 2, 3, 4), \tag{40e}$$

$$\mathbf{u}^i_{bj,xxx} = \mathbf{B}^i_{bj,xxx}\mathbf{d}^i_{bj}, \quad (j = 2, 3, 4), \tag{40f}$$

where $\mathbf{B}^i_{b1}, \mathbf{B}^i_{b1,x}$ are $L \times (L + 4)$ and $\mathbf{B}^i_{bj}, \mathbf{B}^i_{bj,x}, \mathbf{B}^i_{bj,xx}, \mathbf{B}^i_{bj,xxx}$ $(j = 2, 3, 4)$ are $L \times (L + 8)$ known coefficient matrices, respectively.

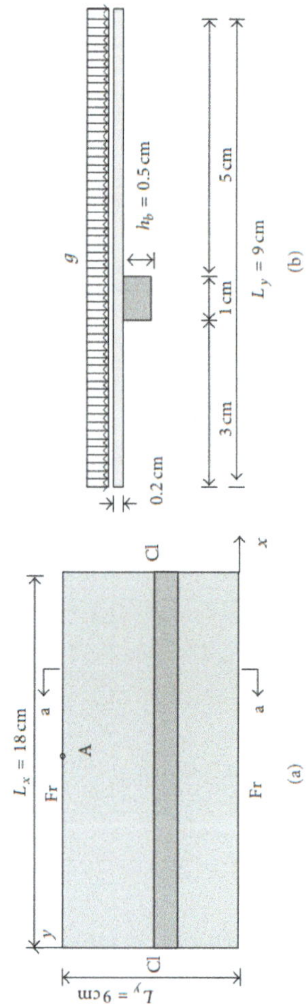

Figure 4: Plan view (a) and section a-a (b) of the stiffened plate of Example 1.

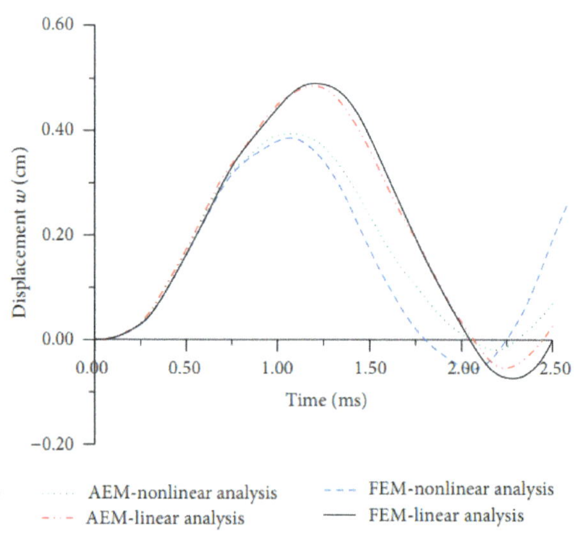

Figure 5: Time history of deflection w (cm) at the middle point A of the free edge of the plate of Example 1.

Applying (17a)–(17d) to the L collocation points and employing (40a)–(40f), $4 \times L$ nonlinear algebraic equations for each (ith) beam are formulated as

$$- E_b^i A_b^i \left[\mathbf{p}_{b1}^i + \left(\mathbf{B}_{b2,x}^i \mathbf{d}_{b2}^i \right)_{dg} \mathbf{B}_{b2,xx}^i \mathbf{d}_{b2}^i + \left(\mathbf{B}_{b3,x}^i \mathbf{d}_{b3}^i \right)_{dg} \mathbf{B}_{b3,xx}^i \mathbf{d}_{b3}^i \right] + \rho_b^i \mathbf{A}_b^i \mathbf{T}_1 \ddot{\mathbf{q}}_1 = \mathbf{q}_{x1}^i + \mathbf{q}_{x2}^i, \tag{41a}$$

$$E_b^i I_z^i \mathbf{p}_{b2}^i - \left(\mathbf{N}_b^i \right)_{dg} \mathbf{B}_{b2,xx}^i \mathbf{d}_{b2}^i - \rho_b^i I_z^i \mathbf{B}_{b2,xx}^i \ddot{\mathbf{d}}_{b2}^i + \rho_b^i \mathbf{A}_b^i \mathbf{B}_{b2}^i \ddot{\mathbf{d}}_{b2}^i - \rho_b^i \mathbf{A}_b^i \left(\mathbf{B}_{b1}^i \ddot{\mathbf{d}}_{b1}^i \right)_{dg} \mathbf{B}_{b2,x}^i \mathbf{d}_{b2}^i = \mathbf{q}_{y1}^i + \mathbf{q}_{y2}^i - \left(\mathbf{B}_{b2,x}^i \mathbf{d}_{b2}^i \right)_{dg} \left(\mathbf{q}_{x1}^i + \mathbf{q}_{x2}^i \right) - \mathbf{X}_{by}^i \left(\mathbf{q}_{x1}^i + \mathbf{q}_{x2}^i \right), \tag{41b}$$

$$E_b^i I_y^i \mathbf{p}_{b3}^i - \left(\mathbf{N}_b^i \right)_{dg} \mathbf{B}_{b3,xx}^i \mathbf{d}_{b3}^i - \rho_b^i I_y^i \mathbf{B}_{b3,xx}^i \ddot{\mathbf{d}}_{b3}^i + \rho_b^i \mathbf{A}_b^i \mathbf{B}_{b3}^i \ddot{\mathbf{d}}_{b3}^i - \rho_b^i \mathbf{A}_b^i \left(\mathbf{B}_{b1}^i \ddot{\mathbf{d}}_{b1}^i \right)_{dg} \mathbf{B}_{b3,x}^i \mathbf{d}_{b3}^i = \mathbf{q}_{z1}^i + \mathbf{q}_{z2}^i - \left(\mathbf{B}_{b3,x}^i \mathbf{d}_{b3}^i \right)_{dg} \left(\mathbf{q}_{x1}^i + \mathbf{q}_{x2}^i \right) + \mathbf{X}_{bz}^i \left(\mathbf{q}_{x1}^i + \mathbf{q}_{x2}^i \right), \tag{41c}$$

$$E_b^i C_S^i \mathbf{p}_{b4}^i - G_b^i I_t^i \mathbf{B}_{b4,xx}^i \mathbf{d}_{b4}^i + \rho_b^i I_p^i \mathbf{B}_{b4}^i \ddot{\mathbf{d}}_{b4}^i - \rho_b^i C_S^i \mathbf{B}_{b4,xx}^i \ddot{\mathbf{d}}_{b4}^i = \mathbf{e}_{y1}^i \mathbf{q}_{z1}^i + \mathbf{e}_{y2}^i \mathbf{q}_{z2}^i - \mathbf{e}_{z1}^i \mathbf{q}_{y1}^i - \mathbf{e}_{z2}^i \mathbf{q}_{y2}^i + \mathbf{X}_{bw}^i \left(\mathbf{q}_{x1}^i + \mathbf{q}_{x2}^i \right), \tag{41d}$$

where $(\mathbf{N}_b^i)_{dg}$ is a diagonal $L\times L$ matrix including the values of the axial forces of the ith beam; the symbol $(\cdot)_{dg}$ indicates a diagonal $L\times L$ matrix with the elements of the included column matrix. The matrices $\mathbf{X}_{bz}^i, \mathbf{X}_{by}^i, \mathbf{X}_{bw}^i$ result after approximating the derivatives of $m_{byj}^i, m_{bzj}^i, m_{bwj}^i$ using appropriately central, backward, or forward differences. Their dimensions are also $L\times L$. Moreover $\mathbf{e}_{y1}^i, \mathbf{e}_{y2}^i, \mathbf{e}_{z1}^i, \mathbf{e}_{z2}^i$ are diagonal $L\times L$ matrices including the values of the eccentricities e_{yj}^i, e_{zj}^i of the components q_{zj}^i, q_{yj}^i with respect to the ith beam shear center axis (coinciding with its centroid), respectively.

Employing (34a)–(34f) and (40a)–(40f), the discretized counterpart of (26a)-(26b), (27a)-(27b) and (28a)-(28b) at the L nodal points of each interface, is written as

$$\mathbf{Y}_1 \mathbf{B}_{p1} \mathbf{d}_{p1} - \mathbf{B}_{b1}^i \mathbf{d}_{b1}^i = \frac{h_p}{2} \mathbf{Y}_1 \mathbf{B}_{p3,x} \mathbf{d}_{p3} + \frac{h_b^i}{2} \mathbf{B}_{b3,x}^i \mathbf{d}_{b3} + \frac{b_f^i}{4} \mathbf{B}_{b2,x}^i \mathbf{d}_{b2} + \left(\varphi_S^{Pi} \right)_{f_1} \mathbf{B}_{b4,x}^i \mathbf{d}_{b4} + \mathbf{K}_{x1}^i \left[1 - \frac{1}{2} \left(\mathbf{B}_{b3,x}^i \mathbf{d}_{b3} \right)_{dg} \left(\mathbf{B}_{b3,x}^i \mathbf{d}_{b3} \right)_{dg} \right] \left(\mathbf{q}_{x1}^i + \mathbf{q}_{x2}^i \right), \tag{42a}$$

$$\mathbf{Y}_2 \mathbf{B}_{p1} \mathbf{d}_{p1} - \mathbf{B}_{b1}^i \mathbf{d}_{b1}^i = \frac{h_p}{2} \mathbf{Y}_2 \mathbf{B}_{p3,x} \mathbf{d}_{p3} + \frac{h_b^i}{2} \mathbf{B}_{b3,x}^i \mathbf{d}_{b3} - \frac{b_f^i}{4} \mathbf{B}_{b2,x}^i \mathbf{d}_{b2} + \left(\varphi_S^{Pi} \right)_{f_2} \mathbf{B}_{b4,x}^i \mathbf{d}_{b4} + \mathbf{K}_{x2}^i \left[1 - \frac{1}{2} \left(\mathbf{B}_{b3,x}^i \mathbf{d}_{b3} \right)_{dg} \left(\mathbf{B}_{b3,x}^i \mathbf{d}_{b3} \right)_{dg} \right] \left(\mathbf{q}_{x1}^i + \mathbf{q}_{x2}^i \right), \tag{42b}$$

$$\mathbf{Y}_1 \mathbf{B}_{p2} \mathbf{d}_{p2} - \mathbf{B}_{b2}^i \mathbf{d}_{b2}^i = \frac{h_p}{2} \mathbf{Y}_1 \mathbf{B}_{p3,y} \mathbf{d}_{p3} + \frac{h_b^i}{2} \mathbf{B}_{b4}^i \mathbf{d}_{b4}^i + \mathbf{K}_{y1}^i \left[1 - \frac{1}{2} \left(\mathbf{B}_{b4,x}^i \mathbf{d}_{b4} \right)_{dg} \left(\mathbf{B}_{b4,x}^i \mathbf{d}_{b4} \right)_{dg} \right] \left(\mathbf{q}_{x1}^i + \mathbf{q}_{x2}^i \right), \tag{43a}$$

$$\mathbf{Y}_2 \mathbf{B}_{p2} \mathbf{d}_{p2} - \mathbf{B}_{b2}^i \mathbf{d}_{b2}^i = \frac{h_p}{2} \mathbf{Y}_2 \mathbf{B}_{p3,y} \mathbf{d}_{p3} + \frac{h_b^i}{2} \mathbf{B}_{b4}^i \mathbf{d}_{b4}^i + \mathbf{K}_{y2}^i \left[1 - \frac{1}{2} \left(\mathbf{B}_{b4,x}^i \mathbf{d}_{b4} \right)_{dg} \left(\mathbf{B}_{b4,x}^i \mathbf{d}_{b4} \right)_{dg} \right] \left(\mathbf{q}_{x1}^i + \mathbf{q}_{x2}^i \right), \tag{43b}$$

$$\mathbf{Y}_1 \mathbf{B}_{p3} \mathbf{d}_{p3} - \mathbf{B}_{b3}^i \mathbf{d}_{b3}^i = -\frac{b_f^i}{4} \mathbf{B}_{b4}^i \mathbf{d}_{b4}^i, \tag{44a}$$

$$\mathbf{Y}_2 \mathbf{B}_{p3} \mathbf{d}_{p3} - \mathbf{B}_{b3}^i \mathbf{d}_{b3}^i = \frac{b_f^i}{4} \mathbf{B}_{b4}^i \mathbf{d}_{b4}^i. \tag{44b}$$

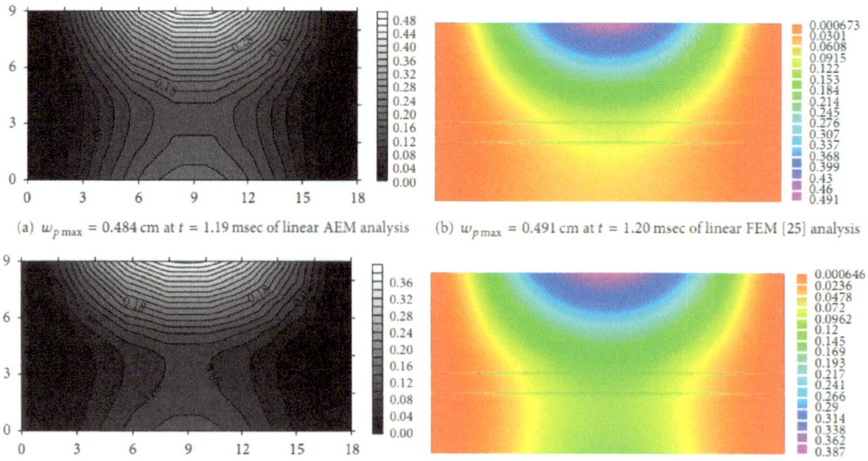

(a) $w_{p\,\max} = 0.484$ cm at $t = 1.19$ msec of linear AEM analysis (b) $w_{p\,\max} = 0.491$ cm at $t = 1.20$ msec of linear FEM [25] analysis

(c) $w_{p\,\max} = 0.394$ cm at $t = 1.08$ msec of nonlinear AEM analysis (d) $w_{p\,\max} = 0.387$ cm at $t = 1.06$ msec of nonlinear FEM [25] analysis

Figure 6: Contour lines of w_p of the stiffened plate of Example 1, employing the present study (a, c) and a FEM [25] solution using solid elements (b, d) at the time of maximum transverse displacement.

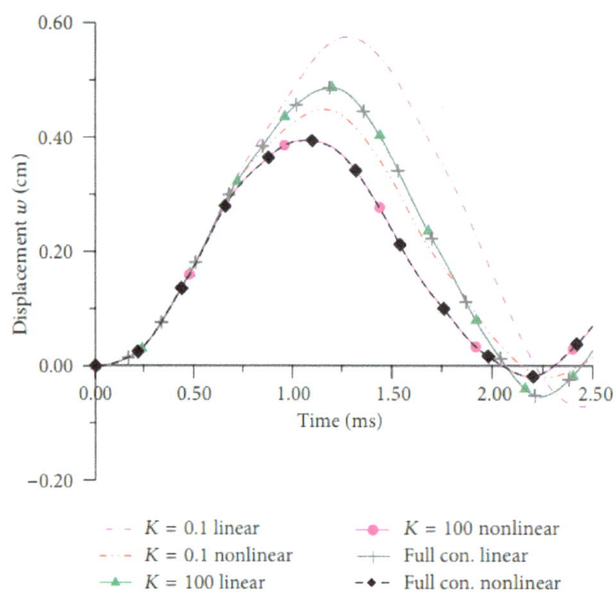

Figure 7: Time history of the deflection w (cm) at the middle point A of the free edge of the stiffened plate of Example 1 for various values of connectors' stiffness.

Equations (30), (35a)–(35c), (37), and (41a)–(41d) together with continuity conditions (42a), (42b), (43a), (43b), (44a), and (44b) constitute a nonlinear system of algebraic equations with respect to q_{x1}, q_{x2}, q_{y1}, q_{y2}, q_{z1}, and q_{z2} (interface forces) and $\mathbf{d}_{pi}$ $(i = 1, 2, 3)$ and $\mathbf{d}_{bj}^{i}$ $(i = 1 \cdots I)$, $(j = 1, 2, 3, 4)$ (generalized unknown vectors of the plate and the beams). This system is solved using iterative numerical methods [52].

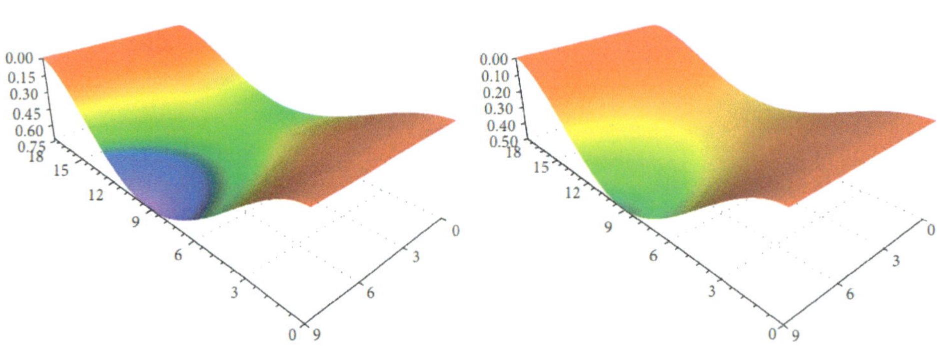

(a) $w_{p\,\max} = 0.725$ cm at $t = 1.47$ msec of linear AEM analysis for $K = 0.005$ (b) $w_{p\,\max} = 0.501$ cm at $t = 1.23$ msec of nonlinear AEM analysis for $K = 0.005$

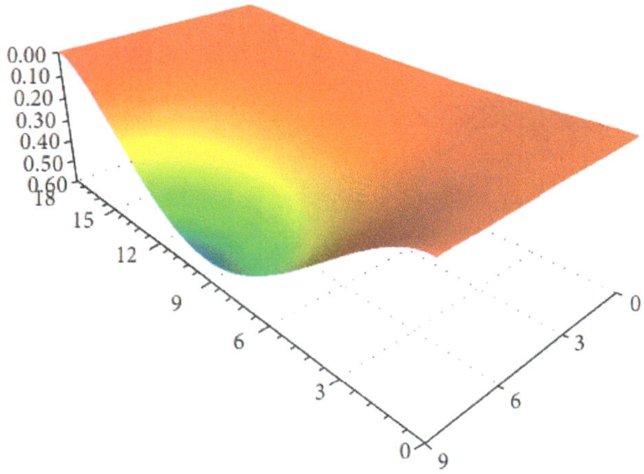

(c) $w_{p\,\max} = 0.521$ cm at $t = 1.22$ msec of linear AEM analysis for $K = 1.0$

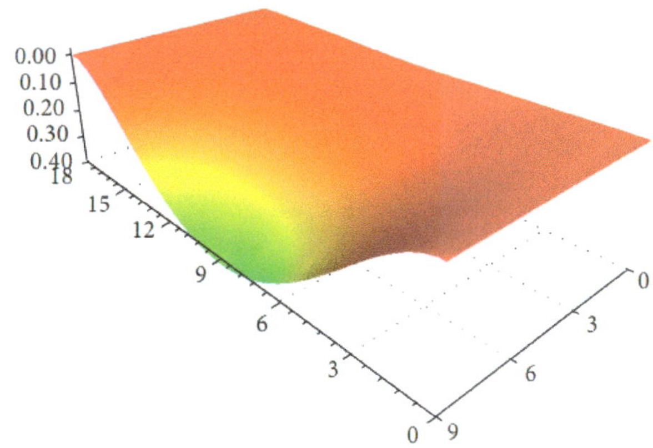

(d) $w_{p\,\max} = 0.419$ cm at $t = 1.11$ msec of nonlinear AEM analysis for $K = 1.0$

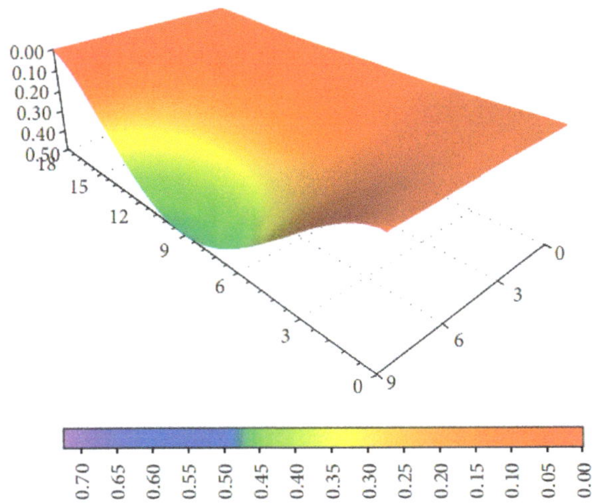

(e) $w_{p\,\max} = 0.495$ cm at $t = 1.20$ msec of linear AEM analysis for $K = 10.0$

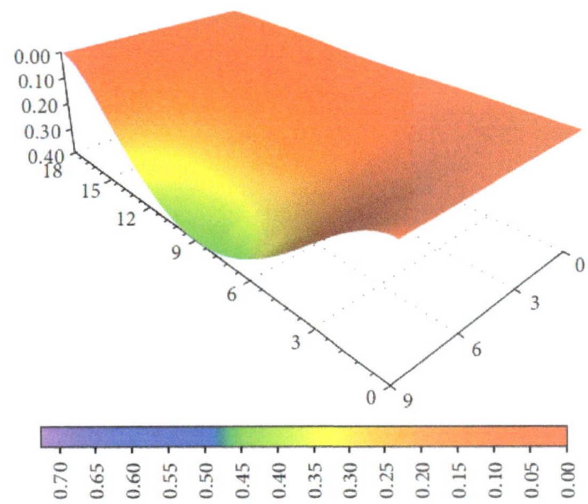

(f) $w_{p\,\max} = 0.401$ cm at $t = 1.08$ msec of nonlinear AEM analysis for $K = 10.0$

Figure 8: Contour lines of w_p of the stiffened plate of Example 1 at the time of maximum transverse deflection ignoring (a, c, e) or taking into account geometrical nonlinearities (b, d, f).

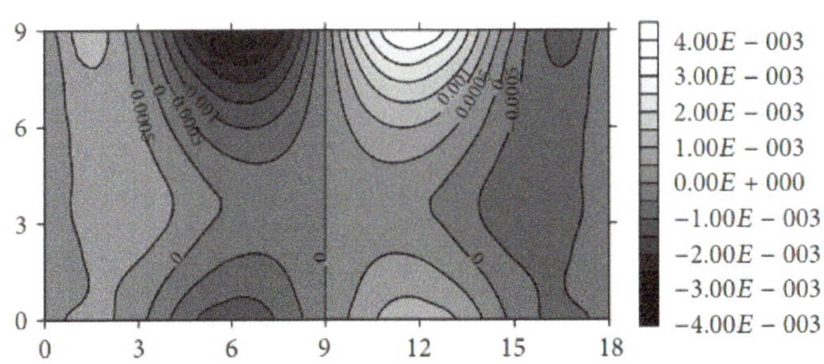

(a) $u_{p\,\max} = 4.5 \times 10^{-3}$ cm at $t = 1.23$ msec for $K = 0.005$

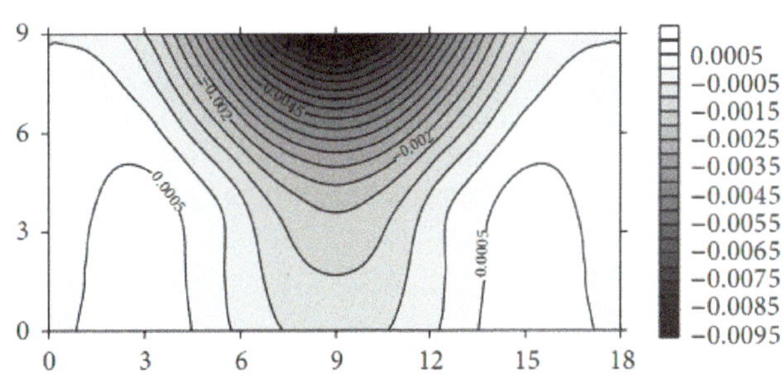

(b) $v_{p\,\max} = 9.3 \times 10^{-3}$ cm at $t = 1.23$ msec for $K = 0.005$

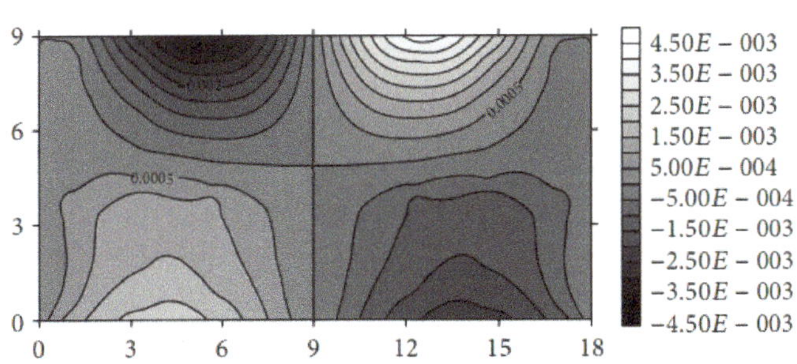

(c) $u_{p\,\max} = 4.3 \times 10^{-3}$ cm at $t = 1.11$ msec for $K = 1.0$

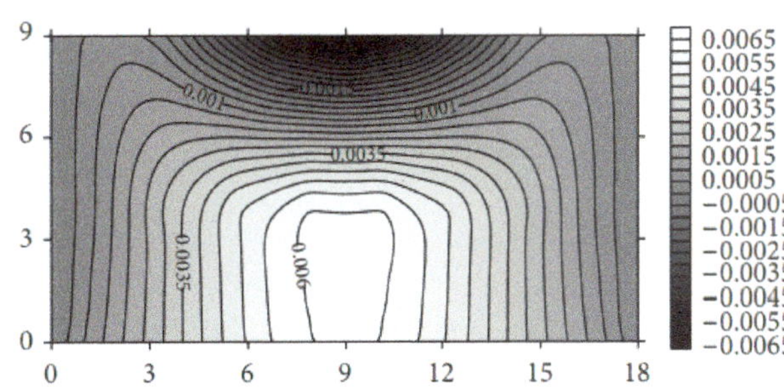

(d) $v_{p\,\max} = 6.4 \times 10^{-3}$ cm at $t = 1.11$ msec for $K = 1.0$

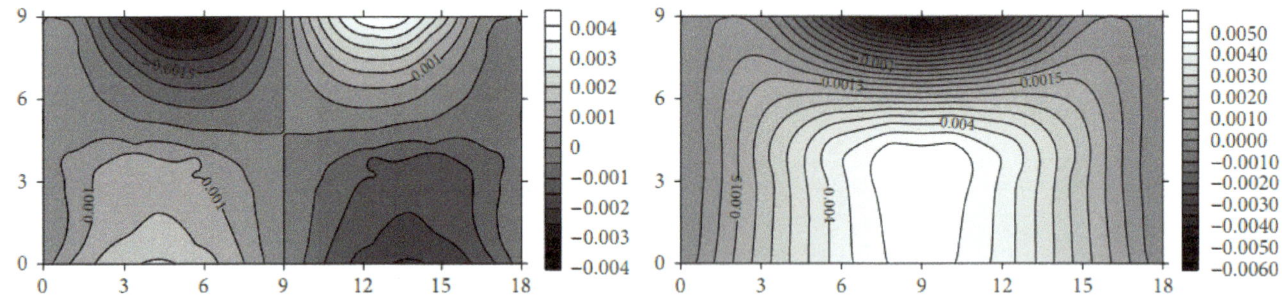

(e) $u_{p\,max} = 4.0 \times 10^{-3}$ cm at $t = 1.08$ msec for $K = 10.0$ (f) $v_{p\,max} = 6.0 \times 10^{-3}$ cm at $t = 1.08$ msec for $K = 10.0$

Figure 9: Contour lines of u_p, V_p of the stiffened plate of Example 1 at the time of maximum transverse deflection w_p taking into account geometrical nonlinearities.

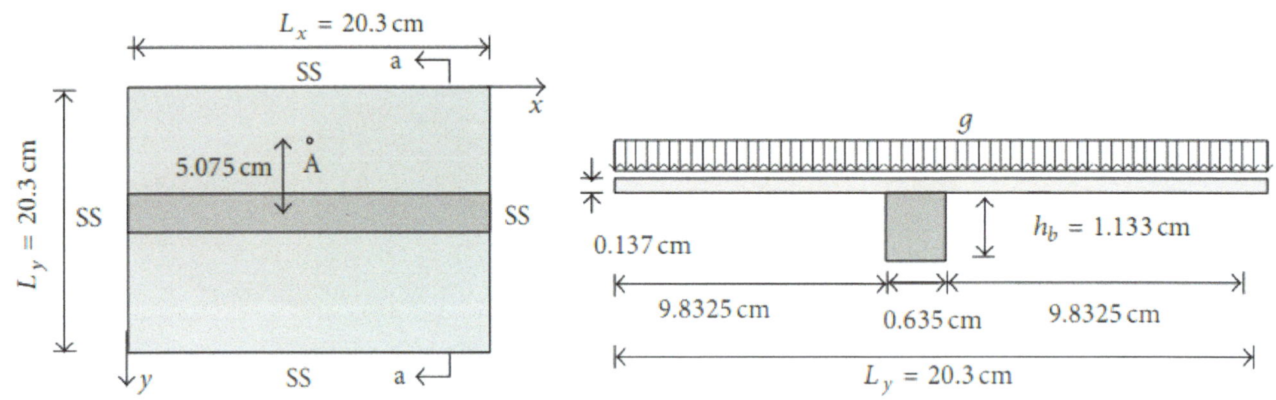

(a) (b)

Figure 10: Plan view (a) and section a-a (b) of the stiffened plate of Example 2.

It is worth noting here that Y_1, Y_2 are position $L{\times}M$ matrices which convert the matrices B_{pi} $(i = 1, 2, 3)$, $\mathbf{B}_{p3,x}$, and $\mathbf{B}_{p3,y}$ into corresponding ones with dimensions $L{\times}M$, appropriately referring to the nodal points of the two interface lines $f^i_{j=1}$, $f^i_{j=2}$, respectively. Moreover, $\mathbf{K}^i_{x1}$, $\mathbf{K}^i_{x2}$, $\mathbf{K}^i_{y1}$, and $\mathbf{K}^i_{y2}$ are diagonal $L{\times}L$ matrices including the value of the flexibilities $1/k^i_{xj}$ and $1/k^i_{yj}$ $(j = 1, 2)$ of the arbitrary distributed shear connectors along the xi and yi directions, respectively.
Finally, it is worth noting that beams placed along the boundary of the plate are treated as every other stiffening beam, since the lines of action of the integrated interface force components q^i_{xj}, q^i_{yj}, and q^i_{zj} $(j = 1, 2)$ will also be internal ones, taking special care during the numerical evaluation of the line integrals in order to avoid their "near singular integral behaviour." According to this, boundary elements that are very close to each other (distance smaller than their length) are divided in subelements, in each of which Gauss integration is applied [51].

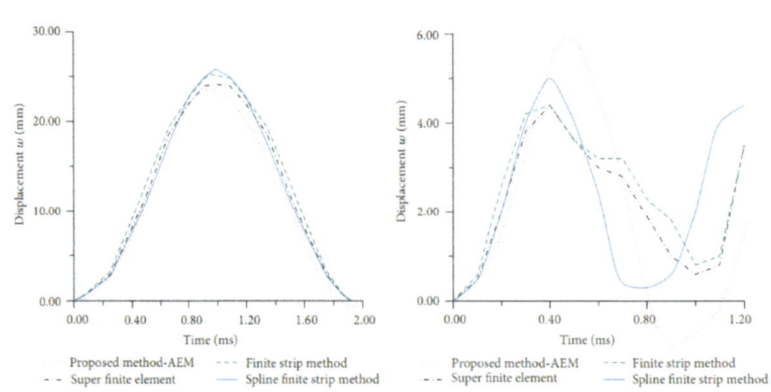

(a) (b)

Figure 11: Time history of linear (a) and nonlinear (b) response of the deflection w (mm) of the half panel center A of the stiffened plate of Example 2.

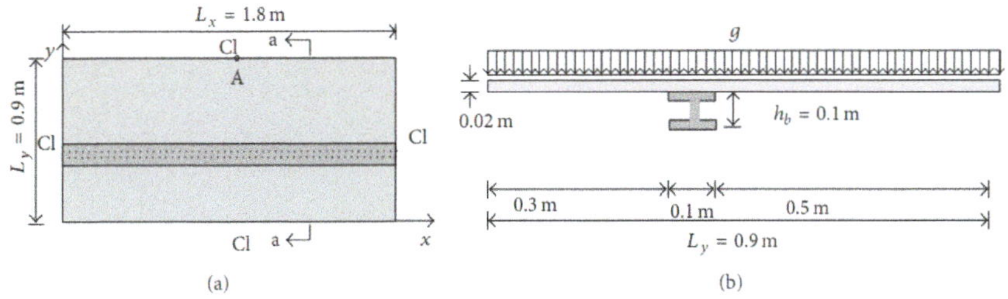

Figure 12: Plan view (a) and section a-a (b) of the stiffened plate of Example 3.

NUMERICAL EXAMPLES

On the basis of the analytical and numerical procedures presented in the previous sections, a FORTRAN program has been written and representative examples have been studied to demonstrate the effectiveness, wherever possible, the accuracy, and the range of applications of the proposed method. It is noted that the term "linear analysis" appearing in all of the following sections refers to the solution of the previously obtained system of equations neglecting all of the nonlinear terms.

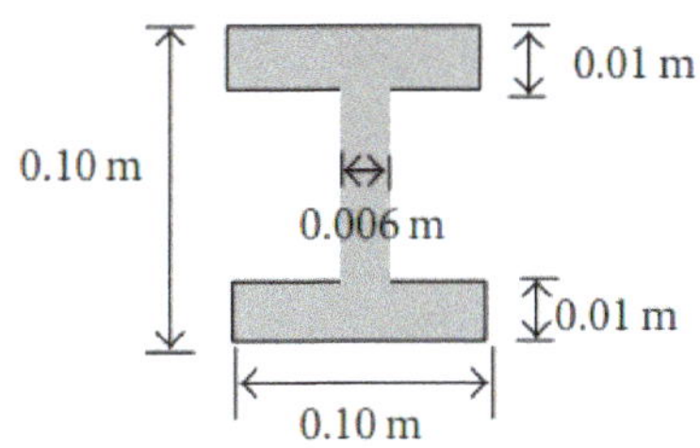

Figure 13: Cross-section of the stiffener of Example 3.

Example 1. A rectangular plate ($h_p = 0.2$ cm, $E_p = E_b = 3 \times 10^7\ \mathrm{kN/m^2}$, $\rho_p = \rho_b = 2.5\ \mathrm{tn/m^3}$, $\nu_p = \nu_b = 0.2$) with dimensions $L_x \times L_y = 18$ cm × 9 cm subjected to a suddenly applied uniform load $g = 50$ kN/m² and stiffened by a rectangular beam of 0.5 cm height and 1.0 cm width, eccentrically placed with respect to the plate center line (Figure 4), has been studied. The plate is clamped along its small edges, while the rest of the edges are free according to the transverse and inplane boundary conditions.

Table 1: Torsion, warping constants, and primary warping function at the interface lines of the stiffening beam of Example 1

h_b (cm)	I_t (cm^4)	C_S (cm^6)	$(\phi_S^P)_{f1}$ (cm^2)	$(\phi_S^P)_{f2}$ (cm^2)
0.5	2.859E − 02	3.175E − 04	−4.979E − 02	4.979E − 02

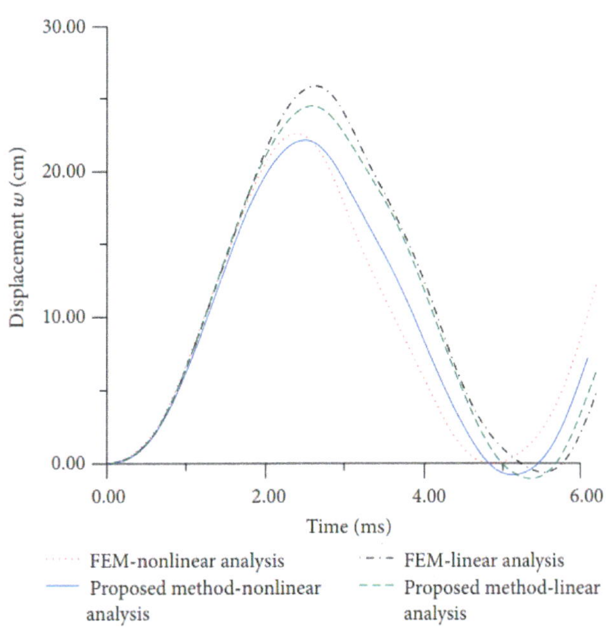

Figure 14: Time history of the maximum displacement w (mm) of the stiffened plate of Example 3.

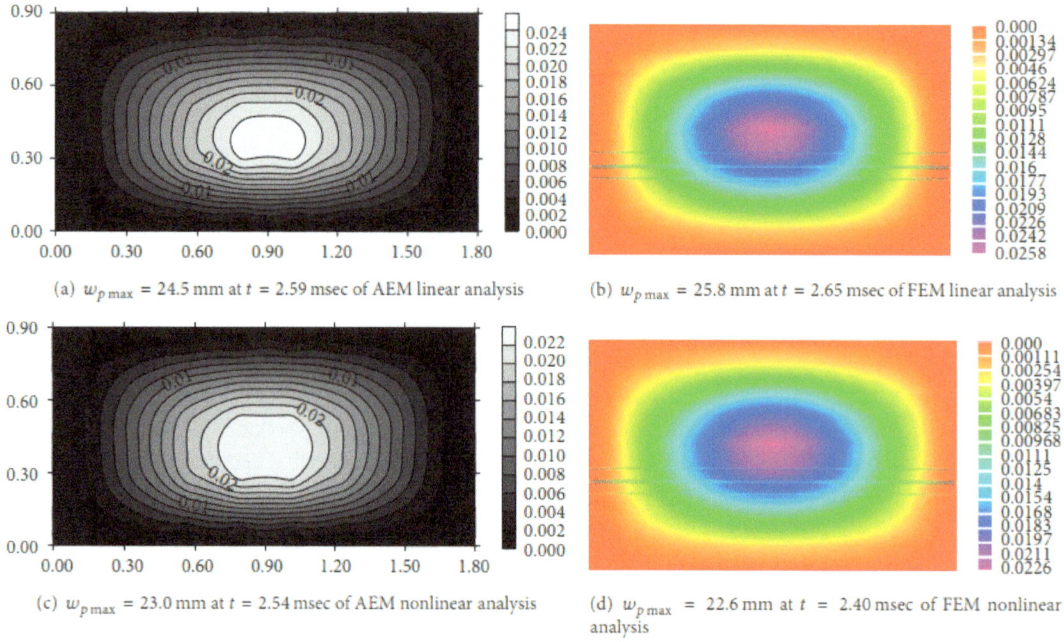

(a) $w_{p\,max}$ = 24.5 mm at t = 2.59 msec of AEM linear analysis

(b) $w_{p\,max}$ = 25.8 mm at t = 2.65 msec of FEM linear analysis

(c) $w_{p\,max}$ = 23.0 mm at t = 2.54 msec of AEM nonlinear analysis

(d) $w_{p\,max}$ = 22.6 mm at t = 2.40 msec of FEM nonlinear analysis

Figure 15: Contour lines of deflection w_p of the stiffened plate of Example 3, employing the present study (a, c) and a FEM [25] solution using solid elements (b, d) at the time of maximum transverse deflection.

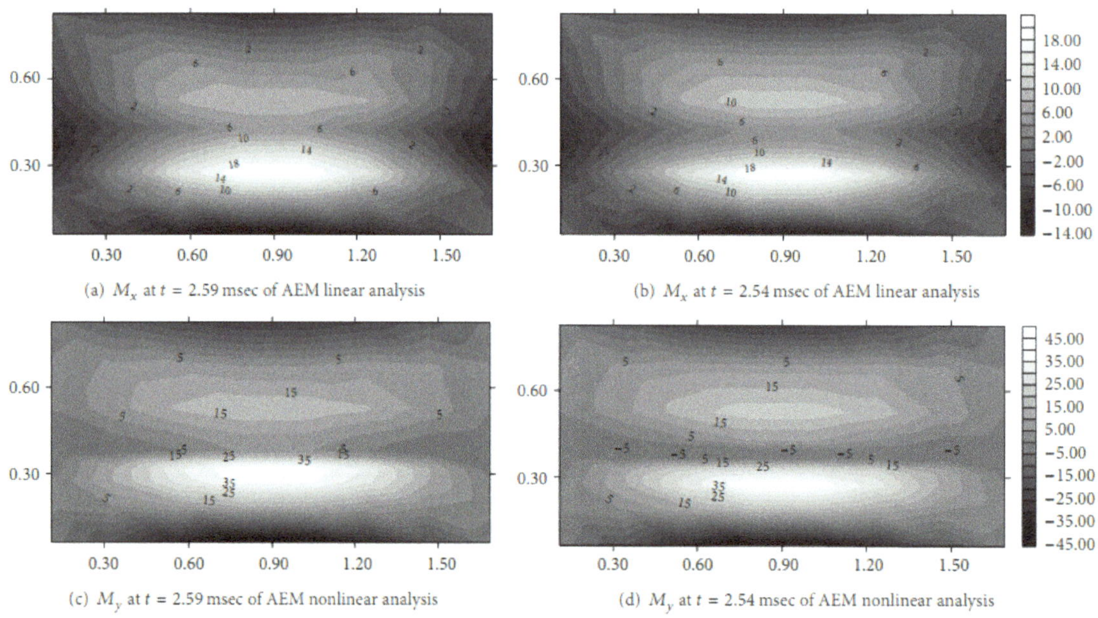

(a) M_x at t = 2.59 msec of AEM linear analysis

(b) M_x at t = 2.54 msec of AEM linear analysis

(c) M_y at t = 2.59 msec of AEM nonlinear analysis

(d) M_y at t = 2.54 msec of AEM nonlinear analysis

Figure 16: Contour lines of moments M_x and M_y (kNm/m) at the time of maximum transverse displacement.

Table 2: Maximum deflection w_p (cm) at point A of the first cycle of motion of the stiffened plate for various values of K of Example 1

K	Linear analysis	Nonlinear analysis
Full connection	0.484	0.391
100.0	0.488	0.395
10.0	0.495	0.401
1.0	0.521	0.419
0.1	0.575	0.449
0.01	0.696	0.494
0.005	0.725	0.501

In Table 1, the torsion It and warping CS constants of the beam cross-section and the values of the primary warping function $(\varphi_S^P)_j$ (j = 1, 2) at the nodes of the two interface lines are presented.

The connection between the slab and the beam is accomplished using a linear distribution of shear connectors along each interface. The adopted relationship for the shear connectors' stiffness is given as

$$k^i_{xj} = k^i_{yj} = 250K \left| x^i - \frac{l^i_{bx}}{2} \right| \text{ kN/m}^2$$

$$(j = 1, 2), \tag{45}$$

where K is a dimensionless magnification factor. In Table 2 the obtained maximum deflections w_p of the first cycle of motion of the stiffened plate, at the middle of the free edge A, are shown for various values of the factor K, performing either a linear or a nonlinear analysis. In Figure 5 the time histories of the deflection $w_p(t)$ at the middle point A of the free edge of the examined stiffened plate, for the case of full connection between the plate and the beam, taking into account or ignoring geometrical nonlinearities, are presented as compared with those obtained from FEM solutions employing 8-noded hexahedral solid finite elements [25].

Table 3: Torsion, warping constants, and primary warping function at the interface lines of the stiffening beam

I_t (m^4)	C_S (m^6)	$(\phi^P_S)_{f1}$ (m^2)	$(\phi^P_S)_{f2}$ (m^2)
$6.984E-08$	$3.374E-09$	$-9.94E-04$	$9.94E-04$

Moreover, in Figure 6 the contour lines of the deflection wp of the stiffened plate (full connection) at the time of maximum transverse displacement are presented as compared with those obtained from the aforementioned FEM solutions, ignoring (Figures 6(a) and 6(b)) or taking into account (Figures 6(c) and 6(d)) geometrical nonlinearities. In order to demonstrate the influence of the shear connectors in the dynamic behaviour of the stiffened plate, in Figure 7, the time histories of the deflection $w(t)$ at the same point A are presented, for various values of the factor K, performing either a linear or a nonlinear analysis. In Figure 8 the contour lines of the deflections w_p of the stiffened plate at the time of maximum displacement are presented, for various cases of connectors' stiffness, ignoring (Figures 8(a), 8(c), and 8(e)) or taking into account (Figures 8(b), 8(d), and 8(f)) geometrical nonlinearities. Moreover, in Figure 9, the contour lines of the displacements u_p, V_p of the stiffened plate at the time of maximum deflection w_p are presented employing nonlinear dynamic analysis.

Example 2. The linear and nonlinear response of a rectangular plate having a central stiffener, as shown in Figure 10, has been studied (E = 68.9 GPa, V = 0.30, ρ = 2.67 tn/m^3). The plate is subjected to a suddenly applied uniform load of g = 300 kPa, while its boundaries are simply supported with restraint against inplane motion. In Figure 11 the linear and nonlinear time history response of the deflection of the half panel center A is depicted. In this figure the obtained results are compared with those presented by Jiang and Olson [53] employing conventional finite strip method, by Koko [54] employing super finite element method, and by Sheikh and Mukhopadhyay [22] employing the spline finite strip method.

As it can easily be observed, there is significant agreement between the results concerning the linear dynamic analysis while deviation between the different types of analysis is observed in nonlinear dynamic analysis, which has been already mentioned by the previous investigators [22, 53, 54].

Example 3. A rectangular plate with dimensions $L_x \times L_y$ = 1.8 m × 0.9 m (Figure 12) stiffened by an I cross-section beam (Figure 13) eccentrically placed with respect to the plate centre line, clamped along all its edges, and subjected to a suddenly applied uniform load g = 750 kN/m^2 has been studied (h_p = 0.02 m, $E_p = E_b$ = 68.9 × 10^6 kN/m^2 , $\rho_p = \rho b$ = 2.67 tn/m^3 , v_p = 0.3). In Table 3, the torsion It and warping C_S constants of the beam cross-section and the values of the primary warping function $(\varphi^P_S)_j$ (j = 1, 2) at the nodes of the two interface lines are presented. In Figure 14 the time histories of the maximum deflection $w(t)$ of the plate for the cases of linear and nonlinear dynamic analysis are presented as compared with those obtained from FEM solutions using 8-noded hexahedral solid finite elements [25].

Moreover, in Figure 15, the contour lines of the displacement w_p of the stiffened plate at the time of maximum transverse displacement are presented as compared with those obtained from the aforementioned FEM solutions, ignoring (Figures 15(a) and 15(b)) or taking into account (Figures 15(c) and 15(d)) geometrical nonlinearities. The convergence of the results between the two methods is noteworthy. Finally, in Figure 16, the contour lines of the corresponded moments M_x, M_y at the time of maximum transverse displacement are presented.

Taking into account the aforementioned convergence of the results between the two methods (AEM, FEM), the importance of the reduction of calculation time when employing the proposed method should be highlighted. More specifically, for the determination of the beam's response to the dynamic loading, a personal computer of 8 Gb RAM and Intel I7 processor of 4 cores with maximum clock speed equal to 3.8 GHz was used. The time step used for the nonlinear analysis is 1 microsecond (μsec) and for a full circle response the calculation time employing the proposed method was 20 minutes as compared with FEM analysis which lasted 50 minutes.

CONCLUDING REMARKS

A general solution for the geometrically nonlinear dynamic analysis of plates stiffened by arbitrarily placed parallel beams of arbitrary doubly symmetric cross-section subjected to arbitrary loading is presented. The proposed model takes into account the nonuniform distribution of the interface shear forces and the nonuniform torsional response of the beams. The main conclusions that can be drawn from this investigation are as follows.(a)The proposed model permits the dynamic response of stiffened plates subjected to arbitrary loading, while both the number and the placement of the parallel beams are also arbitrary (eccentric beams are included). The plate and the beams are supported by the most general boundary conditions including elastic support or restraint.(b)The adopted model permits the evaluation of the longitudinal and transverse inplane shear forces at the interfaces between the plate and the beams in the geometrically nonlinear analysis of the stiffened plate, the knowledge of which is very important in the design of shear connectors in stiffened structures.(c)The nonuniform torsion in which the stiffening beams are subjected is taken into account by solving the corresponding problem and by comprehending the arising twisting and warping in the corresponding displacement continuity conditions. The distributed warping moment arising from the nonuniform distribution of longitudinal inplane forces is also taken into account.(d)The accuracy of the results and the validity of the proposed model are noteworthy.(e)The influence of geometrical nonlinearity on the deformation of the examined stiffened plates is remarkable. In the case of immovable inplane boundary conditions, the displacements decrement can be verified.(f)The increment of the deflection with the decrement of the connectors' stiffness is easily verified, while this decrement results in a more pronounced influence of the geometrical nonlinearity on the response of the stiffened plate. (g)The developed procedure retains most of the advantages of a BEM solution over a FEM approach, although it requires domain discretization.

ACKNOWLEDGMENT

This work has been funded by the State Scholarships Foundation of Greece as a part of postdoctoral research project.

REFERENCES

1. T. Mizusawa, T. Kajita, and M. Naruoka, "Vibration of stiffened skew plates by using B-spline functions," Computers and Structures, vol. 10, no. 5, pp. 821–826, 1979.·
2. R. B. Bhat, "Vibrations of panels with non-uniformly spaced stiffeners," Journal of Sound and Vibration, vol. 84, no. 3, pp. 449–452, 1982.
3. D. W. Fox and V. G. Sigillito, "Bounds for frequencies of rib reinforced plates," Journal of Sound and Vibration, vol. 69, no. 4, pp. 497–507, 1980.
4. D. W. Fox and V. G. Sigillito, "Bounds for eigenfrequencies of a plate with an elastically attached reinforcing rib," International Journal of Solids and Structures, vol. 18, no. 3, pp. 235–247, 1982. ·
5. Y. K. Lin and B. K. Donaldson, "A brief survey of transfer matrix techniques with special reference to the analysis of aircraft panels," Journal of Sound and Vibration, vol. 10, no. 1, pp. 103–143, 1969.
6. G. Aksu and R. Ali, "Free vibration analysis of stiffened plates using finite difference method," Journal of Sound and Vibration, vol. 48, no. 1, pp. 15–25, 1976.
7. G. Aksu, "Free vibration analysis of stiffened plates by including the effect of inplane inertia," Journal of Applied Mechanics, vol. 49, no. 1, pp. 206–212, 1982.
8. M. Mukhopadhyay, "Vibration and stability analysis of stiffened plates by semi-analytic finite difference method, part I: consideration of bending displacements only," Journal of Sound and Vibration, vol. 130, no. 1, pp. 27–39, 1989.
9. M. Mukhopadhyay, "Vibration and stability analysis of stiffened plates by semi-analytic finite difference method, part II: consideration of bending and axial displacements," Journal of Sound and Vibration, vol. 130, no. 1, pp. 41–53, 1989.
10. M. D. Olson and C. R. Hazell, "Vibration studies on some integral rib-stiffened plates," Journal of Sound and Vibration, vol. 50, no. 1, pp. 43–61, 1977.
11. A. Mukherjee and M. Mukhopadhyay, "Finite element free vibration of eccentrically stiffened plates," Computers & Structures, vol. 30, no. 6, pp. 1303–1317, 1988. · ·
12. A. Mukherjee and M. Mukopadhyay, "Recent advances in the dynamic behavior of stiffened plates," The Shock and Vibration Digest, vol. 27, article 6, 1989.
13. R. S. Srinivasan and K. Munaswamy, "Dynamic response analysis of stiffened slab bridges," Computers & Structures, vol. 9, no. 6, pp. 559–566, 1978.
14. E. J. Sapountzakis and J. T. Katsikadelis, "Dynamic analysis of elastic plates reinforced with beams of doubly-symmetrical cross section," Computational Mechanics, vol. 23, no. 5, pp. 430–439, 1999.
15. E. J. Sapountzakis and V. G. Mokos, "An improved model for the dynamic analysis of plates stiffened by parallel beams," Engineering Structures, vol. 30, no. 6, pp. 1720–1733, 2008.
16. E. J. Sapountzakis and V. G. Mokos, "Shear deformation effect in the dynamic analysis of plates stiffened by parallel beams," Acta Mechanica, vol. 204, no. 3-4, pp. 249–272, 2009.
17. R. S. Srinivasan and S. V. Ramachandran, "Linear and nonlinear analysis of stiffened plates," International Journal of Solids and Structures, vol. 13, no. 10, pp. 897–912, 1977.
18. S. R. Rao, A. H. Sheikh, and M. Mukhopadhyay, "Large-amplitude finite element flexural vibration of plates/ stiffened plates," Journal of the Acoustical Society of America, vol. 93, no. 6, pp. 3250–3257, 1993. · ·
19. M. M. Hegaze, "Nonlinear dynamic analysis of stiffened and unstiffened laminated composite plates using a high order element," in Proceedings of the 13th International Conference on Aerospace Sciences & Aviation Technology (ASAT ‹09), vol. 26 28, 2009.
20. M. Kolli and K. Chandrashekhara, "Non-linear static and dynamic analysis of stiffened laminated plates," International Journal of Non-Linear Mechanics, vol. 32, no. 1, pp. 89–101, 1997.
21. Z. Qingjle, L. Shiqi, and Z. Jijia, "Nonlinear dynamic behavior of stiffened plate under instantaneous loading," Computers & Structures, vol. 40, no. 6, pp. 1351–1356, 1991.
22. A. H. Sheikh and M. Mukhopadhyay, "Linear and nonlinear transient vibration analysis of stiffened plate structures," Finite Elements in Analysis and Design, vol. 38, no. 6, pp. 477–502, 2002. · · ·
23. T. Zhang, T.-G. Liu, Y. Zhao, and J.-Z. Luo, "Nonlinear dynamic buckling of stiffened plates under in-plane impact load," Journal of Zhejiang University: Science, vol. 5, no. 5, pp. 609–617, 2004.
24. A. Karimin and M. Belhaq, "Effect of stiffener on nonlinear characteristic behavior of a rectangular plate: a single mode approach," Mechanics Research Communications, vol. 36, no. 6, pp. 699–706, 2009.
25. Siemens PLM Software Inc, "NX Nastran User's Guide," 2008.

26. Y. F. Wu, R. Xu, and W. Chen, "Free vibrations of the partial-interaction composite members with axial force," Journal of Sound and Vibration, vol. 299, no. 4-5, pp. 1074–1093, 2007.
27. R. Xu and Y. Wu, "Static, dynamic, and buckling analysis of partial interaction composite members using Timoshenko›s beam theory," International Journal of Mechanical Sciences, vol. 49, no. 10, pp. 1139–1155, 2007.
28. R. Xu and Y. F. Wu, "Two-dimensional analytical solutions of simply supported composite beams with interlayer slips," International Journal of Solids and Structures, vol. 44, no. 1, pp. 165–175, 2007.
29. R. Q. Xu and Y.-F. Wu, "Free vibration and buckling of composite beams with interlayer slip by two-dimensional theory," Journal of Sound and Vibration, vol. 313, no. 3–5, pp. 875–890, 2008.
30. U. A. Girhammar and D. Pan, "Dynamic analysis of composite members with interlayer slip," International Journal of Solids and Structures, vol. 30, no. 6, pp. 797–823, 1993.
31. C. Adam, R. Heuer, and A. Jeschko, "Flexural vibrations of elastic composite beams with interlayer slip," Acta Mechanica, vol. 125, no. 1–4, pp. 17–30, 1997.
32. U. A. Girhammar, D. H. Pan, and A. Gustafsson, "Exact dynamic analysis of composite beams with partial interaction," International Journal of Mechanical Sciences, vol. 51, no. 8, pp. 565–582, 2009.
33. U. A. Girhammar and V. K. A. Gopu, "Composite beam-columns with interlayer slip—exact analysis," Journal of Structural Engineering, vol. 119, no. 4, pp. 1265–1282, 1993.
34. U. A. Girhammar and D. H. Pan, "Exact static analysis of partially composite beams and beam-columns," International Journal of Mechanical Sciences, vol. 49, no. 2, pp. 239–255, 2007.
35. V. A. Oven, I. W. Burgess, R. J. Plank, and A. A. Abdul Wali, "An analytical model for the analysis of composite beams with partial interaction," Computers and Structures, vol. 62, no. 3, pp. 493–504, 1997. ·
36. S. Schnabl, I. Planinc, M. Saje, B. Čas, and G. Turk, "An analytical model of layered continuous beams with partial interaction," Structural Engineering and Mechanics, vol. 22, no. 3, pp. 263–278, 2006.
37. J. B. M. Sousa Jr., C. E. M. Oliveira, and A. R. da Silva, "Displacement-based nonlinear finite element analysis of composite beam-columns with partial interaction," Journal of Constructional Steel Research, vol. 66, no. 6, pp. 772–779, 2010.
38. G. Ranzi, A. Dall›Asta, L. Ragni, and A. Zona, "A geometric nonlinear model for composite beams with partial interaction," Engineering Structures, vol. 32, no. 5, pp. 1384–1396, 2010.
39. E. J. Sapountzakis and V. G. Mokos, "An improved model for the analysis of plates stiffened by parallel beams with deformable connection," Computers and Structures, vol. 86, no. 23-24, pp. 2166–2181, 2008. · ·
40. J. A. Dourakopoulos and E. J. Sapountzakis, "Nonlinear dynamic analysis of plates stiffened by parallel beams with deformable connection," in Proceedings of the 4th ECCOMAS Thematic Conference on Computational Methods in Structural Dynamics and Earthquake Engineering (COMPDYN ‹13), Kos Island, Greece, June 2013.
41. J. T. Katsikadelis, "The analog equation method. A boundary-only integral equation method for nonlinear static and dynamic problems in general bodies," Theoretical and Applied Mechanics, vol. 27, pp. 13–38, 2002.View at MathSciNet
42. V. Z. Vlasov, Thin-Walled Elastic Beams, Israel Program for Scientific Translations, 1961.
43. E. J. Sapountzakis and V. G. Mokos, "Analysis of plates stiffened by parallel beams," International Journal for Numerical Methods in Engineering, vol. 70, no. 10, pp. 1209–1240, 2007.
44. E. Ramm and T. J. Hofmann, "Stabtragwerke, Der Ingenieurbau," in Band Baustatik/Baudynamik, G. Mehlhorn, Ed., Ernst & Sohn, Berlin, Germany, 1995.
45. H. Rothert and V. Gensichen, Nichtlineare Stabstatik, Springer, Berlin, Germany, 1987.
46. E. J. Sapountzakis and V. G. Mokos, "Warping shear stresses in nonuniform torsion by BEM," Computational Mechanics, vol. 30, no. 2, pp. 131–142, 2003. · ·
47. E. J. Sapountzakis and I. C. Dikaros, "Large deflection analysis of plates stiffened by parallel beams with deformable connection," Journal of Engineering Mechanics, vol. 138, no. 8, pp. 1021–1041, 2012.
48. J. T. Katsikadelis and A. E. Armenakas, "A new boundary equation solution to the plate problem," Journal of Applied Mechanics, vol. 56, no. 2, pp. 364–374, 1989.
49. E. J. Sapountzakis and I. C. Dikaros, "Large deflection analysis of plates stiffened by parallel beams," Engineering Structures, vol. 35, pp. 254–271, 2012.
50. J. T. Katsikadelis and A. E. Armenàkas, "Numerical evaluation of double integrals with a logarithmic or Cauchy-type singularity," Journal of Applied Mechanics, vol. 50, no. 3, pp. 682–684, 1983.View at MathSciNet ·
51. J. T. Katsikadelis, Boundary Elements: Theory and Applications, Elsevier, Amsterdam, The Netherlands, 2002.
52. K. E. Brenan, S. L. Campbell, and L. R. Petzold, Numerical Solution of Initial-Value Problems in Differential-Algebraic Equations, Classics in Applied Mathematics, Society for Industrial and Applied Mathematics, Philadelphia, Pa, USA, 1996.
53. J. Jiang and M. D. Olson, "Nonlinear dynamic analysis of blast loaded cylindrical shell structures," Computers and Structures, vol. 41, no. 1, pp. 41–52, 1991. · ·
54. T. S. Koko, Super finite elements for nonlinear static and dynamic analysis of stiffened plate structures [Ph.D. thesis], Department of Civil Engineering, University of British Columbia, Vancouver, Canada, 1990.

CHAPTER 4

NUMERICAL SIMULATION OF PARTICLE DISTRIBUTION IN CAPILLARY MEMBRANE DURING BACKWASH

Hussam Mansour, Anik Keller, Rolf Gimbel and Wojciech Kowalczyk

Keywords: Capillary membranes, backwash, numerical simulation, multiphase flow, ultrafiltration, particle distribution

INTRODUCTION

The application of ultrafiltration (UF) capillary membranes in water treatment increases exponentially and has become more and more important in industrial fluid separations. It is a promising technology widely employed to enhance the purity and safety of drinking water [1] since it offers the advantages of high efficiency with the use of mainly physical processes, *i.e.*, in many cases without additives of chemicals and relatively low operational costs and energy consumption [2]. However, the process is limited by membrane fouling caused by physical and/or chemical interactions between the membrane and particulate matter existing in the feed water [3]. The accumulation of these particles during filtration forms a cake layer, which creates additional resistance and contributes to the reduction of permeability of the membrane. The formation of this layer was studied in the presence of adhesive forces on the membrane surface and on different distributed membrane pore sizes [4]. A new model was developed for crossflow filtration of polydisperse particles [5] to investigate the factors influencing the cake formation during the filtration process and the cake growth on non-uniform permeable membranes [6,7]. CFD simulations allow better insight into transport mechanisms of particles and consequently offer numerous improvement possibilities on the membrane process including new membrane materials and configurations, determining operation conditions for optimum selectivity and defining the permeate flux to minimize fouling [8]. The total permeate flux of crossflow microfiltration membranes is predicted by 3-dimensional CFD simulations to obtain the pressure distribution upon the membrane surface [9]. The accumulation of particles on porous surfaces during crossflow filtration is described by modeling the colloidal phase transition into condensed phase [10,11]. Coupling Navier-Stokes and Darcy equations in crossflow tubular membrane filtration based on finite difference model provides characterization of the flow behavior and contributes to a better understanding of the growth rate of the polarization boundary layer along the tubular membrane [12–14] and the strength and weakness of this coupling is discussed as well [15]. For 3-dimensional crossflow finite element models for different suspended particles, CFD simulation is developed in order to predict the velocity field, pressure drop and particle concentration along a hollow fiber for an ultrafiltration system and to provide information about the overall resistance of the membrane and shear forces [16]. Many techniques have been extensively investigated and developed to increase the filtration process accuracy and performance. The most common technique is backwash. It is an effective method for removing particle cakes and recovering membrane properties [17]. A periodic backwash lifts off the deposited cake and flushes it out of the capillary membrane for the backwash mode outside-in. Many experimental investigations have been performed to improve the membrane cleaning efficiency and to evaluate the potential of air bubbling injection [18–21]. A new controller system was introduced to achieve an optimal backwash duration [22]. Moreover, the adhesive forces and operating conditions were studied on capillary membranes driven in dead-end mode during backwash [4,23]. Numerical simulation of membrane backwash is a powerful tool to

Citation: Hussam Mansour, Anik Keller, Rolf Gimbel and Wojciech Kowalczyk, Numerical Simulation of Particle Distribution in Capillary Membrane during Backwash, doi:10.3390/membranes3040249.

increase the understanding of this process but until now some aspects are still lacking for elucidation and optimization. From the numerical point of view minor effects have been spent in investigating backwash process and the affecting parameters on its performance. The objective of this study is on the one hand to highlight the function of the exerted forces on the particles inside capillary membranes operated with dead-end outside-in mode during backwash, and on the other hand to analyze the velocity pattern followed by the demonstration of particle enrichment areas according to different initial particle distributions, particle sizes and water flow rates.

The concept of the whole project contains both numerical and experimental aspects. In the current paper we focused on the theoretical approach and first of all on the development and proof of the numerical multiphase model. Since the numerical investigations have a principal character aiming at the analysis of general fluid mechanical phenomena in a capillary, a direct comparison with experiments is not given. Nevertheless, the first observations and evaluations show some qualitative similarities according to the fluid flow and the behavior of particles. Additionally, some further information from the experiment was applied as initial condition in the numerical model, *i.e.*, thickness of the particle deposited layer and packing density as well as the flux at inlet. Moreover, the numerically calculated pressure drop in the capillary membrane was validated by the experiment. Further experimental validation of a numerical model will concentrate on an analysis of a particle distribution that is measured at the outlet of the capillary membrane at different phases of the process. The results of the significant stronger linking of simulations and experiments will be addressed in subsequent studies.

METHODS

Experimental Section

For the filtration experiments both hard (silica powder) (400 mg L^{-1}) and soft (baker's yeast cells) particles (50 mg L^{-1}) are chosen as additives to the feed in order to cause inorganic and organic particulate membrane fouling. Furthermore, spherical polystyrene particles (100 mg L^{-1}) with well-defined size distribution are applied. The filtration device illustrated in Figure 1 (lower part) can generate simultaneously fouling layers in up to three single commercial UF capillary membranes under the same dead-end conditions. The device is operated for 1 liter of the particle suspension at constant flux of 180 L m^{-2} h^{-1} recording temperature (T) and pressure (P). The flow rate and additionally the density of the suspensions are recorded by high sensitive coriolis flowmeters (FL).

A particle analyzer (PA) is implemented in one feed line to monitor the particle shape and size distribution. With this setup the velocity, size and morphology of single particle (bigger than 1 μm) and eventually formed aggregates can be recorded and characterized. The backwash section (upper part of Figure 1) is designed to investigate the removal of particles and fouling layers out of the capillary membranes. During backwash the water is forced through the capillary wall in outside-in direction by applying either constant pressure profiles (1, 2, 3 or 4 bar) or linear pressure function (from 1 to 4 bar). Pressure (P) and backwash flux are monitored continuously. To observe particles and agglomerates flushed out of the capillary the particle analyzer (PA) is implemented. Additionally, the capillary membrane was divided into many segments in which the structure of the deposited layer was analyzed by microscope.

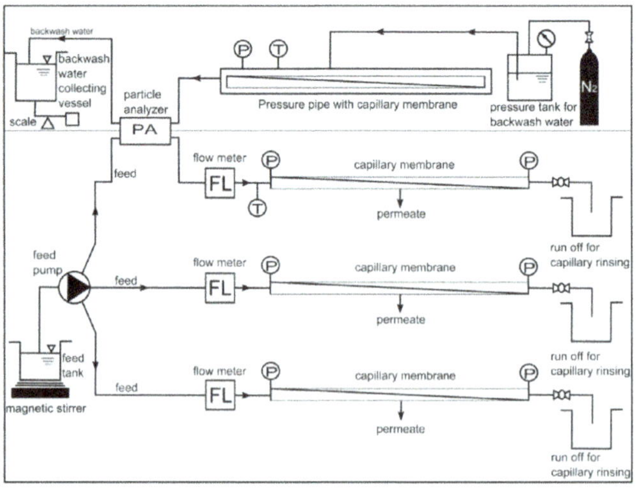

Figure 1: Schematic representation of the filtration and backwash device.

Numerical Simulation

Model Description

Numerical simulation was performed to describe the flow phenomena and motion of particles in the capillary. This study is based on two phase Euler-Euler model for solving the multiphase flow problem. The fluid is simulated as a continuous phase whereas the particles are simulated as a dispersed phase. The continuity (1) and momentum (2) equations are solved with the twoPhaseEulerFoam solver which is based on a finite volume method implemented in free open source CFD framework OpenFOAM-2.0.0 (OpenCFD Ltd.). The continuous phase has the physical properties of pure water, density of 1000 kg m^{-3} and a kinematic viscosity of 1×10^{-6} m^2 s^{-1}. The particles are considered as non-deformable spheres with a constant density of 1050 kg m^{-3}. Reynolds number *Re* of 200–700 related to the capillary diameter, maximum velocity and kinematic viscosity of the fluid indicates a laminar flow inside the capillary.

Governing Equations

The basic equations of the multiphase modeling assume an incompressible fluid with the corresponding conservation of mass and momentum equations [24]. Heat and mass transfer equations are not considered.

- Continuity equation

$$\frac{\partial(\alpha_k)}{\partial t} + \nabla(\alpha_k \overline{u}_k) = 0 \tag{1}$$

- Momentum equation

$$\frac{\partial(\alpha_k \overline{u}_k)}{\partial t} + \nabla(\alpha_k \overline{u}_k \overline{u}_k) + \nabla(\alpha_k \overline{R}_k^{eff}) = -\frac{\alpha_k}{\rho_k}\nabla p + \alpha_k g + \frac{\overline{M}_k}{\rho_k} \tag{2}$$

$$\alpha_k = \frac{V_k}{V} \tag{3}$$

$$\overline{R}_k^{eff} = -\nu_k^{eff}(\nabla\overline{u}_k + \nabla\overline{u}_k^T - \frac{2}{3}I\nabla\overline{u}_k) + \frac{2}{3}Ik_k \tag{4}$$

Modeling of particle motion in the liquid phase includes interfacial momentum exchange between the two phases. The major forces exerted on the particle are responsible for drag, lift and virtual mass force. The sum of momentum rates resulting from these forces sets the total interfacial momentum M_k [24].

$$\overline{M}_k = M^D + M^L + M^V \tag{5}$$

The drag force is caused by the relative motion between phases and defined with:

$$M^D = \frac{3}{4}\frac{\rho_c}{d_k}\alpha_k C_D \left|u_r\right| u_r \tag{6}$$

The drag coefficient calculated according to [25]:

$$C_D = \frac{24}{Re}(1 + 0.15Re^{0.687})$$

The lift force arises due to the shear flow and acts perpendicular to the relative velocity u_r with $C_{L,\alpha} = 0.5$. It is given by [26] as follows:

$$M^L = C_{L,\alpha}\rho_c\alpha_k u_r \times (\nabla \times u_c) \tag{7}$$

The force generated by virtual mass depends on the relative acceleration between the phases.

$$M^V = C_v\rho_c\alpha_k(\frac{Du_c}{Dt} - \frac{Du_{\alpha_k}}{Dt}) \tag{8}$$

D/Dt is the total derivative.

Some other forces like Van der Waals and inter-particle electrostatic forces may arise but their effects on the particles behavior are not considered in this approach. Since it is assumed that the particles are detached from the wall at the beginning of the simulation, adhesive and friction force are also not taken into account.

Two types of capillaries operated in dead-end mode are investigated. The first capillary type has an inner diameter and a length of 1.4 mm, 1500 mm, respectively. It is assumed that the particles are homogeneously suspended within the capillary membrane under initial volume fraction of 10% (Figure 2). This model is descritized in 3 millions structured control volumes generated in ANSYS ICEM CFD (Ansys Ltd) where the elements are uniformly distributed across as well as along the computational domain.

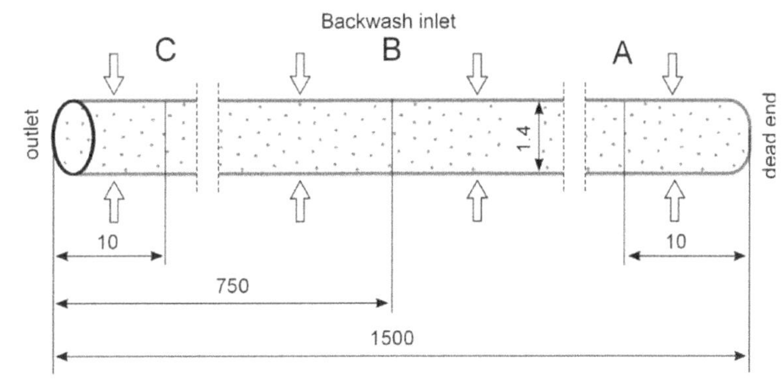

Figure 2: Positions of control lines (A, B, C) for the evaluation of velocity and particle distribution during backwash.

The second capillary type has an inner diameter of 1.4 mm and a length of 980 mm. In this type the particles are deposited on the inner wall of the capillary forming a ring with thickness of 125 μm and a packing density of 0.6 (Figure 3). Here, 10 millions structured elements are generated with a very fine mesh next to the wall which increases towards the capillary center (Figure 4) and the height of the smallest element is 2.3 μm.

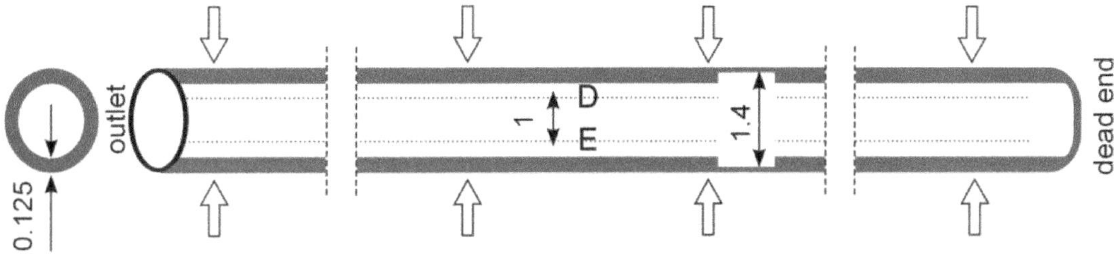

Figure 3: Positions of control lines ($D = 500$ μm, $E = -500$ μm) for the evaluation of velocity and particle distribution during backwash.

Figure 4: Mesh in the calculation domain.

Initial and Boundary Conditions

The initial condition of the particles in the capillary describes (1) a homogeniously distributed suspension at rest and (2) particles deposited in the form of a cake layer. Various boundary conditions have been implemented in numerical simulations in order to examine in details the effects of the operation conditions on performance of backwash process and to improve membrane cleaning efficiency. The diameter of the particles varies between 5 and 20 μm. Moreover, a constant water flux has been chosen between 100 and 400 L m^{-2} h^{-1} and applied at inlet. Dead-end is defined as closed end of the capillary with wall properties and no-slip condition. The ambient pressure is given at outlet. During numerical calculations a convergence criterion of 10^{-6} is set to guarantee a converged solution. The time step control is realized with an adaptive time stepping procedure assumed a maximum Courant number of 0.8. The stable calculations run with the time step in the range of 10^{-5} s. Gaussian linear scheme is implied for the approximation and interpolation. First order, bounded and implicit time discretisation scheme is specified for the solution.

NUMERICAL RESULTS

Transient simulations of the particle accumulation in the membrane capillary are carried out at different operation parameters. The computational approach developed in this study and the evaluation of the results are predicted in term of particle volume fraction in the cell. Moreover, the diagrams and figures presented in dimensionless form defines r^* in the abscissa which denotes the dimensionless diameter of the capillary where zero is on the axis.

The simulation results are evaluated by inserting control lines A, B and C inside the capillary at a distance of 10 mm from the dead-end, in the middle and 10 mm from the outlet across the capillary, respectively (Figure 2) and inserting control lines D and E at a distance of 500 μm from the axis along the capillary (Figure 3).

A parabolic velocity profile is reached in a very short time of the simulation inside the capillary where the flow changes rapidly from radial at dead-end to axial at the outlet. The axial component of the velocity becomes dominant along the capillary compared with the radial velocity component. Due to a constant flux condition at the inlet, the flow rate increases along the membrane and the maximum value is recorded at the outlet.

Analyzing the particle distribution along the three control lines (A, B, C) over the time as illustrated in Figure 5, Figure 6and Figure 7, a heterogeneous particle distribution inside the membrane could be observed. The particles move towards the wall where their accumulation occurs. Thus, the volume fraction increases next to the wall compared with the values around the axis and decreases over the time when the particles are flushed out of the capillary (Figure 7). The main force responsible for the particle distribution in the membrane is the lift force. It determines the flow patterns and particle enrichment areas in the capillary. At the dead-end control line A, where a relatively small axial velocity gradient is observed the lift force is negligible as Figure 5 shows. However, this force gets more significant with increasing the volume flow rate since the axial velocity gradient of the continuous phase becomes larger (Figure 6). The middle control line B illustrates a lateral migration of particles depending on the Reynolds number *Re* and velocity gradient.

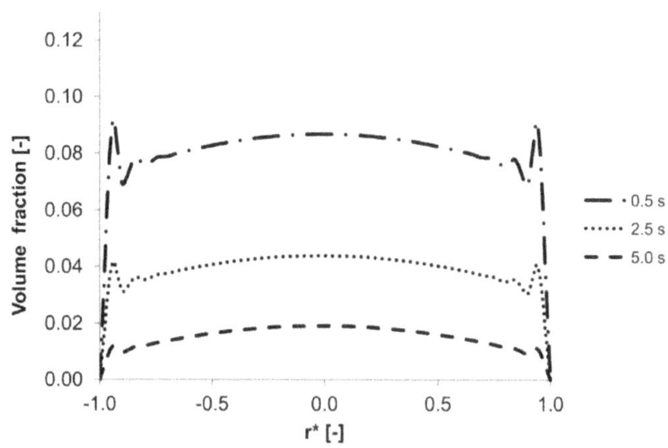

Figure 5: Particle distributions along control line A with backwashing flux of 300 L $m^{-2}h^{-1}$ during 5 s (homogeneous initial distribution, particle diameter of 20 μm).

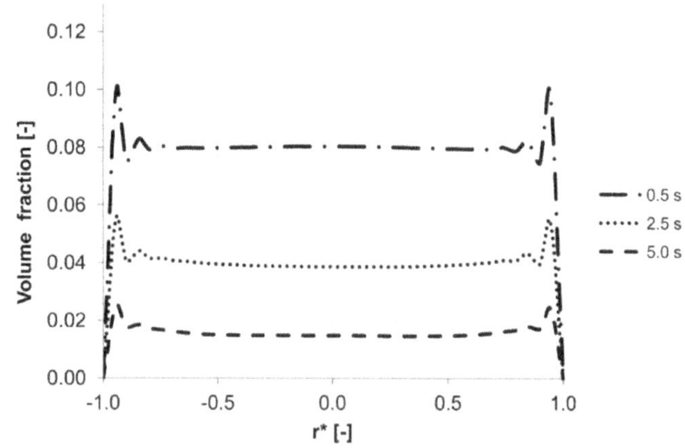

Figure 6: Particle distributions along control line B with backwashing flux of 300 L $m^{-2}h^{-1}$ during 5 s (homogeneous initial distribution, particle diameter of 20 μm).

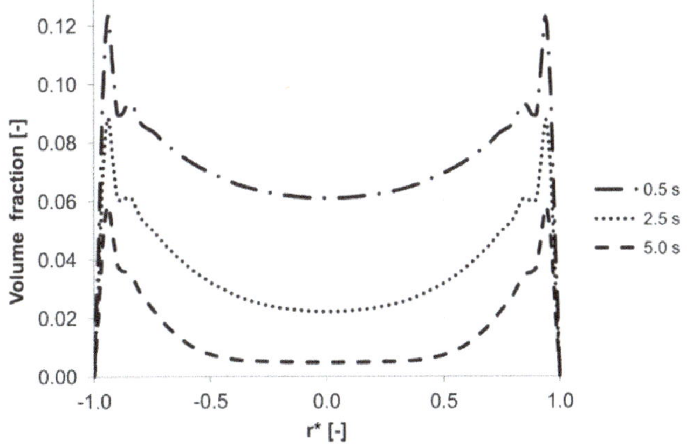

Figure 7: Particle distributions along control line C with backwashing flux of 300 L m^{-2}h^{-1} during 5 s (homogeneous initial distribution, particle diameter of 20 μm).

To emphasize the important role of the lift force, numerical simulation of the backwash process was carried out while the lift force was not considered (Figure 8). Comparing Figure 7 and Figure 8, where the particle distribution at control line C is simulated with and without lift force, respectively, a considerable difference is observed. In contrast to Figure 8, the particles are driven towards the wall due to the effect of the lift force in Figure 7. The distribution of the particles at control line C without lift force is similar to that distribution shown in Figure 5 at control line A where the lift force is very small due to a small axial velocity gradient. This comparison leads to the conclusion that neglecting the lift force causes a similar distribution of the particles at all control lines (A, B, C) in the capillary.

Virtual mass force can be dropped out of the calculation since the relative acceleration between the phases is negligible due to rather small particle diameter (smaller than 20 μm) and almost similar density of both phases.

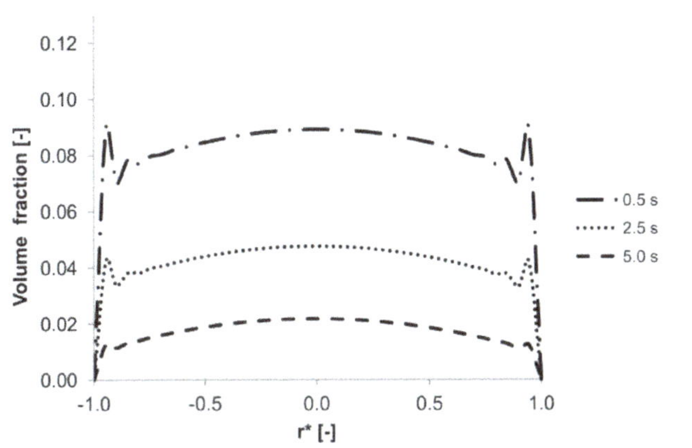

Figure 8: Particle distributions along control line C with not considered lift force and backwashing flux 300 L m^{-2} h^{-1} during 5 s (*cf.* Figure 7).

Considering the second capillary with a deposited layer, Figure 9 and Figure 10 demonstrate the particle distribution at the control lines D and E for different backwash times ranging between 5 and 15 s. The gravity effect is pretty small due to insignificant density difference between the water and considered particles. Nevertheless, the particle distribution in lower and upper areas of the capillary shows a slight deviation from the symmetry on the control lines. As expected, a higher value of volume fraction can be seen in the lower part of the capillary. The particles behavior causes heterogeneity inside the capillary.

This heterogeneity begins next to the capillary outlet and propagates towards dead-end causing a critical increase of the particles volume fraction at certain positions. The tendency to form plugs inside the capillary has a considerable effect on backwash process. For instance, permeability, membrane resistance and energy consumption are influenced by the presence of additional obstacles.

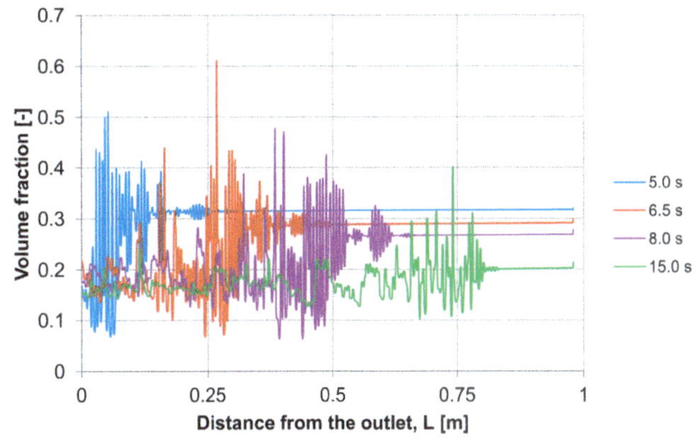

Figure 9: Particle distribution at control line D with backwashing flux of 300 L m^{-2} h^{-1}, (deposited layer as initial distribution, particle diameter of 7 μm).

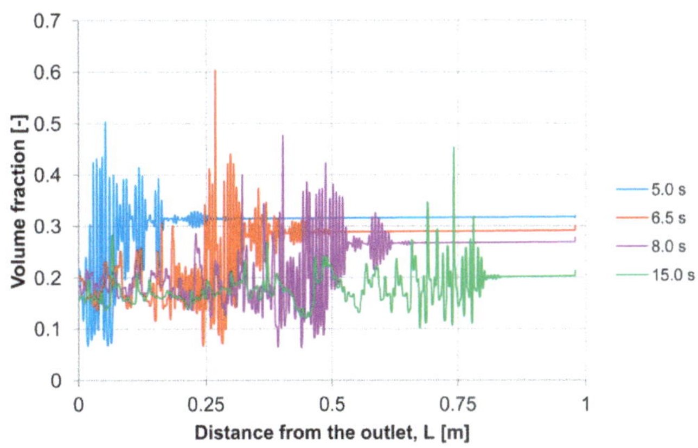

Figure 10: Particle distribution at control line E with backwashing flux of 300 L m^{-2} h^{-1}, (deposited layer as initial distribution, particle diameter of 7 μm).

The dilution of the detached layer depending on the distance from the outlet side for different backwash times (lines D and E in Figure 3) are shown in Figure 9 and Figure 10, respectively.

At both, the upper control line D and the lower line E, consistent particles distribution is observed including some temporary peaks which refer to an enrichment of particles. The volume fraction of the particles at these peaks reaches a similar value of the initial packing density of the deposited layer. These peaks are first formed at the outlet and shifted during continuous operation of the backwash process in dead-end direction. The particle enrichment areas may lead to a probable formation of plugs inside the capillary which are accordingly moving towards dead-end.

Effect of the Particle Size

Further analysis of the backwash process was performed by considering different sizes of the suspended particles. The water flux of 300 L m^{-2} h^{-1} was constant along the inlet. Three different particle sizes were simulated 5 μm, 10 μm and 20 μm which had the same material properties.

The evaluation of the particle distribution along the three control lines (A, B, C) mentioned previously at different backwash times shows different particle behavior in the membrane capillary according to the particle size. For instance, particles with a diameter of 20 μm response quickly to the continuous phase and flushed out without causing accumulation areas inside the membrane. Whereas particles with the size of 5 μm and 10 μm reveal heterogeneous particle enrichment areas in the axial region and close to the wall where accumulation areas are developed. Additionally, a significant enrichment of particles is observed close to the wall of the membrane. The distribution of the volume fraction for different particle sizes along control lines (A, B, C) is illustrated in Figure 11, Figure 12 and Figure 13.

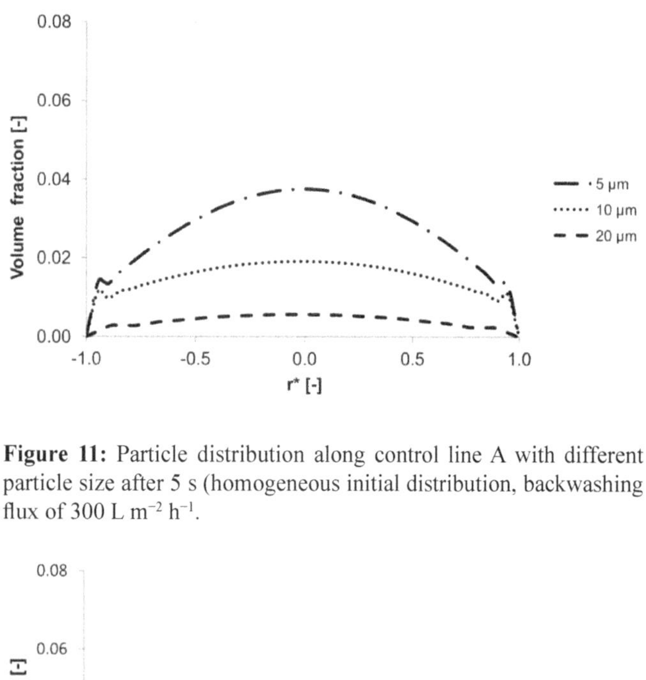

Figure 11: Particle distribution along control line A with different particle size after 5 s (homogeneous initial distribution, backwashing flux of 300 L m^{-2} h^{-1}.

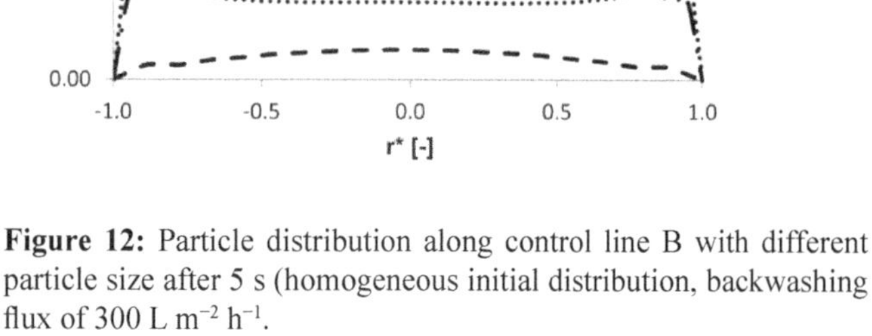

Figure 12: Particle distribution along control line B with different particle size after 5 s (homogeneous initial distribution, backwashing flux of 300 L m^{-2} h^{-1}.

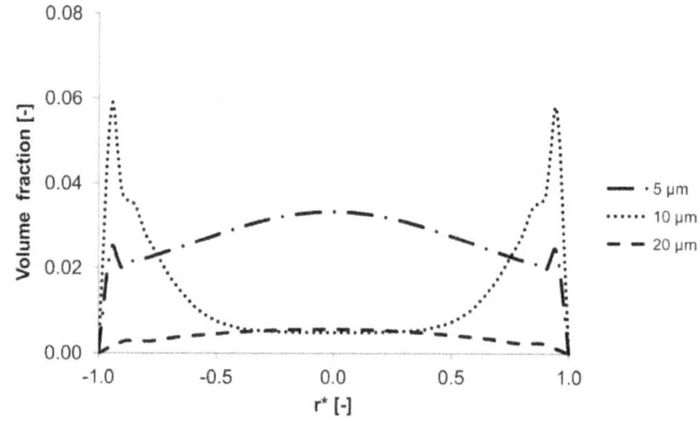

Figure 13: Particle distribution along control line C with different particle size after 5 s (homogeneous initial distribution, backwashing flux of 300 L m^{-2} h^{-1}.

Effect of the Backwash Flux

Different backwash fluxes have been investigated in order to cover a wide range of operation parameters implemented in industrial water treatment plants. An effective cleaning process is accompanied by an optimal and easy adjustable flux, which flushes off the particles and prevents the formation of plugs. Based on homogeneous particle distribution shown in Figure 2, the effect of backwash flux was investigated while other operation conditions were kept constant. The simulation results highlight the fact that backwash flux is a significant parameter in determining the particle accumulation areas inside the membrane. The analysis of the diagrams Figure 14, Figure 15 and Figure 16 at the control positions A, B and C shows a decrease of the volume fraction of particles with increasing mass flow rate. Moreover, for a flux of 100 L m^{-2} h^{-1} the particles are driven to the outlet without considerable lateral migration. The motion across the streamlines and the corresponding behavior of the suspended particles are caused by the relative velocity which at low Reynolds number points parallel to capillary axis. Thus, the lift force tends to drift away accumulated particles to the capillary outlet.

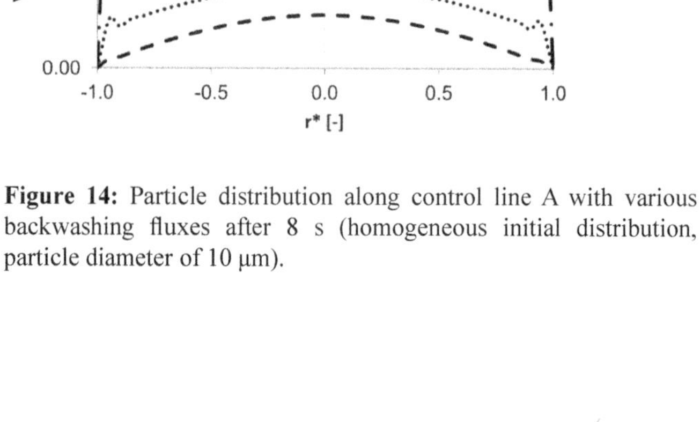

Figure 14: Particle distribution along control line A with various backwashing fluxes after 8 s (homogeneous initial distribution, particle diameter of 10 μm).

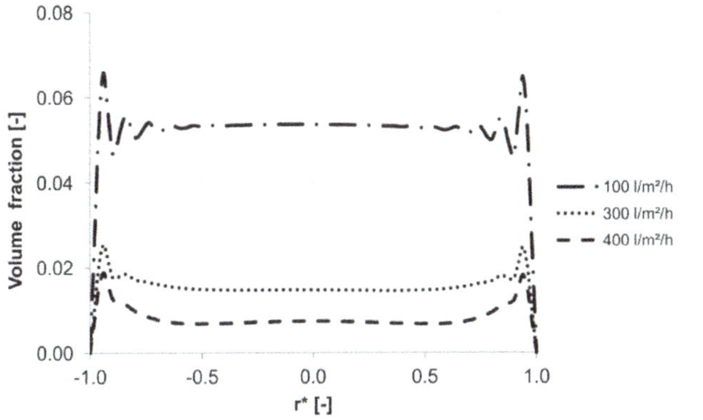

Figure 15: Particle distribution along control line B with various backwashing fluxes after 8 s (homogeneous initial distribution, particle diameter of 10 μm).

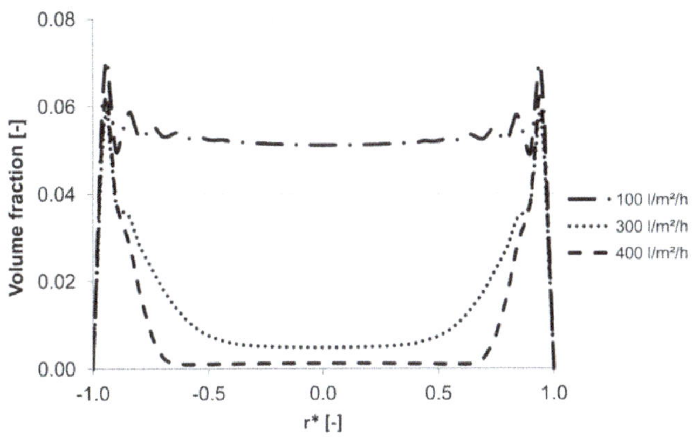

Figure 16: Particle distribution along control line C with various backwashing fluxes after 8 s (homogeneous initial distribution, particle diameter of 10 μm).

Figure 14, Figure 15 and Figure 16 illustrate the volume fraction of particles across the capillary at the three control lines (A, B, C in Figure 2) over backwash time. The same figures also demonstrate a faster washing out of particles from the capillary axis when the capillary is operated with a backwash flux of 300 L m^{-2} h^{-1} and 400 L m^{-2} h^{-1}.

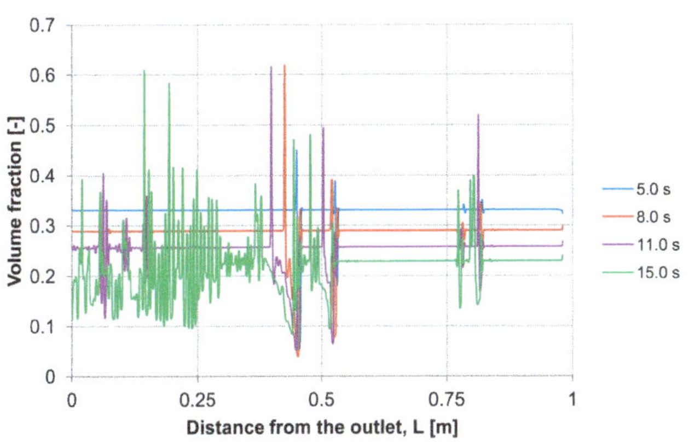

Figure 17: Particle distribution at control line D with backwashing flux of 100 L m^{-2} h^{-1}, (deposited layer as initial distribution).

In order to investigate the real conditions for the backwash process, initial particle distribution delivered by the experimental data (Figure 3) was adjusted and correlated to different fluxes. The evaluation of the simulation results confirms a relatively smooth motion of the particles with less peaks and turbulent behavior inside the capillary at low flux (Figure 17 and Figure 18).

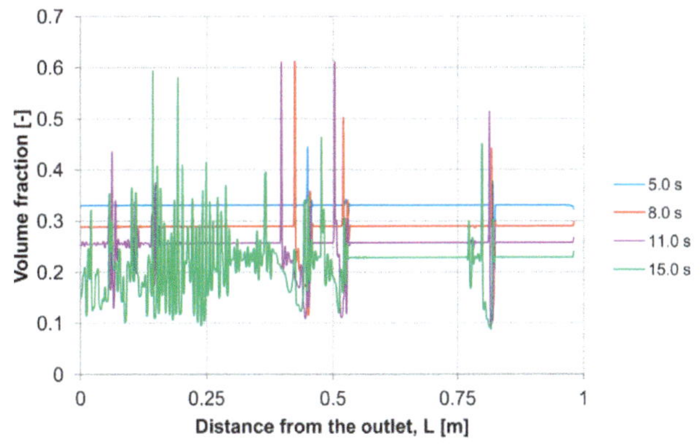

Figure 18: Particle distribution at control line E with backwashing flux of 100 L m^{-2} h^{-1}, (deposited layer as initial distribution).

DISCUSSION AND CONCLUSION

Numerical simulations were performed in order to find operation parameters for membrane backwash suitable for quantitative process description, which reflects on a better membrane cleaning performance. This study considered the effect of the lift force on the behavior of the particles as it shifts the suspended particles into wall direction due to the large velocity gradient close to the wall. It was found that this force has the dominant influence in enhancing the lateral migration of the particles in wall direction and in determining the enrichment areas and possibly the formation of plugs inside the membrane. It was proven that the lateral migration of particles and the location of its maximum volume fraction are strongly influenced by the particle size and the inlet water flux. For large particle size, the backwash process maybe favorable because these particles are distributed more homogeneously over the cross section inside the capillary and are washed out of the capillary relatively early. In contrast to the small particles which tend to cause enrichment areas where the axial velocity component is getting larger compared to the radial velocity component. Increasing backwash inlet flux provides fast destruction of the deposited layer and more heterogeneity of particle distribution inside the capillary whereas, decreasing the inlet flux causes a smooth motion of particles, which may avoid the high possibility of the membrane clogging. For the studied process, the virtual mass force becomes insignificant when the particle size is in microscale.

ACKNOWLEDGMENTS

Financial support by the Deutsche Forschungsgemeinschaft (DFG) is gratefully acknowledged.

REFERENCES

1. Gimbel, R.; Panglisch, S.; Loi-Bruegger, A.; Hobby, R.; Lerch, A.; Strugholtz, S. New Approaches in Particle Separation with UF/MF Membranes. Proceedings of the IWA International Conference on Particle Separation, Toulouse, France, 9–11 July 2007.
2. Bowen, W.R.; Jenner, F. Theoretical descriptions of

membrane filtration of colloids and fine particles: An assessment and review. *Adv. Colloid Interface Sci.* 1995, *56*, 141–200.

3. Charcosset, C. Principles on Membrane and Membrane Processes. In *Membrane Processes in Biotechnologies and Pharmaceutics*; Elsevier: Amsterdam, The Netherlands, 2012; pp. pp. 1–41.
4. Broeckmann, A.; Busch, J.; Wintgens, T.; Marquardt, W. Modeling of pore blocking and cake layer formation in membrane filtration for wastewater treatment. *Desalination* 2006, *189*, 97–109.
5. Foley, G.; Malone, D.M.; MacLoughlin, F. Modelling the effects of particle polydispersity in crossflow filtration. *J. Membr. Sci.* 1995, *99*, 77–88.
6. Lee, K.J.; Wu, R.M. Simulation of resistance of cross-flow microfiltration and force analysis on membrane surface. *Desalination* 2008, *233*, 239–246.
7. Fu, L.F.; Dempsey, B.A. Modeling the effect of particle size and charge on the structure of the filter cake in ultrafiltration. *J. Membr. Sci.* 1998, *149*, 221–240.
8. Ghidossi, R.; Veyret, D.; Moulin, P. Computational fluid dynamics applied to membranes: State of the art and opportunities. *Chem. Eng. Process. Process Intensif.* 2006, *45*, 437–454.
9. Rahimi, M.; Madaeni, S.S.; Abbasi, K. CFD modeling of permeate flux in cross-flow microfiltration membrane. *J. Membr. Sci.* 2005, *255*, 23–31.
10. Bacchin, P.; Espinasse, B.; Bessiere, Y.; Fletcher, D.F.; Aimar, P. Numerical simulation of colloidal dispersion filtration: Description of critical flux and comparison with experimental results. *Desalination* 2006, *192*, 74–81.
11. Hwang, K.; Wang, Y. Numerical simulation of particles deposition in cross-flow microfiltration of binary particles. *Tamkang J. Sci. Eng.* 2001, *2*, 119–125.
12. Damak, K.; Ayadi, A.; Zeghmati, B.; Schmitz, P. A new Navier-Stokes and Darcy's law combined model for fluid flow in crossflow filtration tubular membranes. *Desalination* 2004, *161*, 67–77.
13. Pak, A.; Mohammadi, T.; Hosseinalipour, S.M.; Allahdini, V. CFD modeling of porous membranes. *Desalination* 2008, *222*, 482–488.
14. Rosén, C.; Trägårdh, C. Computer simulations of mass transfer in the concentration boundary layer over ultrafiltration membranes. *J. Membr. Sci.* 1993, *85*, 139–156.
15. Nassehi, V. Modelling of combined Navier-Stokes and Darcy flows in crossflow membrane filtration. *Chem. Eng. Sci.* 1998, *53*, 1253–1265.
16. Marcos, B.; Moresoli, C.; Skorepova, J.; Vaughan, B. CFD modeling of a transient hollow fiber ultrafiltration system for protein concentration. *J. Membr. Sci.* 2009, *337*, 136–144.
17. Hillis, P.; Padley, M.B.; Powell, N.I.; Gallagher, P.M. Effects of backwash conditions on out-to-in membrane microfiltration. *Desalination* 1998, *118*, 197–204.
18. Qaisrani, T.M.; Samhaber, W.M. Impact of gas bubbling and backflushing on fouling control and membrane cleaning. *Desalination* 2011, *266*, 154–161.
19. Remize, P.J.; Guigui, C.; Cabassud, C. Evaluation of backwash efficiency, definition of remaining fouling and characterisation of its contribution in irreversible fouling: Case of drinking water production by air-assisted ultra-filtration. *J. Membr. Sci.* 2010, *355*, 104–111.
20. Serra, C.; Durand-Bourlier, L.; Clifton, M.J.; Moulin, P.; Rouch, J.C.; Aptel, P. Use of air sparging to improve backwash efficiency in hollow-fiber modules. *J. Membr. Sci.* 1999, *161*, 95–113.
21. Van de Ven, W.J.C.; Puent, I.G.M.; Zwijnenburg, A.; Kemperman, A.J.B.; van der Meer, W.G.J.; Wessling, M. Hollow fiber ultrafiltration: The concept of partial backwashing. *J. Membr. Sci.* 2008, *320*, 319–324.
22. Smith, P.J.; Vigneswaran, S.; Ngo, H.H.; Ben-Aim, R.; Nguyen, H. Design of a generic control system for optimising back flush durations in a submerged membrane hybrid reactor. *J. Membr. Sci.* 2005, *255*, 99–106.
23. Serra, C.; Clifton, M.J.; Moulin, P.; Rouch, J.C.; Aptel, P. Dead-end ultrafiltration in hollow fiber modules: Module design and process simulation. *J. Membr. Sci.* 1998, *145*, 159–172.
24. Rusche, H. Computational Fluid Dynamics of Dispersed Two-Phase Flows at High Phase Fractions. Ph.D. Thesis, Imperial College London, London, UK, December 2002.
25. Schiller, L.; Naumann, A.Z. Ueber die grundlegenden Berechnungen bei der Schwerkraftaufbereitung. *Ver. Deut. Ing.* 1933, *77*, 318–320.
26. Auton, T.R. The lift force on a spherical body in a rotational flow. *J. Fluid Mech.* 1987, *183*, 199–218.

CHAPTER 5

AUTOMATIC METHOD FOR BUILDING INDOOR BOUNDARY MODELS FROM DENSE POINT CLOUDS COLLECTED BY LASER SCANNERS

Enrique Valero , Antonio Adán and Carlos Cerrada

Keywords: 3D modeling, B-rep models, laser scanners, 3D data processing

INTRODUCTION

During the last decade laser scanners have gained popularity in architecture, engineering, construction and facility management (AEC/FM). Other measurement methods such as total stations, measuring tapes and based-stereo-camera prototypes are too time-consuming or inaccurate compared to scanners, particularly in large-scale environments. Moreover the high density provided by a single scan (which could be over several million points) makes this technology very suitable for use in the 3D modeling of facilities.

Most of the time, the dense raw data provided from scanners are manipulated by a designer or processed by an engineer in order to create a simplified model of the scenario. This well known reverse engineering process is applied to 3D model creation. Thus, a simplified model provides a high-level representation of the scenario that ranges from a single CAD model, in which a wall is represented as a set of independent planar surfaces, to a Building Information Model (BIM), in which a wall is thought as a volumetric object composed by multiple surfaces with several relevant properties like color, material, cost, etc. Much of the emphasis in previous works has been on creating visually realistic non-parametric models rather than accurate parametric ones. Some examples in this category include methods by El-Hakim et al.[1], which is focused on indoor environments and Früh et al.[2] and Remondino et al.[3], which are focused on outdoor environments. In these cases, a great part of the modeling process is supervised by the modeler so that it cannot be said that models are obtained in an automatic manner.

In this paper we introduce a method that automatically yields Boundary Representation Models (B-rep) for indoors after processing dense point clouds collected by laser scanners from key locations through an existing facility. The 3D model then represents the current state of the building, what is named as the "as-is condition". This model does not necessarily have to be coincident with either the designed ("as-designer condition") or the built model ("as-built condition"). In fact, the facility could have lightly been modified or restored from its initial design. In other occasions, we cannot access to the design drawings or they merely do not exist. Thus, the automatic creation of "as-is models" of inhabited scenarios is a challenging research field which is gaining attention from different applications including architecture, engineering, robotics, etc.[4].

Although there are not many publications dealing with the automatic creation of 3D models, in the last years several interesting works related with some of the stages of this process can be found in the literature. A review for automatic reconstruction of as-built building information models can be found in [4]. The process of creating a single-semantic model can vary depending on the input and expected output. Generally, the automatic modeling process can be divided into three steps: data acquisition, data processing and data modeling, and most of the published papers concentrate on the second stage: data processing.

In this framework, 3D data processing means processing millions of unstructured 3D data in order to obtain higher level

Citation: Enrique Valero, Antonio Adán and Carlos Cerrada, Automatic Method for Building Indoor Boundary Models from Dense Point Clouds Collected by Laser Scanners, doi:10.3390/s121216099.

data structures. Several approaches that convert 3D data into high-level representations in buildings context can be found over the last years. We can here distinguish between proposals to detect and model single objects or particular parts of large scenarios and those than completely model indoors and outdoors.

As regards the first sort of proposals, one of the earlier works that obtain 3D models from laser scanner in a local space context is the one of Kwon et al.[5]. They introduce a set of algorithms for fitting sparse point clouds to a set of single volumetric primitives (cuboids, cylinders and spheres) which can be extended to groups of primitives belonging to the same object. An automated recognition/retrieval approach for 3D CAD objects in the construction context is presented in [6]. In [7] a semiautomatic method to match 3D existing models to data of industrial building steel structures is proposed. The author develops a variant of the ICP algorithm for data registration in order to recognize CAD models objects in large site laser scans. The method is semiautomatic because a coarse registration needs to be performed manually. The same author presents in [8] a plane based registration system for coarse registration of laser scanners data with 3D models in the context of AEC/FM industry. The approach is based on finding planes from the point cloud and matching them with the ones extracted from the 3D models. Nevertheless, the matching process is performed by hand. Other authors [9] propose the combination of point clouds, acquired by means of laser scanners, and photogrammetry in order to generate 3D models of a building under construction. The work of Rusu et al.[10] correctly recognizes and localizes relevant kitchen objects including cupboards, kitchen appliances, and tables. They interpret the improved point clouds in terms of rectangular planes and 3D geometric shapes. One of the innovations consists of including a novel multi-dimensional tuple representation for point clouds and robust, efficient and accurate techniques for computing these representations, which facilitate the creation of hierarchical object models.

Detailed models of part of the walls and building façades are obtained in [2,11–13]. In [11,12] the data processing goes from detecting windows through low data density regions to discover other data patterns in the façade. In [13] important façade elements such as walls and roofs are distinguished as features. Then, the knowledge about the features' sizes, positions, orientations, and topology is introduced to recognize these features in a segmented laser point cloud. Früh et al.[2] develop a set of data processing algorithms for generating textured façade meshes of cities from a series of scans, corresponding to vertical 2D surfaces, obtained by a laser scanner. Thrun et al. developed a plane extraction method based on the expectation-maximization algorithm [14]. Other researchers have proposed plane sweep approaches to find planar regions [15,16]. Valero et al.[17] focus on the modeling of those linear moldings that typically surround doorways, windows, and divide ceilings from walls and walls from floors.

With respect to the automatic creation of complete indoor models, geometric surfaces or volumetric primitives can be fitted to a 3D point cloud to model walls, doors, ceilings, columns, beams, and other structures of interest. In its most simple format, the modeled primitives are annotated with labels (e.g., "wall"). In a higher modeling degree, spatial and functional relationships between nearby structures and spaces are established. Interesting developments can be seen in [18] and [19]. In these works, Adan et al. identify and model the main structural components of an indoor environment (walls, floors, ceilings, windows, and doorways) despite the presence of significant clutter and occlusion, which occur frequently in natural indoor environments, but they deal with rectangular rooms. Furthermore, in the present work we present an approach to identify walls in rooms with more complicated geometries. Okorn et al.[20] present an automated method for creating accurate 2D floor plan models of building interiors. They project the points onto a 2D ground plane and create a histogram of points density, from which line segments corresponding to walls are extracted using a Hough Transform. The first steps of our proposal seem to be inspired in the same strategy since we also project the points to extract the approximate 2D wall's location. Nevertheless, this is only used to later retrieve the corresponding 3D data and delimitate with precision the points belonging to the walls.

This work aims to make progress in the creation of automatic 3D models from point clouds provided by scanners [21]. The contributions of this work lie in three points: accuracy of the model, coherent representation and evaluation of the method. First, we propose a voxelization (discretization) of the space with which to accurately define the planes fitted to the points belonging to the ceiling, floor and walls of the facility. Second, we generate a complete boundary representation model of the indoor in which faces, edges and vertices are coherently connected. Third, once the method is designed and implemented, we make a performance evaluation by measuring the geometric modeling accuracy, the recognition accuracy and the relationship modeling accuracy. The following sections explain the main stages to achieve a complete B-rep model of interiors and present the results obtained.

3D DATA SEGMENTATION

Floor and Ceiling Segmentation

The first stage in the 3D model creation process consists of efficiently segmenting the data obtained by the scanner from different positions. Here we assume that preprocessing stages like outliers/noise filtering and data registration are done, so that our input is an unstructured large cloud of points.

One of the problems which arise in the automatic walls detection is the voxelization of the space in the sense of which the best voxels' size is and where the origin of the space is set. In [18,19] this aspect is not tackled. In the present work, we deal with the optimization of the voxels' size, allowing us to adjust more precisely and at the same time, the ceiling, floor and walls of rectangular rooms into the voxel planes. Thus, a voxel plane contains the majority of the data of each wall.

First of all, we face up the identification and segmentation of the floor and the ceiling of the room. The method proposed in this section assumes that, as it is usual in construction, floor and ceiling are parallel structures. From here on the word "wall" will be indistinctly used to refer any flat indoor structure (ceiling, floor or wall). Our approach is based on creating an optimum discretization of the space (from here on called "voxel space") and then accurately defining the planes which contain the maximum number of points which lie in the floor and ceiling.

Formally, the voxel space can be defined in a universal coordinate system (UCS) by means of the voxel size (ε, δ, σ) and the coordinates of centroid's voxel (v_x, v_y, v_z), v_z being voxel's height, according to the construction context.

Assuming cubic voxels, our objective is then addressed on determining the minimum voxel size, characterized by the parameter ε, and the first coordinate of the voxel plane v_z which contains the maximum number of the points belonging to a wall. Once the voxelization of the space is carried out, most of the points belonging, for example, to the floor are contained in a narrow parallelepiped M whose height is ε (see Figure 1(a)). The uncertainty in ε can be limited by means of the flatness specifications provided in the construction standards.

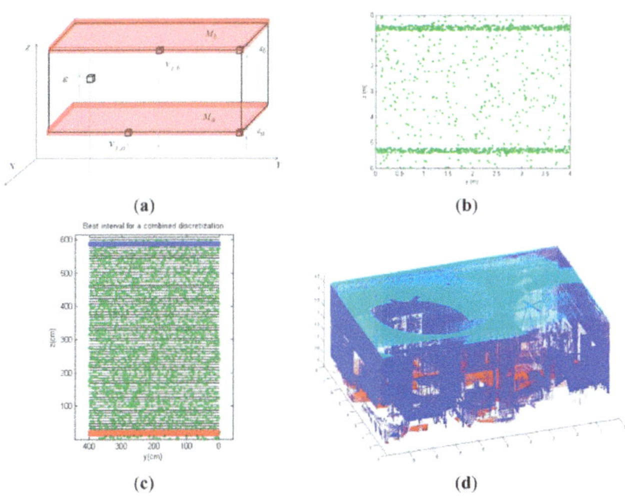

Figure 1: (**a**) Representation of the voxelized space. (**b**) and (**c**) Simulated data of the interior of a room and final discretization of the space. Blue and red planes of voxels containing the majority of the points of floor and ceiling. (**d**) Real segmentation. Point cloud sensed in a room and segmented points belonging to ceiling (cyan) and floor (blue).

The trade-off between the size of the voxels and the number of points contained in M is regulated by the objective function (1), in which $p(\varepsilon, v_z)$ is the percentage of the data contained in M. Therefore, this function evaluates the percentage of the wall's data contained in a voxel plane versus the voxel's size. Figure 1 shows the voxelization of the space according to the definition of the volume M for the floor (M_a) and the ceiling (M_b):

$$F(\varepsilon, v_z) = \frac{p^2(\varepsilon, v_z)}{\varepsilon} \qquad (1)$$

In order to obtain the maximum value of function F, attaining the maximum number of the wall's data inside a voxel plane, different restricted ranges are defined for variables ε and v_z. On the one hand, ε is delimited by the range [ε_1, ε_2] in which ε_1 corresponds to the precision of the scanner (in our case, 1 cm) andε_2 is determined by the flatness tolerance, allowed in the international specification DIN 18202.

On the other hand, v_z is initialized in the position m, which corresponds to the maximum value of the data distribution around the wall. For each value of ε in the range [ε_1, ε_2] the value of v_z is evaluated for the range [m − ε, m + ε] and function F is then calculated. Once this process is finished, the maximum value of F (max F in Algorithm 1) provides the optimum values ε' and v'_z respectively. The algorithm is shown below.

When the algorithm is separately applied for ceiling and floor, two different voxel space configurations are generated. Let ($v'_{z,a}$, ε'_a) and ($v'_{z,b}$, ε'_b) be the position and voxel size parameters calculated for the floor and the ceiling of the room. We propose a new function G which integrates both voxelization proposals.

```
F(ε,v_z)← 3D data distribution in axis Z
m← fitted Gaussian function
max F← F(ε_1,m)
for each ε←ε_1 to ε_2 do
    for each z ←m-ε to m+ε do
        F(ε,z)
        if F(ε,z) > max F then
            max F← F(ε,z); ε' ← ε ; v'_z← z
        end
    end
end
```

Algorithm 1: Calculation of optimum values ε' and v'_z.

Equation (2) imposes one unique voxel size E and the positions of the planes of voxels $V'_{z,a}$ and $V'_{z,b}$, attaining the maximum value of G which provide us the best simultaneous planes of voxels for ceiling and floor. The new proposed function to optimize is as follows:

$$\arg\max_{\Omega}\{G(E, V_{z,a}, V_{z,b})\}$$

$$\Omega = E \in [\varepsilon'_1, \varepsilon'_1], V_{z,a} \in [v_{z,a} - \frac{\varepsilon'_a}{2}, v_{z,a} + \frac{\varepsilon'_a}{2}], V_{z,b} \in [v'_{z,b} - \frac{\varepsilon'_a}{2}, v'_{z,b} + \frac{\varepsilon'_a}{2}],$$

$$G\{E, V_{z,a}, V_{z,b}\} = p_a(E, V_{z,a}) + p_b(E, V_{z,b}) \qquad (2)$$

p_a and p_b are the occupation percentages for the parameters ε'_a and ε'_b. The pseudocode of the algorithm which obtains G is detailed in Algorithm 2.

```
for each ε ← ε_1 to ε_2 do
    for each z ← z_a-ε/2 to z_a+ε/2 do
        calculate ceiling voxel centroid b
        calculate number of points in intervals a and b
        calculate percentage of points p_a and p_b
    end
    for each z ← z_b-ε/2 to z_b+ε/2 do
        calculate ceiling voxel centroid a
        calculate number of points in intervals a and b
        calculate percentage of points p_a and p_b
    end
end

G = maximum(sum(p_a,p_b))
```

Algorithm 2: Obtaining the function G.

Performance Tests

The approach detailed above has been tested in simulated and real data. Figure 1 shows an example of the results obtained under simulation. In Figure 1(b) a front view of a simulated point cloud of a room is depicted. It can be seen two dense point regions, which correspond to the floor and ceiling, and sparse regions which simulate the rest of the sensed points in the room. The maximum values of two Gaussian functions fitted to the data distribution determine the initial values of $v_{z,a}$ and $v_{z,b}$. Figure 1(c) shows the voxels planes projected over the plane YZ, the 3D data and two slices in red and blue that contain the majority of the sensed points. The red voxels plane contains points of the floor and the blue one contains points of the ceiling. We tested the algorithm over twenty simulated rooms. The occupation percentage averages for ceiling and floor were 96.9% and 92.1% respectively.

Figure 1(d) presents the result for a real case. We illustrate the 3D data sensed by a laser scanner from five positions of an inhabited classroom. The data segments corresponding to the ceiling and floor are painted in cyan and red. The total point cloud was composed by 1.5 million points, and the size of the segmented regions was 187,000 points for the ceiling and 95,000 points for the floor.

Walls Segmentation

Rectangular Indoor Plans

In this section we present an approach to segment 3D data corresponding to each one of the walls of a rectangular indoor plan. A rectangular plan is the easiest case to be dealt with. The strategy explained in Section 2.1 can here be extended by considering three pairs of parallel voxels planes, so that the Equation (2) can be easily extended. The objective is to find six parallelepipeds (slices) with centers c_i, $i = 1,\ldots 6$ and with a common width ε which contain the maximum number of points belonging to the walls of the room. Formally, the objective is:

$$\arg\max_{\varepsilon, c_i}(G)$$

$$G(\varepsilon, c_1, c_2, \ldots, c_6) = \sum_{i=1}^{6} p_i(\varepsilon, c_i) \quad \varepsilon \in [\varepsilon'_{min}, \varepsilon'_{max}], c_i \in \left[z_i - \frac{\varepsilon'_i}{2}, z_i + \frac{\varepsilon'_i}{2}\right], i = 1\ldots 6 \tag{3}$$

Arbitrary Indoor Plans

Assuming that floor and ceiling lie in parallel planes, the approach proposed in Section 2.1 can always be used to detect the floor and the ceiling in any non-rectangular room of a building. Figure 2(b) illustrates the extraction of the points belonging to the floor and ceiling of an arbitrary plan. However, the identification of the walls is a more complex task.

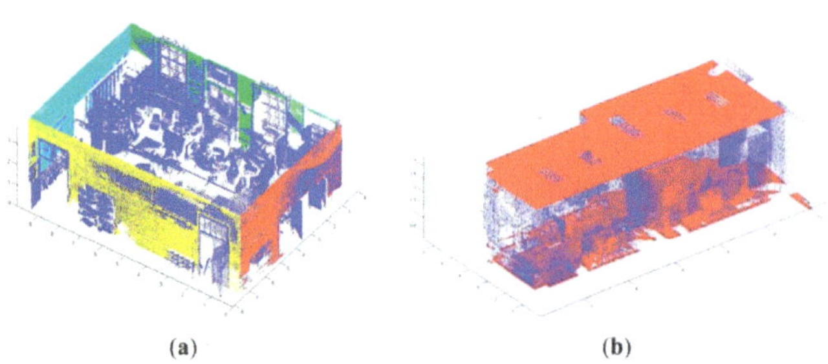

Figure 2: Segmentation of floor and ceiling in rectangular (**a**) and arbitrary (**b**) plans. (a) shows the segments of the four walls of the rectangular indoor presented in Figure 1(d).

The projection of the 3D data from a specific viewpoint allows us to obtain a normalized binary image in which each pixel can be occupied by one or more 3D points (white pixels in Figure 3(b)) or not. From here on, we will denote I the projected image of the data from a top view. This image will help us to obtain a coarse location and position of the walls which will be later refined. The segmentation process is as follows.

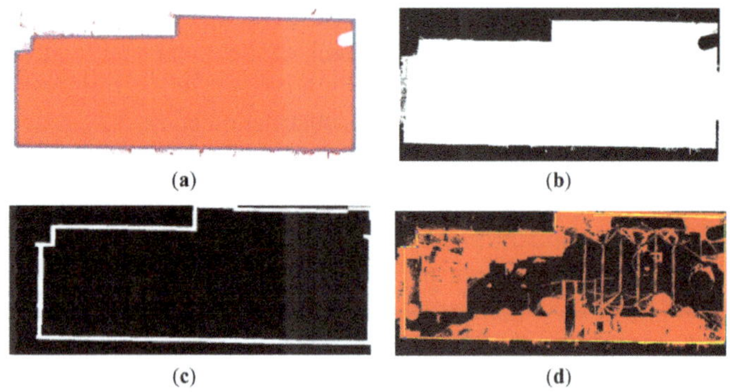

Figure 3: Stages in a wall segmentation process. (**a**) Visualization of the point cloud from a zenital viewpoint. (**b**) Binary image generated after discretization. (**c**) Boundary detection. (**d**) Defining edges and corners in the image.

After creating the image I, the boundary of the room is extracted and, through a Hough Transform algorithm, the set of edges corresponding to the walls in the projected image are detected in a 2D context. As the reader may suppose, if a wall or the connectivity between two walls is completely occluded by a piece of furniture or a constructive component, the boundary of the room does not fit to the walls but these components. Once the room boundaries are demarcated, the points out of the boundaries are removed automatically. Afterwards, we figure out the intersections between edges and obtain the corners in the image. Figure 3 shows the steps of the segmentation process: (a) 3D point cloud viewed from the top of the room; (b) Discretization of the view and generation of binary image I; (c) Boundary extraction in I. (d) Edge and corner detection.

Assuming vertical walls, the segments and corners in the image signify planes and edges in the 3D context. And the planes will be used to segment the 3D points which lie into each wall. Thus, we calculate the mean square distance of the point cloud to the walls and classify each point into a wall. Figure 4 shows the segmentation of points belonging to the walls for two different rooms.

Figure 4: Retrieval of 3D points corresponding to the walls.

CREATION OF B-REP MODELS

The segmentation stage provides the set of points belonging to each wall (including floor and ceiling) of the indoor scene. The following step consists of converting this raw information into high level surface representation. Within the 3D representation models universe we have chosen the boundary representation (B-rep) model [22]. In B-rep, a shape is described by a set of surface elements along with the connectivity information which describe the topological relationship between the elements. The process to achieve a B-rep of an interior space starts calculating the planes which best fit every set of the segments associated to the walls. To do this, we have used the Singular Value Decomposition (SVD) technique [23]. Through SVD, the closeness between the plane and the segmented points is easily calculated. The steps to calculate the plane equation that best fits a generic set of points are shown below.

Each point cloud $P = (p_1, p_2, \ldots, p_n)$, corresponding to a wall, can be fit to a plane defined by the equation:

$$\Pi : Ax + By + Cz + D = 0 \tag{4}$$

The best fit plane is that which minimizes the sum of the distances between every point p_i and the plane Π. Therefore, we can calculate each fit plane by minimizing the expression:

$$\sum_{i=1}^{n} \frac{\left|Ax_i + By_i + Cz_i + D\right|^2}{A^2 + B^2 + C^2} \tag{5}$$

Setting the partial derivative with respect to D equal to zero, we obtain:

$$D = -\left(Ax_0 + By_0 + Cz_0\right) \tag{6}$$

in which $p_0 = (x_0, y_0 z_0)$ is the centroid of P. Replacing (6) in (5):

$$\sum_{i=1}^{n} \frac{\left|A(x_i - x_0) + B(y_i - y_0) + C(z_i - z_0)\right|^2}{A^2 + B^2 + C^2} \tag{7}$$

Let us introduce the matrix $M = [p_1 - p_0\; p_2 - p_0 \ldots p_n - p_0]^T$, in which $p_i = (x_i, y_i, z_i)$ and $p_0 = (x_0, y_0, z_0)$, and the vector $v = [A\; B\; C]^T$. We can show the problem over a matrix representation:

$$\frac{\left(v^T M^T\right)\left(Mv\right)}{v^T v} \tag{8}$$

This expression is called a Rayleigh Quotient and is minimized by the eigenvector of M^TM that corresponds to its smallest eigenvalue. Next, we use the singular decomposition of $M = USV^T$, in which the columns of V are the singular vectors of Mand the eigenvectors of M^TM. Therefore, the solution of Equation (5), provide us the normal of the plane Π, $\overrightarrow{n_\Pi} = [A \quad B \quad C]$. The parameter D is calculated from Equation (6).

The last stage consists of calculating the intersections between connected planes. Note that the topological relationship between the walls is established as the edges and corners are extracted in figure I. So, we know which planes have to be themselves intersected and, therefore, we can find the 3D edges and corners of the room. For instance, the ceiling and floor's planes pairwise intersect with each wall's plane and define the edges at the top and down of the room; and vertices are extracted after intersecting three planes.

Figure 5 illustrates an example with the results of our approach. Part (a) represents the planes fitted to the walls and part (b) shows the set of labeled vertices of the room. Figure 4(c) contains the relationship graph in which adjacent faces in the diagram share one edge.

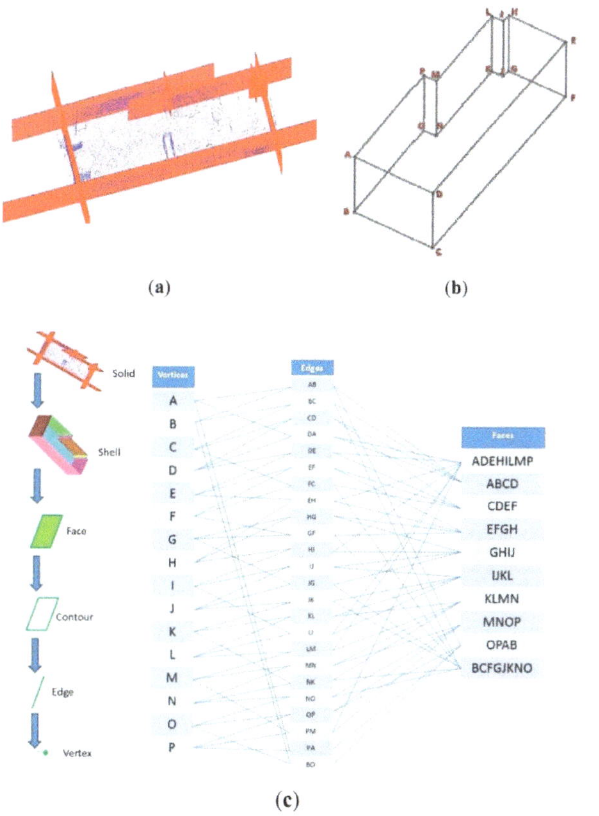

Figure 5: (**a**) Planes fitting the walls. Note that the planes (in red) do not represent the walls but the planes which fit the walls. They are merely used to illustrate the intersections of such planes. (**b**) Labeled vertices of the room. (**c**) Decomposition of a solid in simple objects. Relationship between topological elements in the test room.

RESULTS

Our approach was tested on panoramic range data appertaining to inhabited interiors. Note that we are dealing with very complex scenarios, in which furniture and other objects contribute to clutter and occlusions. They contain a wide variety of objects that occlude not only the walls, but also the ceiling and floor. Three to six laser scans were taken per room. A FARO Photon laser scanner provided 38 million points per room. Figures 7 and 8 show the resulting 3D B-rep models for two rooms.

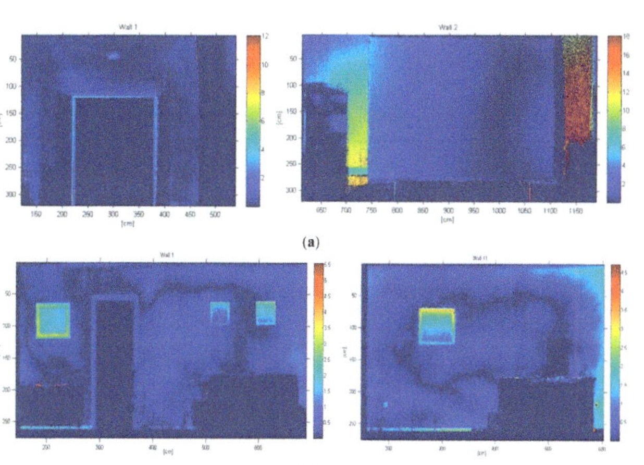

Figure 6: Deviation of the data to the fitted planes for two walls of the lab (**a**) and two walls of the living room. (**b**) Colorbars are coded in centimeters.

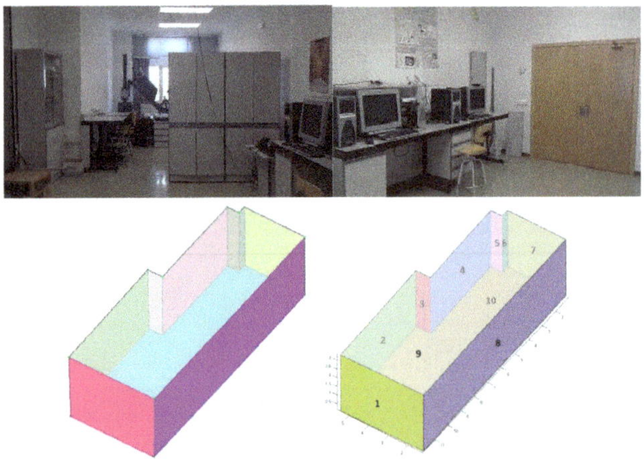

Figure 7: Images of the tested lab (**top**). Our model and the ground truth model (**bottom**).

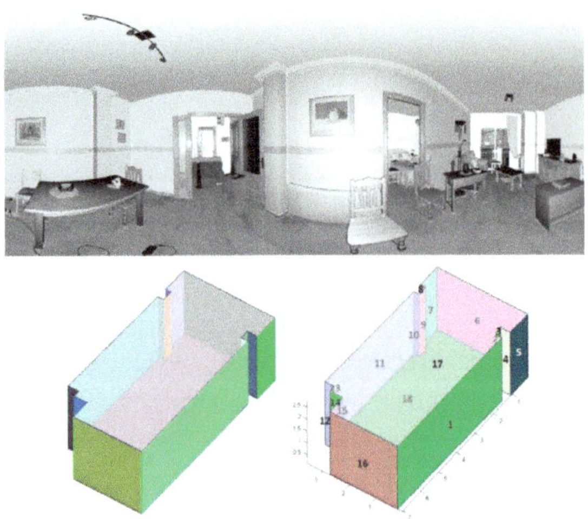

Figure 8: Planar image of the living room (**top**). Our model and the ground truth model (**bottom**).

In this section we present the results obtained for two different inhabited indoors. These interiors do not have rectangular but arbitrary plans. The first room corresponds to the Virtual Reality Lab at the Escuela Técnica Superior de Ingenieros Industriales (UCLM). The second one is the living room of a private flat.

After defining the B-rep model we aim to investigate the accuracy of the obtained models. Firstly, we determine the error committed when we represent the walls of the rooms by means of planes. We thus measure the quadratic distance of every point of the walls to the corresponding plane. Distances between the sensed points and the planes are represented in colormaps in Figure 6(a) for different walls of the lab and (b) for two walls belonging to the living room. It can be seen different regions where an important variation of color is produced. These areas can be owing to typical objects hanging on the wall (pictures, posters, and so on), moldings and doorframes. Of course, some of the errors come from the fact that the walls, ceiling and floor are not totally flat. The corresponding error means, for both inhabited interiors, are presented in Table 1, in which $\bar{d}$ represents the mean value for the distances between each 3D point and its corresponding plane.

The mean error of the fitted planes ranged between 0.59 and 4.64 cm for the lab and between 0.19 and 6.5 cm for the room. In any case, for the majority of the walls the mean error is around below 2 centimeters, despite being severely occluded by tables and chairs. On the other hand, the wall which fitted worse was number 2 of the lab. In this case, there was a big projection panel which largely occluded the wall. Most points sensed in this part of the room corresponded to the panel, so that the calculated plane fitted the panel instead of fitting the wall.

We focused most of our analysis on understanding the performance of our modeling results, since this aspect of the algorithm is considerably less studied than planar wall modeling. We considered two aspects of the performance: first, how reliably the walls can be detected, and second, how accurately they are modeled. To answer the first question, we compared the detected walls with walls of the ground truth model. No fails were reported in this aspect and all existing walls were correctly detected. Failed detections mainly might occur in severe occlusion circumstances.

The second question was tackled first generating the geometrical ground truth of the scenes. We constructed by hand the ground truth models of the rooms with the help of a Leica DISTOTM A6 laser tape measure, which provides 1 millimeter accuracy. In order to assess the error committed, the ground truth models were then compared with our 3D models (see Figures 7 and 8). The results are summarized in Table 2.

Table 1: Mean deviation of the scanned data to fitted walls

Lab	Wall 1	Wall 2	Wall 3	Wall 4	Wall 5	Wall 6	Wall 7	Wall 8	Mean
$\bar{d}$ [cm]	1.13	4.64	1.71	0.59	0.81	1.44	1.23	1.16	**1.58**

Living room	Wall 1	Wall 2	Wall 3	Wall 4	Wall 5	Wall 6	Wall 7	Wall 8	Wall 9	Wall 10	Wall 11	Wall 12	Wall 13	Wall 14	Wall 15	Wall 16	Mean
$\bar{d}$ [cm]	0.55	1.11	0.82	0.19	6.52	4.60	0.24	1.86	0.33	0.57	0.49	4.11	0.44	0.4	0.28	1.08	**1.47**

Table 2: Parameters calculated for each pair of walls

Lab	Wall 1	Wall 2	Wall 3	Wall 4	Wall 5	Wall 6	Wall 7	Wall 8	Wall 9	Wall 10	Mean
α (°)	1.09	0.09	1.12	1.58	6.62	4.83	0.08	0.45	-	-	**1.59**
$d_v(cm)$	0.81	0.80	0.89	0.90	0.90	0.89	0.89	0.90	-	-	**0.87**
$d_g(cm)$	0.52	3.21	1.83	1.96	1.12	0.60	0.98	4.86	-	-	**1.88**

Living room	Wall 1	Wall 2	Wall 3	Wall 4	Wall 5	Wall 6	Wall 7	Wall 8	Wall 9	Wall 10	Wall 11	Wall 12	Wall 13	Wall 14	Wall 15	Wall 16	Wall 17	Wall 18	Mean
α (°)	2.25	10.3	0.95	2.49	0.12	0.72	6.62	0.16	3.86	2.55	1.12	0.89	0.58	0.20	0.11	0.56	-	-	**1.85**
$d_v(cm)$	0.80	0.79	0.79	0.80	0.79	0.79	0.80	0.80	0.80	0.80	0.80	0.80	0.80	0.80	0.80	0.80	-	-	**0.80**
$d_g(cm)$	0.74	0.39	0.09	0.99	0.15	1.80	0.11	2.05	0.32	1.46	2.18	1.31	1.29	2.15	1.17	4.45	-	-	**1.29**

We first compared the lengths of the vertical and horizontal edges of each face of the ground truth models with ours and the difference between are denoted as d_v and d_h in Table 2. The value of d_v was similar for all pairs of walls; around 0.87 cm for the lab and 0.80 cm for the living room. The smallest value of d_h in the living room was for number 3. In this case, the mean value was less than 2 cm in both rooms.

In order to compare the accuracy of the orientation of the faces, we calculated the difference between the respective normal vectors (α in Table 2). The smallest faces yielded a high rate in α, which distorted the average value. Thus, although the mean value of α were 1.59° and 1.85° for the lab and the living room respectively, in the majority of the faces α was less than this value.

Once we have presented the results for these two inhabited interiors, we can establish some difference between them. The living room has more and smaller walls than the laboratory, what might lead the reader to believe that the segmentation process is particularly complicated in this case. However, if we compare the mean values for the different parameters in Tables 1 and 2, the deviations with the ground truth are lower in the living room's case. This result can be due to the fact that a multitude of pieces of furniture occlude the walls of the room and, consequently, the segmentation process may yield more imprecise segments.

CONCLUSIONS

Automatic creation of "as-is models" of inhabited scenarios is a challenge research field which is gaining attention from different applications in AEC/FM contexts. In this paper an approach for automatically creating Boundary Representation Models (B-rep) from dense point clouds collected by laser scanners is presented.

The method here proposed is based on segmenting the data by optimizing the discretization of the space. We make a 2D projection of the data in the voxel space and coarsely determine the segments in which the points lie. The voxels' size is adjusted to the walls, taking into account the flatness tolerances in construction, improving the approach of previous works [18,19] in which predefined voxels' sizes were used. Then the boundary representation model is generated by intersecting the planes that contains the 3D segments, calculating the faces, edges and vertices, and establishing the relationship between components. This idea has been tested in arbitrary shape plans providing excellent results.

The results lead us to state that the majority of the wall's points were correctly segmented. We have also evaluated the performance and accuracy of our method comparing the ground truth and the query B-rep models. Overall modeling accuracy for the AEC domain typically needs to be at least within 2.5 cm of the ground truth, so that the precision of our model is clearly above the standard (between 0.8 cm and 1.88 cm for vertical and horizontal edges).

This research is a part of a larger project which aims to obtain automated reverse engineering of buildings including more complex semantic models. Future improvements to the method will be addressed in two lines: extend the method to non-flat walls and generate B-rep models including details (paintings, panels, etc.) and parts of the walls (moldings, doorframes, windows frames, etc.) in inhabited buildings. As other potential applications, these 3D models can be helpful to create, in an automatic manner, virtual scenes in which digitized objects are introduced in order to generate virtual exhibitions as the ones shown in [24–26].

ACKNOWLEDGMENTS

This research has been carried out under contract with the Spanish CICYT through the DPI-2008-05444, DPI 2009-14024-C02-01 and DPI2011-26094 projects. It also belongs to the activities carried out in the frame of the RoboCity2030-II excellence research network of the CAM (ref. S2009/DPI-1559).

REFERENCES

1. El-Hakim, S.F.; Boulanger, P.; Blais, F.; Beraldin, J.A. A system for indoor 3D mapping and virtual environments. Proc. SPIE 1997, 3174, 21–35.
2. Früh, C.; Jain, S.; Zakhor, A. Data processing algorithms for generating textured 3D building facade meshes from laser scans and camera images. Int. J. Comput. Vis 2005, 61, 159–184.
3. Remondino, F.; El-Hakim, S.; Gonzo, L. 3D Virtual Reconstruction and Visualization of Complex Architectures. Proceedings of 3rd International Workshop on 3D Virtual Reconstruction and Visualization of Complex Architectures (3D-Arch), Trento, Italy, 25–28 February 2009.
4. Tang, P.; Huber, D.; Akinci, B.; Lipman, R.; Lytle, A. Automatic reconstruction of as-built building information models from laser-scanned point clouds: A review of related techniques. Automat. Constr 2010, 19, 829–843.
5. Kwon, S.; Kim, F.; Haas, C.; Liapi, K.A. Fitting range data to primitives for rapid local 3D modeling using sparse point range clouds. Automat. Constr 2004, 13, 67–81.
6. Bosche, F.; Haas, C. Automated retrieval of 3D CAD model objects in construction range images. Automat. Constr 2008, 17, 499–512.
7. Bosche, F. Automated recognition of 3D CAD model objects in Laser scans and calculation of as-built dimensions for dimensional compliance control in construction. Automat. Constr 2010, 24, 107–118.
8. Bosche, F. Plane-based registration of construction laser scans with 3D/4D building models. Adv. Eng. Informat 2012, 26, 90–102.
9. El-Omari, S.; Moselhi, O. Integrating automated data acquisition technologies for progress reporting of construction projects. Automat. Constr 2011, 20, 699–705.
10. Rusu, R.B.; Marton, Z.C.; Blodow, N.; Dolha, M.; Beetz, M. Towards 3D point cloud based object maps for household environments. Rob. Auton. Syst 2008, 56, 927–941.
11. Bohm, J.; Becker, S.; Haala, N. Model Refinement by Integrated Processing of Laser Scanning and Photogrammetry. Proceedings of 2nd International

Workshop on 3D Virtual Reconstruction and Visualization of Complex Architectures (3D-Arch), Zurich, Switzerland, 12–13 July 2007.

12. Bohm, J. Façade Detail from Incomplete Range Data. Proceedings of the ISPRS Congress, Beijing, China, 3–11 July 2008.
13. Pu, S.; Vosselman, G. Knowledge based reconstruction of building models from terrestrial laser scanning data. ISPRS J. Photogramm 2009, 64, 575–584.
14. Thrun, S.; Martin, C.; Liu, Y.; Hähnel, D.; Emery-Montemerlo, R.; Chakrabarti, D.; Burgard, W. A realtime expectation-maximization algorithm for acquiring multiplanar maps of indoor environments with mobile robots. IEEE Trans. Robot 2004, 20, 433–443.
15. Hähnel, D.; Burgard, W.; Thrun, S. Learning compact 3D models of indoor and outdoor environments with a mobile robot. Rob. Auton. Syst 2003, 44, 15–27.
16. Budroni, A.; Böhm, J. Toward Automatic Reconstruction of Interiors from Laser Data. Proceedings of Virtual Reconstruction and Visualization of Complex Architectures (3D-Arch), Venice, Italy, 22–24 August 2005.
17. Valero, E.; Adan, A.; Huber, D.; Cerrada, C. Detection, Modeling, and Classification of Moldings for Automated Reverse Engineering of Buildings from 3D Data. Proceedings of International Symposium on Automation and Robotics in Construction (ISARC), Seoul, Korea, 2–4 March 2011; pp. 546–551.
18. Adan, A.; Xiong, X.; Akinci, B.; Huber, D. Automatic Creation of Semantically Rich 3D Building Models from Laser Scanner Data. Proceedings of the International Symposium on Automation and Robotics in Construction (ISARC), Seoul, Korea, 2–4 March 2011; pp. 342–347.
19. Adan, A.; Huber, D. 3D Reconstruction of Interior Wall Surfaces under Occlusion and Clutter. Proceedings of International Conference on 3D Imaging, Modeling, Processing, Visualization and Transmission (3DIMPVT), Hangzhou, China, 16–20 May 2011; pp. 275–281.
20. Okorn, B.E.; Xiong, X.; Akinci, B.; Huber, D. Toward Automated Modeling of Floor Plans. Proceedings of the Fourth International Symposium on 3D Data Processing, Visualization and Transmission, Paris, France, 18–20 May 2010.
21. Valero, E.; Adan, A.; Cerrada, C. Automatic construction of 3D basic-semantic models of inhabited interiors using laser scanners and RFID sensors. Sensors 2012, 12, 5705–5724.
22. Mortenson, M. Geometric Modeling, 1st ed; John Wiley and Sons: New York, NY, USA, 1985.
23. Golub, G.; Reinsch, C. Singular value decomposition and least square solutions. Numer. Math 1970, 14, 403–420.
24. Corcoran, F.; Demaine, J.; Picard, M.; Dicaire, L.; Taylor, J. Inuit3D: An Interactive Virtual 3D Web Exhibition. Proceedings of the Conference on Museums and the Web, Boston, MA, USA, 17–20 April 2002.
25. Walczak, K.; Cellary, W.; White, M. Virtual museum exhibitions. IEEE Computer 2006, 39, 93–95.
26. Monnerat, M.; Romano, P.; Grillo, O.; Haguenauer, C.; Azevedo, S.; Cunha, G. The Dinos Virtuais Project: A virtual approach of a real exhibition. IADIS Int. J. WWW/Internet 2010, 8, 136–150.

CHAPTER 6 AN AERIAL–GROUND ROBOTIC SYSTEM FOR NAVIGATION AND OBSTACLE MAPPING IN LARGE OUTDOOR AREAS

Mario Garzón , João Valente, David Zapata and Antonio Barrientos

Keywords: Autonomous navigation, cooperative perception, obstacle avoidance, outdoor mapping, collaborative robotics, UGV, UAV

INTRODUCTION

The navigation of a mobile robot can be described as the problem of finding a suitable and collision-free motion between the two poses (position and orientation) of this robot, while obstacle mapping consists of using the sensing capabilities to obtain the representation of unknown obstacles in such a manner that it is useful for navigation [1]. This work presents an alternative to solve these two problems by using a system that combines an aerial and a ground robot, using only the localization system of the terrestrial robot (UGV) and the image from a single camera on board the aerial robot (UAV).

The localization and obstacle detection techniques that are needed for ground robot navigation systems can use relative or absolute sensors. Relative sensors refers to those that are usually internal to the robot, and its measurement is given with respect to another position or state. Absolute sensors provides a measurement with respect to an external or global reference frame [2]. The idea presented here is to provide a vision system on board the aerial robot as a relative sensor, and fuse this information with the absolute position of the ground robot in order to achieve an absolute measurement of the position of the obstacles. By doing this it is possible to obtain a mechanism of collaborative navigation and obstacle avoidance, that takes advantage of the heterogeneity of a mixed robotic system.

There are two main objectives for this system: The first is to obtain the relative distance from the UGV to any unknown obstacle surrounding it, and in that way, guarantee the safe navigation towards the goal or way-point. The second one is to obtain the absolute position of all obstacles detected during the navigation and build a map with their localization; this means having a list of the geo-referenced position of the obstacles. Some previous works have used a combination of aerial and ground robots to approach similar objectives [3–5], but there is no previous work that would have used only one on-board aerial camera for obstacle detection and avoidance; also, to our knowledge, no previous obstacle mapping was performed with fusion of an aerial image and the localization of the UGV.

One of the advantages of using an aerial visual navigation system is that the UGV field of view (FOV) is dynamic. This means that by controlling the UAV height and relative position, the UGV at some stage can perform a pseudo-zoom on an obstacle or any interesting object, as well as reconnaissance of areas outside the UGV's FOV. Moreover, it is also possible to identify rugged terrain, floor openings or negative obstacles, and other unexpected navigation obstructions in the UGV surrounds. This collaboration ensures the UGV's safety while it performs other inspection tasks. Finally, the system does not require previous knowledge from the environment and it can cover large navigation perimeters.

The collision-free navigation system is being developed in several phases. In the first step, the UGV and the obstacles in the robot pathway are identified by processing the aerial image, then techniques based in potential fields are applied to enable simple navigation and obstacle avoidance. After that, the proposed architecture will allow a local map to be built and obtain a geo-referenced position of the obstacle seen by the UAV. Also, each obstacle found can be memorized, and a global map with all the obstacles can be built. This will enable

Citation: Mario Garzón, João Valente, David Zapata and Antonio Barrientos, An Aerial-Ground Robotic System for Navigation and Obstacle Mapping in Large Outdoor Areas, doi:10.3390/s130101247.

the system to merge reactive and deliberative methods. The outline of this article is as follows: After this brief introduction, Section 2 reviews some related works. Section 3 introduces the addressed challenges. The techniques used to estimate the UGV's pose and the obstacle position are described in Section 4. Then, in Section 5, the UGV navigation and control systems used are explained. After that, the experiments performed and the results obtained both in simulations and with real robots are presented in Section 6. Finally, the conclusions are presented.

RELATED WORK

Mobile robotic systems, both aerial and terrestrial, have been studied and developed over the years for several civil and military purposes. Some of those applications are focused on using mobile robots to help or substitute humans in tedious or dangerous tasks, as well as to survey and patrol large unstructured environments. This task is one of the objectives of the ROTOS project in which this work is framed.

In order to perform perimeter surveillance, a robot must be able to generate a trajectory to explore, or to navigate from, an initial point to a final point without colliding with other vehicles or obstacles. This autonomous navigation is one of the most ambitious issues in robotics research.

Concerning visual navigation, many reactive and deliberative navigation approaches have been presented up to now, e.g., in structured environments using white line recognition [6], in corridor navigation using View-Sequenced Route Representation [7], or more complex techniques combining visual localization with the extraction of valid planar region [8], or visual and navigation techniques to perform visual navigation and obstacle avoidance [9]. Some works integrate and fuse vision data from the UAV and UGV for best target tracking [3], another work [10] presents uncertainty modeling and observation–fusion approaches that produce considerable improvement in geo-location accuracy. Also, a different work [11] presents a comparative study of several target estimation and motion-planning techniques and remarks on the importance (and the difficulty) of a single UAV at maintaining consistent view of moving targets.

By merging all those capacities and characteristics together, it is possible to develop a unique sensing and perception of a collaborative system. Work [4] focus their work on a multi-robot system based on a vision-guide autonomy quad-rotor. They describe a way to take off, land and track over the UGV, where the UGV is equipped with two LEDs and a flat pattern in the surface. However, the quad-rotor does not provide information to the UGV about the environment. On the other hand, work [12] presents a motion-planning and control system based on visual servoing (i.e., the use of feedback from a camera or a network of cameras) in a UGV without cameras on board, but not specifically with a UAV. Even work [5] proposes a hierarchy in which several UAVs with aerial cameras can be used not only to monitor, but also to command a swarm of UGVs.

Some of those studies have been tested in several different contexts, such as environment monitoring [13], pursuit-evasion games [14], fire detection and fighting [15], multi-robot localization in highly cluttered urban environments [16], de-mining [17], and other multi-purpose collaborative tasks [18].

PROBLEM FORMULATION AND SOLVING

In this section, we introduce the problem from a theoretical perspective. In the first place, a geometric description of the problem, as well as the relationships between the different reference frames, are given. Then, a solution is proposed based on the aerial–ground system kinematics and some assumptions. As mentioned before, there are two main objectives of the aerial–ground system:

- **Obstacle avoidance:** provide collision-free navigation for the UGV during the execution of its mission through visual feedback from a low cost mini UAV.
- **Obstacle mapping:** obtaining the global position of the obstacles and built a geo-referenced map from it.

Coordinate Frames and Definitions

In the first place, the coordinate systems are defined using the right-hand rule. This step is important and should not be avoided since those definitions are used in the development of the guidance, navigation and control (GNC) sub-system. The coordinate frames from all physical bodies within the workspace are defined with the right-hand rule. The reference frames adopted by the aerial–ground robotic system are shown in Figure 1.

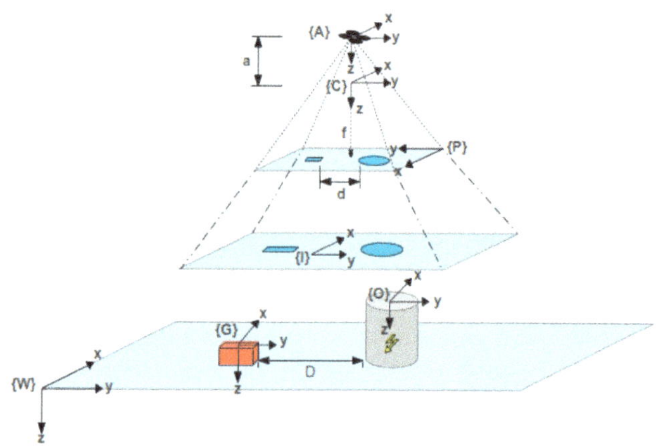

Figure 1: Coordinate frames.

The nomenclature used in the coordinate systems is the follow: W (world frame), G (ground frame), O (Obstacle frame), I (image plane), P (pixels plane), C (camera frame), and A (aerial frame). All the coordinate frames are defined as three dimensional (3D), with the exception of the image plane and the pixels plane, which are two-dimensional (2D).

In this work, some assumptions have been considered. The goal is that the UAV follows the UGV and hovers at a minimum height above it. This minimum height must ensure that UGV is always within the camera FOV (field of view (FOV)), likewise the UGV surroundings. In this way, any obstacle nearby can easily be identified. Therefore, it is assumed that the UAV has a GNC (guidance, navigation and control) sub-system that enables it to follow and hover above UGV. Moreover, one single obstacle is identified at a time in the camera FOV.

Sensor Geometric Model

The sensor model can be mapped by a physical model (more complex, by considering highly non-linear equations) or analytical (approximated, polynomial functions). Additionally, the sensor model, i.e., the digital camera, can be expressed either forward (image to ground) or inversely (ground to image). In this work, the forward physical model is considered. Let us start from the pinhole camera model from Equation (1), where $(u, \upsilon, 1)^T$ is a 2D point in the pixel frame and $(X, Y, Z, 1)^T$ a 3D point in the workspace of the robot. The perspective projection is scaled by a factor s.

$$\underbrace{s \times \begin{bmatrix} u \\ v \\ 1 \end{bmatrix}}_{s\times p} = \underbrace{\begin{bmatrix} f_x & 0 & c_x \\ 0 & f_y & c_y \\ 0 & 0 & 1 \end{bmatrix}}_{K} \underbrace{\begin{bmatrix} r_{11} & r_{12} & r_{13} & t1 \\ r_{21} & r_{22} & r_{23} & t2 \\ r_{31} & r_{32} & r_{33} & t3 \end{bmatrix}}_{[R|T]} \underbrace{\begin{bmatrix} X \\ Y \\ Z \\ 1 \end{bmatrix}}_{P} \quad (1)$$

Intrinsic Parameters

The intrinsic parameters of the camera are addressed by the first matrix from Equation (1), denoted as K. With these parameters, a relationship can be established between the pixel coordinates in the image and its position in the camera reference frame. The intrinsic parameters of interest (known a priori from the camera specifications) are: f, focal length of the camera; c_x and c_y, which are the principal points of the image (i.e., pixel coordinates definition from the center) and lens distortion coefficients (explained later ahead in Section 3.2.3.).

Extrinsic Parameters

The extrinsic parameters, on the other hand, define the camera position and orientation with reference to the world reference system. The obstacles position within the world coordinate frame can be obtained in two ways:

- Obtaining the obstacle position in the world coordinate frame and apply all geometric transformations (i.e., forward camera model) from the pixel plane to the world reference frame.
- Obtain the distance from the UGV to the obstacle in meters in the image plane, and compute the obstacle position in the world reference frame from the UGV position and orientation in the world reference frame (given by a GPS). Herein, there is no need to translate the coordinates to the world frame.

The proposed approach employs the methodologies described in the item 2. Thus, the only extrinsic parameter employed and necessary is the orientation from the camera which ensures geometric corrections on the image acquired by the camera shipped on the UAV.

From Figure 1, it can be noticed that the camera reference frame is displaced from the camera reference frame over the z-axis. This displacement represents the distance from the camera to the UAV center of gravity (CoG). The camera position with relation to the UAV can be expressed as

$$\begin{bmatrix} x_C^A \\ y_C^A \\ z_C^A \end{bmatrix} = \begin{bmatrix} x_A \\ y_A \\ z_A \end{bmatrix} + \begin{bmatrix} 0 \\ 0 \\ a \end{bmatrix} \quad (2)$$

Furthermore, the UAV position is already given in world coordinates (i.e., provided by the GPS). Thus, the camera position in the world reference frame is given again by Equation (2).

The position and heading from the UGV and the obstacle position in the image reference frame are extracted and projected in the pixel frame by applying computer vision algorithms and techniques. The obstacle position or the UGV position can be obtained by applying this simple perspective projection in Equation (3), its relative position to each other, obtained by a simple subtraction, can be used when the navigation is purely reactive.

$$\begin{bmatrix} x_O^I \\ y_O^I \end{bmatrix} = \begin{bmatrix} f_x & 0 & c_x \\ 0 & f_y & c_y \end{bmatrix} \begin{bmatrix} cos(\theta) & sin(\phi)sin(\theta) & cos(\phi)sin(\theta) \\ 0 & cos(\phi) & -sen(\phi) \\ -sin(\theta) & sin(\phi)cos(\theta) & cos(\phi)cos(\theta) \end{bmatrix} \begin{bmatrix} x_p z_A \\ y_p z_A \\ 1 \end{bmatrix} \quad (3)$$

The angle between the obstacle and the UGV in the UGV's reference frame, denoted as γ, can be obtained by subtracting the UGV heading in the camera plane denoted as α from the angle of the obstacle, also on the camera plane and denoted as θ. Both of those angles can be obtained by using Equation (3). This is better illustrated in Figure 2.

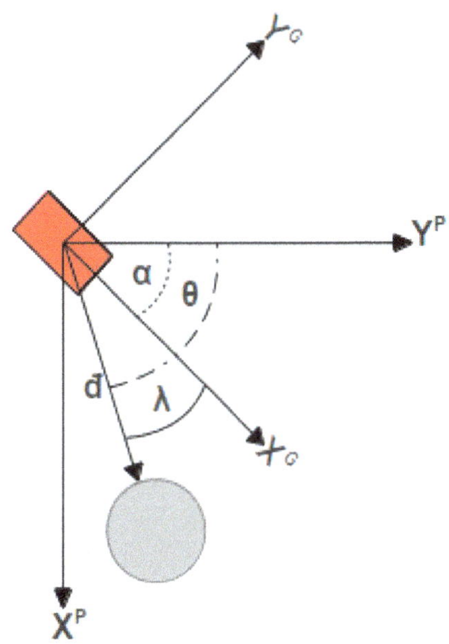

Figure 2: UGV heading geometry. The angle of interest is given by $\gamma = \theta - \alpha$.

Finally, the last objective of the system is to build a global map with all obstacles found in the path of the UGV. Each obstacle identified by the UAV camera is geo-referenced and targeted in a geo-referenced map. Thus, the obstacle position in the world coordinate frame is given by

$$\begin{bmatrix} x_O^W \\ y_O^W \\ z_O^W \end{bmatrix} = \begin{bmatrix} cos(\gamma) & -sin(\gamma) & 0 \\ sin(\gamma) & cos(\gamma) & 0 \\ 0 & 0 & 1 \end{bmatrix} \begin{bmatrix} x_O^I - x_G^I \\ y_O^I - y_G^I \\ z_O^I - z_G^I \end{bmatrix} + \begin{bmatrix} x_G^W \\ y_G^W \\ z_G^W \end{bmatrix} \quad (4)$$

Lens Distortions

Additionally, the Brown-Conrady model is applied to rectify the lens distortion. The lens distortion model is given by two distortion coefficients: the radial $(k_1,\ldots, k_n)$ and the tangential $(d_1,\ldots, d_n)$. Thus, considering $(x'_p, y'_p)^T$, an undistorted image point, and $(x_p, y_p)^T$, a distorted image point, the rectified equations are given by,

$$\begin{bmatrix} x_p \\ y_p \end{bmatrix} = \begin{bmatrix} x'_p(1 + k_1 r^2 + k_2 r^4) + 2d_1 + 2p_1 x'_p y'_p + p_2(r^2 + 2x'^2_p) \\ y'_p(1 + k_1 r^2 + k_2 r^4) + p_1(r^2 + 2x'^2_p) + 2p_2 x'_p y'_p \end{bmatrix} \tag{5}$$

Features Extraction

As was already mentioned, the features extracted with the camera shipped on the UAV were obtained by applying computer vision techniques. The UGV position and orientation in the pixel plane is obtained through the pseudo-code shown in Algorithm 1. In this case, a polygon is defined as a geometric shape with more than three vertices. This procedure is accomplished by finding in the image processed a set of contours addressing the ground robot shape or any good polygonal feature addressing the robot. The image processing sequence is exemplified in Figure 3. A threshold is applied to a gray scale image, followed by a canny edge detector. Once the polygon is identified, its centroid can be easily obtained. Moreover, the UGV orientation relative to the image is obtained through the contour principal moments of inertia.

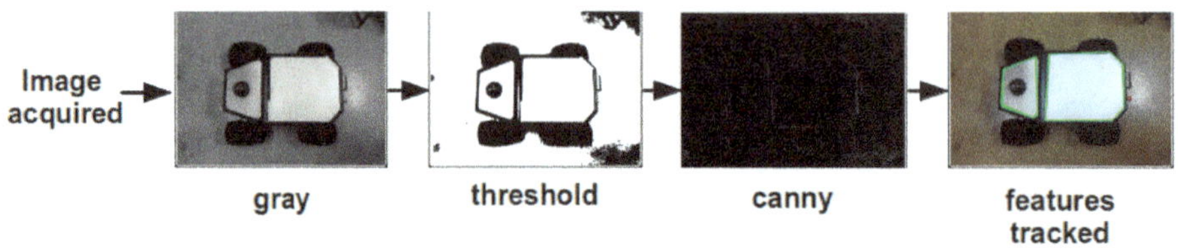

Figure 3: Features extraction procedure example for a ground robot.

$Contors \leftarrow FindCountours(Image_{<gray>})$
if $3 < length(Contors)$ **then**
  $P \leftarrow GetPolygn(Contours)$
  $Point_{<p_x, p_y>} \leftarrow Centroid(P)$
end if
$M \leftarrow getMomentums(P)$
$\alpha \leftarrow Angle(M)$
$(x^P_G, y^P_G) \leftarrow getCentroid(P)$

Algorithm 1: UGV pose extraction.

Concerning obstacles, those with round and square shape (e.g., silos, water or gas tanks, containers) have been considered for identification in this work. The square features are extracted similarly to the aforementioned procedure. However, the round top shapes are featured, employing Hough transformations. It should be also highlighted that the obstacle is identified from its top view, since the camera is pointing downwards (see pseudo-code in Algorithm 2).

$C \leftarrow HoughCircles(Image_{<gray>})$
$(x^P_O, y^P_O) \leftarrow Centroid(C)$
$r \leftarrow Radius(C)$

Algorithm 2: Obstacle position extraction.

Before the experiments, some parameters must be tuned manually, as, for example, the threshold, which must be regulated according to weather conditions, in particular, light exposure.

POSITION ESTIMATION

This section describes the techniques used to estimate the UGV's pose and the obstacle position. First the UGV's pose in a global coordinate frame is estimated by fusing the odometry, IMU and GPS readings using an Extended Kalman Filter (EKF), after that, another Kalman Filter is used to estimate the position of the obstacles using the results of the image processing algorithm.

Ground Robot Pose Estimation

The vehicle pose estimation problem can be defined as the calculations necessary to estimate the vehicle state based on the readings from several sensors. This problem is solved by using an EKF, which produces at time k a minimum mean squared error estimate $\hat{\mathbf{s}}(k \mid k)$ of a state vector $\mathbf{s}(k)$. This estimation is the result of fusing a prediction of the state estimate $\hat{\mathbf{s}}(k \mid k-1)$ with an observation $\mathbf{z}(k)$ of the state vector $\mathbf{s}(k)$.

- Vehicle Model

The vehicle model represents its three-dimensional pose (position and orientation). This pose can be parametrized as $\mathbf{s} = [t, \psi]^T = [x, y, z, \varphi, \theta, \psi]^T$, where $t = [x, y, z]^T$ are the Universal Transverse Mercator (UTM) coordinates and the relative height of the vehicle, and $\boldsymbol{\Psi} = [\varphi, \theta, \psi]^T$ are the Euler angles on the X-Y-Z axis also known as Roll, Pitch, Yaw.

In order to use an EKF to estimate the pose of the robot, it is necessary to express it as Multivariate Gaussian Distribution $\mathbf{s} \sim N(\mu, \Sigma)$. This distribution is defined by a six-element column vector μ, representing the mean values. Moreover, a six-by-six symmetric matrix Σ represents the covariance.

$$\mu = [E[x], E[y], E[z], E[\varphi], E[\theta], E[\psi]]^T$$

$$\Sigma = \begin{bmatrix} Cov[x,x] & Cov[x,y] & \dots & Cov[x,\psi] \\ Cov[y,x] & Cov[y,y] & \dots & Cov[y,\psi] \\ \vdots & \vdots & \ddots & \vdots \\ Cov[\psi,x] & Cov[\psi,y] & \dots & Cov[\psi,\psi] \end{bmatrix} \quad (6)$$

Now it is necessary to define a non-linear conditional probability density function f (**s** (k), **u** (k + 1)) which represents the probability of the predicted position given the current vector state **s**(k) and the control vector **u**(k + 1).

$$\mathbf{s}(k+1) = f\left(\mathbf{s}\left(k\right), \mathbf{u}\left(k+1\right)\right) \quad (7)$$

It is also very useful to define a global UGV transformation T_{UGV} consisting of a rotation matrix R(Ψ) obtained from the Euler angles, and a translation t in the global reference frame.

$$T_{\text{UGV}} = \begin{bmatrix} R(\Psi) & t \end{bmatrix} \quad (8)$$

- Measurement Models

The vehicle pose is updated according to the readings from three different sensors: The internal odometry of the UGV, a IMU sensor and a GPS. Thus, it is necessary to create a model that links the measurements of each sensor with the global position of the mobile robot. As was done with the pose of the vehicle, the observations of the sensors need to be expressed as Gaussian Distribution; this means that they should provide a vector of mean values and a symmetrical matrix of covariances.

As it was done for the estimated UGV pose, a transformation T_i is defined for each measurement. These transformations are:

Odometry

The odometry provides a relative position and orientation constraint. To be able to use these measurements in a global reference frame, the transformation between two successive readings of the odometry ($T_{odom-read}(k) \rightarrow T_{odom-read}(k+1)$) is calculated, and the resulting transformation is applied to the previous estimated position of the UGV.

$$T_{odom}(k+1) = T_{\text{UGV}}(k) * T_{odom-read}(k)^{-1} * T_{odom-read}(k+1) \quad (9)$$

–

GPS.

The readings from the GPS are converted to UTM using the equations form USGS Bulletin 1532 [19], and they are used to provide a global position constraint. However, the GPS does not provide information for the orientation of the robot, so only the position should be taken into account.

$$T_{\text{GPS}}(k+1) = \begin{bmatrix} R(\Psi_o) & t_{\text{GPS-READ}}(k+1) \end{bmatrix} \quad (10)$$

–

IMU.

The IMU readings are pre-processed to fuse the gyroscopes, accelerometers and magnetometers. In order to provide a global constraint of the orientation of the UGV, no position or translation constraints are obtained from this sensor.

$$T_{\text{IMU}}(k+1) = \begin{bmatrix} R(\Psi_{\text{IMU-READ}}) & t_o \end{bmatrix} \quad (11)$$

With the transformations defined, it is possible to define a probability density function for each one of them. In all cases, the H matrix (necessary to compute the estimation of the EKF) will be an identity matrix of dimension six. For the GPS and the IMU, the covariance associated with the rotation or translation, which are not provided, will be replaced by very high values.

These models and the Kalman Filter itself are implemented using the Bayesian Filter Library [20], which is fully integrated in the ROS environment. The estimated pose is updated at a pre-defined frequency with the data available at that time; this is an estimation of the pose to be made even if one sensor stops sending information. Also, if the information is received after a time-out, it is disregarded.

Figure 4 shows a test trajectory performed with the robot in a simulated environment. The positions were translated to a common reference frame in order to be able to compare them. It can be observed that the EKF performs a correction of the position from the odometry, thereby reducing the mean square error in comparison with the real position obtained directly from the simulator.

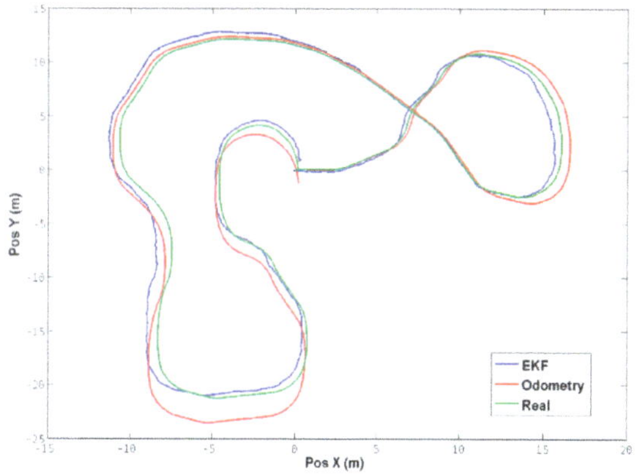

Figure 4: Test Trajectory for EKF. A trajectory was performed in order to test the performance of the Extended Kalman Filter.

Transformations and Obstacle Pose Estimation

Once the pose of the UGV is estimated, it is possible to apply the processing and transformations described in Section 3 to obtain the global position of the obstacle. For each image processed, a transformation like the one described on Equation (4)is obtained, and it is applied to the estimated position of the robot. This produces several measurements of the position of each obstacle detected. Since the global position of the UGV is obtained in UTM coordinates, the global position of the obstacles is given in the same reference frame. All of the positions are stored and then the mean value and the standard deviation are calculated. Finally. a table with the mean values and standard deviations is produced.

NAVIGATION AND CONTROL

This section describes the algorithms and control laws used for the navigation of both the ground and aerial robots. Since this work is not oriented to find new control theories, a well-studied potential fields navigation is used to control the ground robot, and the aerial robot works in teleoperated mode.

Aerial Robot

The main objective of the aerial robot is to have the ground robot and as much as possible of its pathway in the field of view of its downward camera. This is assured by involving a human pilot who controls the air robot in teleoperated mode. However, there is a mid-level control system working to facilitate the operation of the air robot. The mid-level control system provides a height control based on the readings of a ultra sound range meter, it also provides a pose stabilization algorithm that attempts to keep the robot as steady as possible in order to obtain a better image, this stabilization technique is based on a fusion of optical flow and IMU readings. Since the dynamics of the aerial robot are much faster than the ground robot, there are several time lapses where the aerial robot is hovering; in those moments, the stabilization algorithm proves its utility.

Alternatively, it is possible to control the aerial robot using a ground control station (GCS) that receives the telemetry and images from the quad-rotor, before using a position control algorithm to maintain the target (UGV) in the center of the image, and sending commands to control the aerial robot flight motions (Roll ψ and Pitch θ angles). These control commands are sent to the aerial robot usually 30 times per second to obtain smooth movements. Additionally, the GCS is used for configuring and monitoring the aerial robot (see [21] for a detailed architecture).

Ground Robot

The ground robot is controlled by using a ROS implementation of the Steering Behaviors [22]. This potential fields navigation technique is based on the definition of a simple vehicle model, which is a point mass approximation. The vehicle is described by its position, orientation and velocity. Its velocity is modified by applying forces or accelerations that are limited by a maximum force parameter. This allows for a simple computational model to operate and is independent from the locomotion schema.

The pose of the vehicle is defined by four vectors, the first one is for the position of the robot, and the other three are ortho-normal vectors that define orientation (Forward, Side and Up), having the X axis as the normalized direction of the velocity or Forward vector. Since only planar movement is considered the Z axis will be constant, and the Y will be the cross product of X and Z (or Forward and Up). The control signal obtained from the algorithm is one vector that represents the steering force resulting from all the behaviors (i.e., evasion, seeking, alignment). This vector is added to the current velocity and then is mapped in order to be translated to the vehicle reference frame. It is subsequently converted into two control signals—linear and angular velocity—that are sent to the ground robot control system.

The use of the steering behaviors library for the navigation of the UGV has two main advantages: it allows the implementation of several behaviors for a single UGV or a set of them, and it is possible to dynamically add or remove obstacles, which is very convenient in this case. For this initial effort, three steering behaviors were implemented: one for seeking the target point, another for arriving at it, and the third one to avoid obstacles.

The seeking behavior tries to steer the vehicle towards a pre-defined position in the global coordinate system. The behavior adjusts the UGV velocity so it is aligned towards the target; the desired velocity is a vector defined from the vehicle position to the target at the maximum speed, and the steering vector is the difference between this desired velocity and the UGV current velocity. If only seeking behavior is used, the UGV will pass through the target and then turn back. The arriving behavior is similar to the seeking, but it implements a braking force when the vehicle is within a pre-defined distance to the target, thus making the UGV stop when it arrives at it. Figure 5 shows the Seek and Avoid behaviors, with their resulting steering vectors and the auxiliar items used to calculate them.

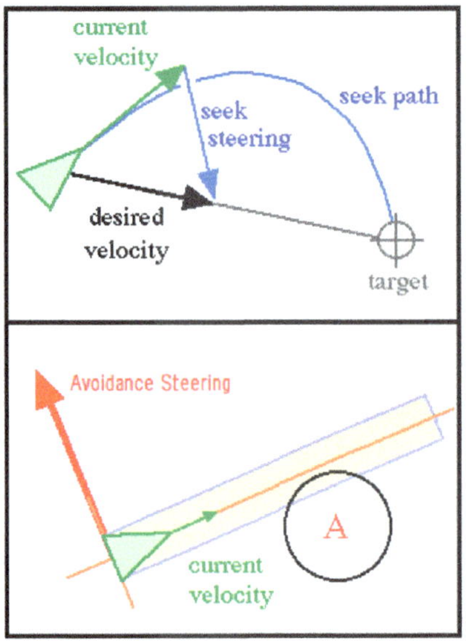

Figure 5: Steering Behaviors. Two Behaviors (Seek and Avoid) and their corresponding vectors are shown.

The avoidance behavior is intended to provide the vehicle with the ability to maneuver in an environment where obstacles may appear; it takes action only when an obstacle is detected in front to the UGV, meaning that if it is moving parallel to a wall, the avoidance will take no action. The basic algorithm assumes that both the vehicle and the obstacles are spheres, although variations can be made to take into account the shape, and then extends the bounding sphere of the UGV to create an imaginary cylinder lying along the forward axis of the UGV, the length of the cylinder depends on the speed of the vehicle and its ability to steer. The algorithm then calculates if any of the given obstacles intersect with the UGV›s cylinder. If there is no collision, a zero vector is returned; if a collision is found, the center of the obstacle is projected on the side axis of the vehicle and a steer in the opposite direction is generated. If two or more obstacles are found, the one nearest to the vehicle prevails over the rest. The two behaviors (seek and avoid) can be combined by adding the steering forces resulting for each one. It is also possible to give more priority to one or another by performing a weighted addition.

EXPERIMENTS

This section presents the results obtained with the proposed aerial–ground system. The approaches described in the previous sections have been implemented first on a simulation environment, and then with real robots. The software architecture used allows to use the same implementation in both environments, but it was necessary to use a realistic simulation of the entire system. First, a quad-rotor aerial robot model was used, as well as a skid-steering ground robot. Also, the sensors simulations that include typical error sources were added to the both robot models, as well as the camera model on board the UAV. By doing this, we ensure that the simulations and the real results are consistent, and that the algorithms can be translated to the real robots. The simulated and real environments are shown in Figure 6.

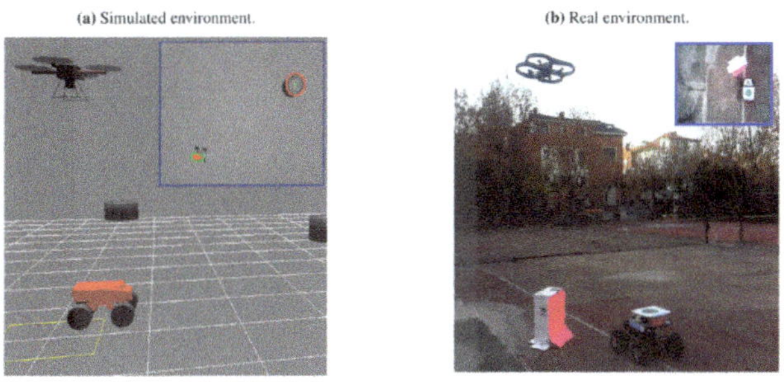

Figure 6: UAV and UGV in the proposed environments (**a**) Simulated environment. (**b**) Real environment. The blue squares in each figure represent the image acquired from the camera on board the aerial robot.

Simulations

Two main tests have been performed on the simulated environment. The first one consists of a single obstacle detection and avoidance maneuver, it is done to check each part of the solution proposed, and thus validate it.

Figure 7 shows the trajectory performed for the first test. The position of the UGV in UTMs is obtained from the Extended Kalman Filter. The position of the obstacle is calculated from the aerial image and the corresponding transformations, as was described in Section 3. The total distance covered was 16.9 meters and was executed in 25 seconds.

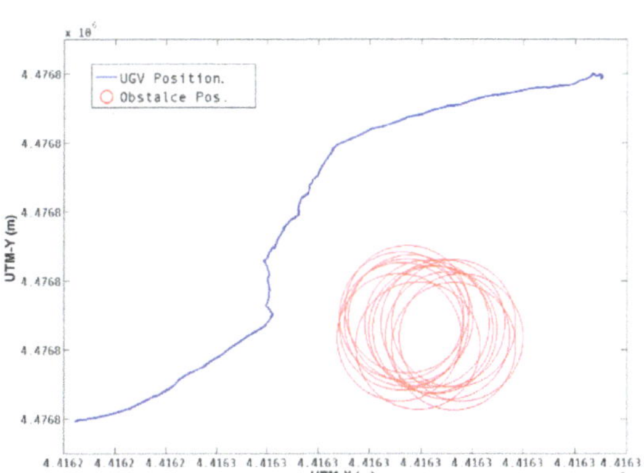

Figure 7: Single obstacle Avoidance. The trajectory of the UGV while performing the avoidance maneuver, and the detected obstacle positions are shown.

Figure 8 shows how each position was obtained, a transformation from the UGV position to the obstacle position is obtained according to Equation (4) and is represented by blue arrows. The resulting obstacle position in the world coordinate frame is represented by red circles. Once the obstacle is left behind, its position is no longer of interest, nor is it taken into account. It can be observed that there is a dispersal of the observations of the obstacle position, so it is of interest to have a characterization of the resulting positions of the obstacle.

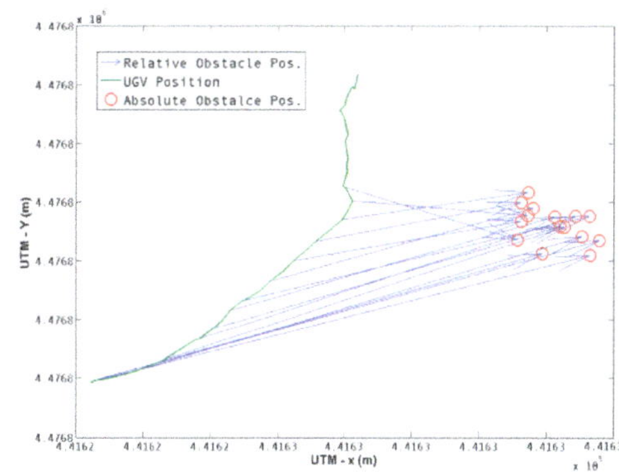

Figure 8: Obstacle Position Estimation. The green line denotes the UGV position. The blue arrows determine the transformation from each robot position to the center of the obstacle marked with red circles.

In Figure 9 the real obstacle position and the positions obtained with the system are translated to a common reference frame and compared. The mean square error and the standard deviation were calculated and are shown in Table 1. It is possible to see that both values are in good error ranges. The mean square error is less than 20% of the diameter of the obstacle, which is 1 meter, in the X axis and 30% on the Y axis.

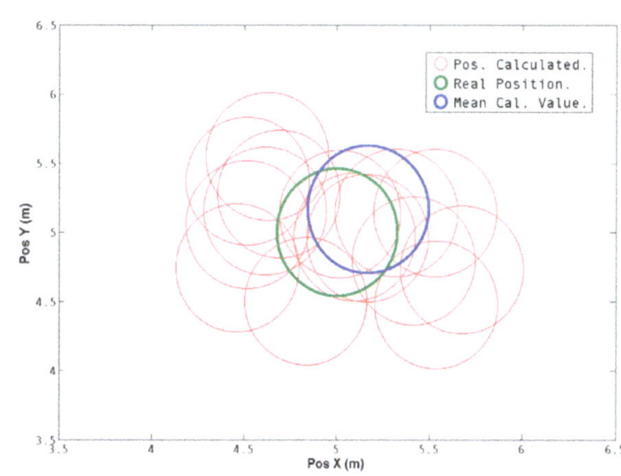

Figure 9: Obstacle Position (Real vs. Estimated). The red circles denote the estimated obstacle position. The green one shows the real position of the obstacle and the blue is the mean value of the calculated positions. All positions were translated to a common reference frame.

Table 1: Mean Square Error and Standard Deviation of the calculated obstacle position

	Mean Square Error (m)	*Standard Deviation*
X Axis	0.1954	0.4236
Y Axis	0.3333	0.3053

Once the obstacle position is obtained, the navigation algorithm uses this information to perform the avoidance maneuver according to the seek and avoid behaviors described in Section 5.2. The outcome of this navigation algorithm is the velocity command in each time step. The command is translated to the UGV reference frame, converted to linear and angular velocity, and sent to the robot's control system. Figure 10 shows the velocity commands generated for this test, together with the UGV's trajectory and the obstacle position.

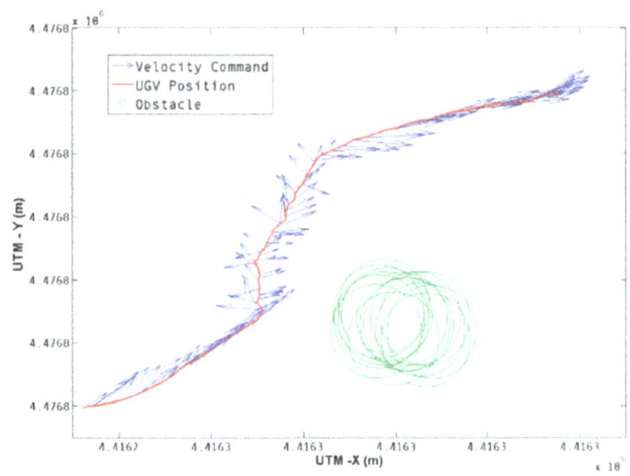

Figure 10.:Velocity Commands. The red line describes the UGV trajectory, and the blue arrows represent the velocity commands at each position.

For the second test, three way-points were defined, in an environment with six cylindrical obstacles. While visiting the way-points, some obstacles were detected, and the corresponding avoidance was executed if necessary. Figure 11 shows the full trajectory, as well as the obstacles detected. Five of the six obstacles were detected, and all three way-points were reached successfully.

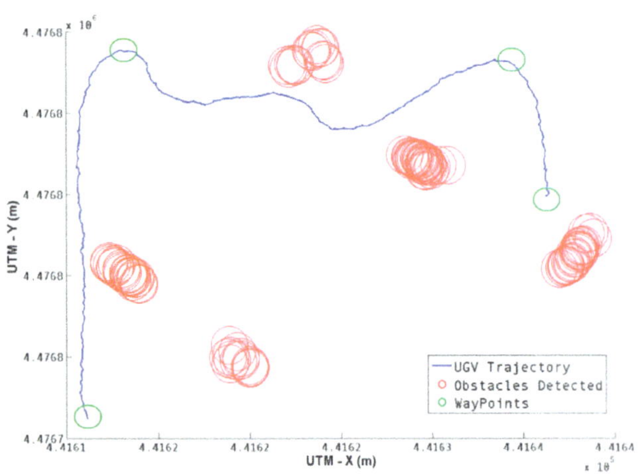

Figure 11: Second Test Trajectory. The blue line describes the UGV trajectory. The red circles represent the obstacles found in the pathway and the green circles are the way-points.

The position is given in UTM coordinates, as was explained in Section 4.2. Since the values obtained for the standard deviation are in all cases less than one meter, it can be concluded that this data gives a set of obstacle positions can be used for the navigation system. Table 2 shows the mean value and standard deviation of the five obstacles detected during the test.

Table 2: Mean Value and Standard Deviation for all the obstacles found in the Test Trajectory

	Mean Pos X (m)	*Std. Dev. X*	*Mean Pos Y (m)*	*Std. Dev. Y*
Obstacle 1	441637.8	0.4743	4476756.3	0.6360
Obstacle 2	441623.3	0.7489	4476768.4	0.5152
Obstacle 3	441629.4	0.4939	4476761.9	0.2612
Obstacle 4	441613.1	0.5246	4476755.1	0.4076
Obstacle 5	441619.3	0.4924	4476749.7	0.3820

Real Environment

A set of tests were designed and implemented in order to check the feasibility of the system outside of the simulated environment. The first test was oriented to check the inter-process communications. Therefore, a set of hovering flights were performed and data both from the UGV and the UAV was acquired in real-time using the middle-ware and software architecture developed for this system [21].

The second test was carried out in an outdoor environment. Its objectives was to obtain real images from the UAV and test the feature extraction algorithms. To facilitate the tests, a platform with a geometric shape pattern was shipped in the UGV. The platform designed have two main capabilities: it can be used for take-off and emergency landings, and it can also transport the UAV in case it ran out of battery. An additional advantage is that the platform design made it easier to track the UGV and distinguish it from other features on the ground. Finally, the navigation algorithm for the UGV was tested using a fixed target and a virtual obstacle position. A set of images from those previous tests are shown in Figure 12.

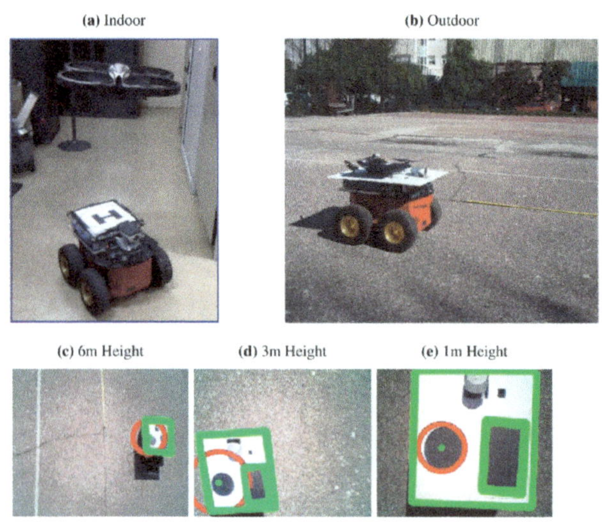

Figure 12: Initial Tests with real robots: **(a,b)** Communication and hovering. **(c)**, **(d)** and **(e)** Aerial Images captured at different heights.

During the tests described previously, the image acquisition rate and processing times, likewise the telemetry data from both the UGV and the UAV, were tested and measured. At last, all the data processing and the commands sent to the navigation system were also measured in terms of its time period. The results of all those measurements are shown in Table 3.

Table 3: Frequencies and Time Consumption for image and data streaming and processing

	Average Frequency (Hz)	*Max. Period (s)*	*Min. Period (s)*
Image Streaming	10	0.446	0.003
UAV Telemetry	13	0.404	0.001
UGV Telemetry	20	0.051	0.049
UGV Navigation System	10	0.103	0.096
Image and Data Processing	37	0.102	0.010

It can be observed that the image streaming runs with an average frequency of 10Hz, and that all the other times or frequencies are faster than that value. In accordance with the software architecture (defined in [21]), the UGV's control and navigation systems run in the UGV on-board computer, and the image acquisition and processing is done on the base-station PC. Also, the inter-process communication core is handled by the ROS middle-ware framework, which enables the system to work at the same frequency as the image streaming. The results from those previous tests shown that the communications and image processing algorithms are feasible enough to perform obstacle-free navigation in real time.

In order to accomplish the last experiment with the full aerial–ground system. A set of obstacles were placed in different positions of the robot workspace. Then, a set of pre-defined and fixed targets were given to the UGV ensuring that to arrive there it must avoid the obstacles previously defined.

The obstacle-avoidance maneuver was executed using the steering behaviors algorithm. The UAV was manually controlled in hovering mode over the UGV, and images from the aerial vehicle were acquired and processed using the feature extraction algorithm. The position of the UGV and the obstacle were extracted from the image and the information of the obstacles position sent to the UGV navigation system. Figure 13 shows a sequence of aerial images obtained during a trajectory, with the objects identified in each frame, it also shows the trajectory obtained from the odometry of the UGV The variables measured during the experiment are shown in Table 4.

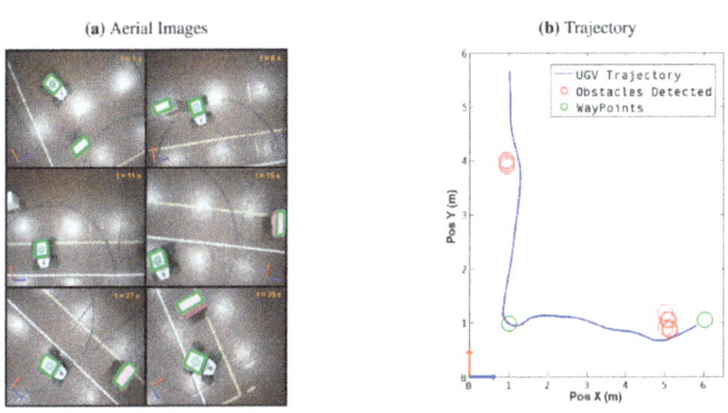

Figure 13: Field tests: (**a**) Sequence of images obtained from the UAV while performing an obstacle-avoidance maneuver. The orientation is represented with a blue and red arrow for the X and Y axes. (**b**) The UGV trajectory read from the EKF is shown in blue, additional marks for the target and the obstacle have been added.

Table 4: Trajectory Parameters and Additional Information

Trajectory Time:	$39\ s$
UGV Max Speed	$0.3\ m/s$
UGV Mean Speed	$0.2048\ m/s$
UAV Max altitude	$4.812\ m$
UAV Mean altitude	$4.51\ m$
Total Trajectory Distance:	$10.1999\ m$

The obstacles were successfully detected and avoided using the proposed system; the mean position and standard deviation of the observations are shown in Table 5. It should be emphasized that the error in the estimated position of the obstacles was less than 0.15meters in the last experiment. Moreover, the UGV maintained a mean safe distance from the obstacles of 0.3 meters. Therefore, despite the obstacles' position error, the UGV is unlikely to collide due to the safe distance between the obstacles.

Table 5: Mean Value and Standard Deviation for the obstacles found in the trajectory

	Mean Pos X (m)	*Std. Dev. X*	*Mean Pos Y (m)*	*Std. Dev. Y*
Obstacle 1	0.9896	0.0194	4.1373	0.0468
Obstacle 2	5.0930	0.1143	1.1158	0.0492

CONCLUDING REMARKS

A hybrid robotic system has been designed and implemented in order to provide a safe navigation system for a UGV, using the aerial image from a camera on board the UAV as the only source of information about the environment. The system is able to perform local real-time navigation and exploration in large semi-structured environments; it can also build near-accurate maps with the absolute position of the obstacles found it its path. These maps can be used to generate local path planning or be fed back to other robots or mission planners. The geographic positions from the obstacles are obtained through an original fusion method employing both the real-time aerial image and the UGV absolute position provided by the GPS.

The robustness of the system was checked according to the mission requirements. Therefore, a set of experiments were done both in a simulation environment and an outdoor scenario with real robots. The ground robot moved around unknown and cluttered environments without colliding.

The results obtained are encouraging to continue researching and extend the potential of this collaborative robotic system.

ACKNOWLEDGMENTS

This work was supported by the Robotics and Cybernetics Group at Technical University of Madrid (Spain), and funded under the projects ROTOS: Multi-robot system for outdoor infrastructures protection, sponsored by Spanish Ministry of Education and Science (DPI2010-17998), and ROBOCITY 2030, sponsored by the Community of Madrid (S-0505/DPI/000235).

REFERENCES

1. Choset, H.; Lynch, K.; Hutchinson, S.; Kantor, G.; Burgard, W.; Kavraki, L.; Thrun, S. Principles of Robot Motion: Theory, Algorithms, and Implementations; MIT Press: Boston, MA, USA, 2005.
2. Krotkov, E. Position Estimation and Autonomous Travel by Mobile Robots in Natural Terrain. Kent Forum Book, 1997. Available online: http://www.ri.cmu.edu/pub_files/pub3/krofkov_eric_1997_2/krotkov_eric_1997_2.pdf (accessed on 3 January 2013).
3. Moseley, M.B.; Grocholsky, B.P.; Cheung, C.; Singh, S. Integrated long-range UAV/UGV collaborative target tracking. Proc. SPIE 2009, 7332.
4. Li, W.; Zhang, T.; Klihnlenz, K. A vision-guided autonomous quadrotor in an air-ground multi-robot system. Proceedings of 2011 IEEE International Conference on Robotics and Automation (ICRA), Shanghai, China, 9–13 May 2011; pp. 2980–2985.
5. Chaimowicz, L.; Kumar, V. Aerial shepherds: Coordination among uavs and swarms of robots. In Distributed Autonomous Robotic Systems 6; Springer: Tokyo, Japan, 2007; pp. 243–252.
6. Ishikawa, S.; Kuwamoto, H.; Ozawa, S. Visual navigation of an autonomous vehicle using white line recognition. IEEE Trans. Patt. Anal. Mach. Intell. 1988, 10, 743–749.
7. Matsumoto, Y.; Inaba, M.; Inoue, H. Visual navigation using view-sequenced route representation. Proceedings of 1996 IEEE International Conference on Robotics and Automation, Minneapolis, MN, USA, 22–28 April 1996; pp. 83–88.
8. Dao, N.X.; You, B.J.; Oh, S.R. Visual navigation for indoor mobile robots using a single camera. Proceedings of 2005 IEEE/RSJ International Conference on Intelligent Robots and Systems (IROS 2005), Edmonton, AB, Canada, 2–6 August 2005; pp. 1992–1997.
9. Cherubini, A.; Chaumette, F. Visual navigation with obstacle avoidance. Proceedings of 2011 IEEE/RSJ International Conference on Intelligent Robots and Systems (IROS), San Francisco, CA, USA, 25–30 September 2011; pp. 1593–1598.
10. Grocholsky, B.; Dille, M.; Nuske, S. Efficient target geolocation by highly uncertain small air vehicles. Proceedings of 2011 IEEE/RSJ International Conference on Intelligent Robots and Systems (IROS), San Francisco, CA, USA, 25–30 September 2011; pp. 4947–4952.
11. Dille, M.; Grocholsky, B.; Nuske, S. Persistent Visual Tracking and Accurate Geo-Location of Moving Ground Targets by Small Air Vehicles. Proceedings of AIAA Infotech@Aerospace Conference, St. Louis, MO, USA, 29–31 March 2011.
12. Rao, R.; Kumar, V.; Taylor, C. Visual servoing of a UGV from a UAV using differential flatness. Proceedings of IEEE/RSJ International Conference on Intelligent Robots and Systems (IROS 2003), Las Vegas, NV, USA, 27 October–1 November 2003; pp. 743–748.
13. Elfes, A.; Bergerman, M.; Carvalho, J.R.H.; de Paiva, E.C.; Ramos, J.J.G.; Bueno, S.S. Air-ground robotic ensembles for cooperative applications: Concepts and preliminary results. Proceedings of 2nd International Conference on Field and Service Robotics, Pittsburgh, PA, USA, 29–31 August 1999; pp. 75–80.
14. Vidal, R.; Rashid, S.; Sharp, C.; Shakernia, O.; Kim, J.; Sastry, S. Pursuit-evasion games with unmanned ground and aerial vehicles. Proceedings of IEEE International Conference on Robotics and Automation, Seoul, Korea, 21–26 May 2001; pp. 2948–2955.
15. Phan, C.; Liu, H. A cooperative UAV/UGV platform for wildfire detection and fighting. Proceedings of Asia Simulation Conference: 7th International Conference on System Simulation and Scientific Computing, Beijing, China, 10–12 October 2008; pp. 494–498.
16. Chaimowicz, L.; Grocholsky, B.; Keller, J.F.; Kumar, V.; Taylor, C.J. Experiments in Multirobot Air-Ground Coordination. Proceedings of the 2004 International Conference on Robotics and Automation, Barcelona, Spain, 18–22 April 2004; pp. 4053–4058.
17. MacArthur, E.Z.; MacArthur, D.; Crane, C. Use of cooperative unmanned air and ground vehicles for detection and disposal of mines. Proc. SPIE 2005, 5999, 94–101.
18. Valente, J.; Barrientos, A.; Martinez, A.; Fiederling, C. Field tests with an aerial–ground convoy system for collaborative tasks. Proceedings of 8th Workshop de RoboCity2030-II: Robots Exteriores, Madrid, Spain, 2 December 2010; pp. 233–248.
19. Snyder, J.P. Map Projections: A Working Manual; Supersedes USGS Bulletin 1532; U.S. Geological Survey, U.S. Government Printing Office: Washington, DC, USA, 1987.
20. Gadeyne, K.; BFL: Bayesian Filtering Library. 2001. Available online: http://www.orocos.org/bfl (accessed on 3 January 2012).
21. Garzón, M.; Valente, J.; Zapata, D.; Chil, R.; Barrientos, A. Towards a ground navigation system based in visual feedback provided by a mini UAV. Proceedings of the IEEE Intelligent Vehicles Symposium Workshops, Alcal de Henares, Spain, 3–7 June 2012.
22. Reynolds, C.W. Steering Behaviors For Autonomous Characters. Proceedings of Game Developers Conference, San Jose, CA, USA, 15–19 March 1999; pp. 763–782.

CHAPTER 7

SET UP OF AN AUTOMATED MULTI-COLOUR SYSTEM FOR INTERIOR WALL PAINTING

Berardo Naticchia, Alberto Giretti and Alessandro Carbonari

Keywords: Miniature painting robot, scaled down laboratory, multi-color spraying end tool.

INTRODUCTION AND PROBLEM STATEMENT

The main advantages that can justify the adoption of robots in the construction industries are generally grouped as follows (Kahane, B. and Rosenfeld, Y., 2004):

- improving productivity and save human labor;
- allowing higher quality standards for buildings.

When construction workers and robots are properly integrated in building tasks, the whole construction process can be better managed and savings in human labor and timing are obtained as a consequence. In addition, it would offer the opportunity to reduce or eliminate human exposure to difficult and hazardous environments, which would solve most of the problems connected with safety when many activities occur at the same time. Finally, automated machines are able to produce higher quality products, as they would promote the passage from craft production to industrial production, with consequent higher accuracy and lower levels of defects.

As listed in the next section 2, there are a lot of contributions pertaining to the improvement of efficiency and safety in construction, which are brought about by automation, provided it will be able to produce highly autonomous robots for building sites. The main requirements presently pursued are: developing dedicated robots to execute the same activities that workers are usually in charge of; equipping robots with the sensors and control systems required to make them autonomously move throughout every job site. However we think that the other objective, namely that of quality improvement, should be reconsidered to be not only connected with better accuracy in production, but also strongly linked to the possibility to adopt new technologies and ergonomics for the execution of building tasks. As a consequence, when we think about building automation we should take the opportunity to reconsider new procedures and new tools to accomplish every task.

Recent research in the building sector has been mainly devoted to automating interior finishing tasks. Among them we have chosen to start working on interior wall painting. Thus far, some very interesting robots for interior wall painting have been manufactured, like in (Kahane, B. and Rosenfeld, Y., 2004). They have been conceived to be able to make wall spraying in such a way that reproduces what performed by human workers. Big efforts have been made to improve their efficiency and the robustness of their control systems, too. We think that such very valuable research in construction can be further improved by exploiting the possibility to use automatic systems to re-think new procedures for executing building tasks. In the case of wall painting, we are studying the feasibility of a new approach for interior wall painting, which gives users the possibility to paint with a multicolour device, that would be able to reproduce artworks similarly to what is done by printing machines: this is also intended as an example of how known industrial technologies can be adjusted for convenient applications in construction. Moreover, architects will be free to design new environments, even because this system can be used as well for standard spraying on interior walls. In this paper we have tested one four-color spraying system. Another big advance that can be brought about by automation in the building sector, is the one of rethinking new

Citation: Berardo Naticchia, Alberto Giretti and Alessandro Carbonari, Set Up of an Automated Multi-Colour System for Interior Wall Painting, ISSN 1729-8806.

ergonomics. Robots are able to work at different scales than human workers: (Bach, Fr. W., Rachkov, M., Seevers, J. and Hahn, M., 1995) maintain that there are some tasks, including painting of big construction areas, which are difficult and expensive to perform using standard ergonomics. They suggest the use of climbing robot, equipped with many sensors to coordinate one another.

However there are a lot of other situations where human sized ergonomics is again the best approach: hence we have set up two parallel laboratories, one for testing at full size and another one which is thought to test miniature robotic systems, as the two solutions will complement each other. In this paper all the experiments have been carried out at the smaller scale, because it is considered to be the most critical. Our successful experimental campaign shows how convenient the use of such new technologies can be: this way it is possible to widen the set of performances allowed for the painting tasks, without renouncing to the efficiency and the other advantages traditionally obtained.

This paper is organized as follows: section 2 summarizes the state of the art in this field of research; section 3 outlines the research that we have already performed, whose tests are shown and commented in section 4. Conslusions and future research are summed up in section 5. All the tests take advantage of the extensions listed in section 7 and references in section 8.

STATE OF THE ART

Interesting examples of automated construction processes are the CIC (Yasuyoshi M. and Roozbeh K., 1993) and the SMART systems (Yamazaki Y. and Maeda J., 1998). CIC is the acronym for Computer Integrated Construction: it requires the use of a mobile contruction system moving throughout the building site. Even if it requires the adoption of ready made products and it would be difficultly applied for different construction technologies, however it laid the foundation of automation in the building sector with wide use of robots and control systems. Another similar example is represented by the S.M.A.R.T. system, developed by Shimizu Corporation, which encompasses also the design and management phases in the automated construction process.

The very interesting contribution given by the aforementioned (Kahane, B. and Rosenfeld, Y., 2004) and(Warszawsky, A. and Rosenfeld, Y. 1994) are more specific for the execution of building tasks. They are developing the multi-purpose TAMIR robot, which has been used to perform a number of experiments on wall painting using a spray gun with its on/off control nozzle operated by connections of the robot's controller. Through full scale experiments it was shown that, according to the break-even point technique, robots are always more profitable than human work when highly autonomous machines are adopted. This holds because one operator may supervise also two different robots; in this way it is convenient also for cheap labor markets. This is due to the fact that human workers are released to the simultaneous performance of complementary subtasks, which might improve the profitability of the whole system. TAMIR was tested also in the execution of other tasks like plastering, as performed in (Forsberg, J., Aarenstrup, R., Wernersson, A., 1997), about which there is however less information available. The main purpose of these experiments was connected with the necessity to develop highly autonomous machines for building sites: even if that problem is not completely solved, however first good results have been obtained, which perorate the feasibility of this technique. All these experiments follow the common approach of testing human ergonomics prototypes.

At the same time, another school of thought is working on miniature robots, which can be used for a number of purposes. Some of them are climbng robots for inspection and diagnostic (Bach, Fr. W., Rachkov, M., Seevers, J. and Hahn, M., 1995); climbing robot for curtain wall cleaning (Zhang, H., Zhang, J., Zong, G. 2006), for miniature navigation (Yilma, B., Seif, M. A. 1999). These robots have been developed because they are more lightweight and move at high speed, thus allowing operations that would be very cumbersome, and sometimes impossibe, in case of use of heavy and slow automatic machines. Among the wide variety of applications where they can be used, there is also the possibility to adopt them in construction: they could be used to access high rise building walls or dangerous environments and perform many operations, including wall painting. They can constitute also one valid alternative to the painting robots thus far presented, because their lightness enables them to quickly move throughout every job site, despite the presence of obstacles or other operators.

Other applications of miniature robots in different fields of research strengthen the validity of our proposal and the interest in this approach. In (Stoeter, S. A., 2005) the development of a control system for miniature robots is investigated. The author observs that untethered miniature robots with physical dimensions around or lower than 10 cm do not permit powerful onboard computers because of their size and powerful constraints. This limit has reduced in the past their functionality to that of remote controlled robots. In this contribution control is performed by a larger companion robot, that closes the loop using a video-based approach system. Such a framework was experimentally validated by performing the task of staircase climbing, which resulted to be feasible. Another robotic miniature system has been proposed for reconnaissance and surveillance by (Hogen, D.F. et al., 2000). It has been conceived to be able to carry out a wide range of missions, thanks to its multiple mobile modes and sensing capabilites. Data are then transmitted to a "ranger" robot or to a human leader. Both the "ranger" and miniature robots are able to accomplish together very complex operations. The other aspect that will be studied in the next paragraphs is concerrned with multi colour painting. This concept is not new in the field of manufacturing, like the ones offered for purchasing in many websites, e.g. (Goldspray, S.L., 2001). But this concept is original for construction. Manufacturing industry presents some examples of industrial plants for color mixing, that are aimed at reproducing every type of hue, saturation and brightness, starting from some primary colors. However those plants are very heavy and dangerous, as they store big amount of paints and work in airless mode, hence at high pressure. Moreover they are simply connected to a spraying gun, with no control system. The step forward that we would like to pursue by integrating a new light multicolor device within robots for contruction is given by the possibility to reproduce the functioning of printing machines for interior walls of buildings. It can be accomplished in case the multicolor device is opportunely coupled with robot arm's movements, as it will be widely explained in the next section 3.

RESEARCH COURSE

The new experimental approach

Our approach is focusing on two main concepts: allowing new ergonomics to better perform construction works and exploiting all the features uniquely provided by automation to produce high quality works, otherwise not feasible by human workers.

The development of new ergonomics is necessary just to improve the feasibility of automation, which is otherwise limited when it tries to reproduce human ergonomics. In order to better express our concern, we could think about what happened with painting: some prototypes have been manufactured reproducing the same working procedures presently adopted in building sites: they require the same tools and size of working area. At this stage the researchers have obtained very good results: they have set very good automatic machines that can work as humans do, hence with the same limits. To cite some of them we can notice that robots are generally quite heavy and cannot access every place in building sites; they would need strong and long braces to hold robots when painting external walls of high rise buildings. That means that human size robots must be combined with other nimbler machines in order to provide a complete service for building painting. When both technologies are available, robots will be able to completely replace human work.

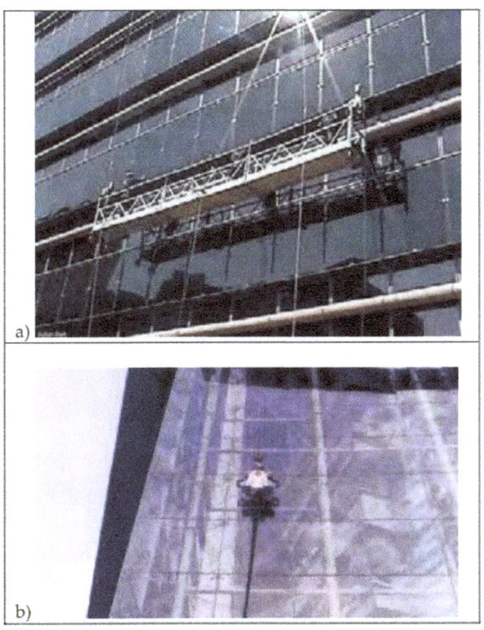

Figure 1: Human ergonomics (a) and automated ergonomics (b) adopted for glazed envelope cleaning of high rise buildings.

We have also one example of this kind, which is fully described in (Zhang, H., Zhang, J., Zong, G., 2006) and summed up in Fig. 1: even if wall cleaning has historically been performed by men working on suspended platforms like in Fig. 1-a, the coming of new technologies helped inventors think about the new and more efficient technology depicted in Fig. 1-b. The second case is more feasible and efficient because it does not ask for the use of strong and heavy supporting devices to lift operators up to very high levels; hence equipment costs are lowered and a smaller area of work is required. In addition, no worker is required to access high level and dangerous platforms like in the first case, eliminating risks of injuries. The same solution would not have been pursued in case heavy cleaning robot had been developed for such a purpose.

The same concept holds for painting robots moving through building sites or close to external walls: heavy machines can reproduce human ergonomics and work properly just in case they have the required room for their movements and stay on strong floors. In case accessing some rooms to be painted is very difficult or dangerous due to the conditions pertaining the specific building site, then heavy machines would be not suitable and the use of different products would be required. Hence our research is aimed at developing two kinds of automatic technologies for painting. The first is relative to a reduced ergonomics for painting, where some small and lightweight robots are coordinated one another. The second is based on a human-like ergonomics, reproducing the same movements and adopting the same tools as it is done by human workers. In the first case a multitask control system will be developed, which is thought to be able to manage the movements of every single robot and to coordinate their relative positions: it will require also the use of many sensors, some of them aimed at recording the extension of painted surface (Fig. 2). This second part of the control system is not necessary in case of robots working according to human ergonomics.

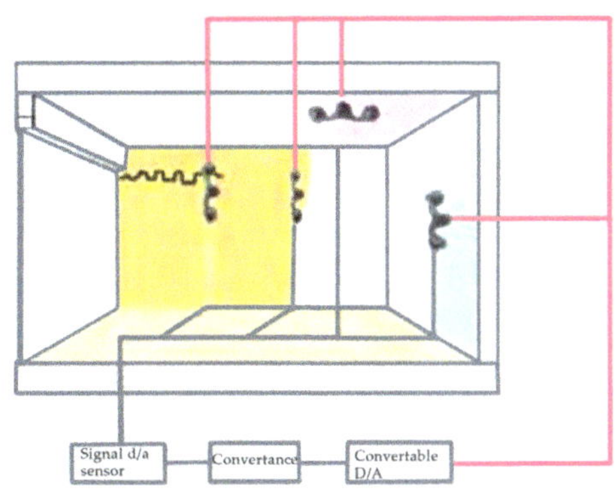

Figure 2: Multitask control for miniature robots.

Figure 3: Robots for scaled down (a) and full scale (b) experiments.

The two solutions will be complementing each other, meaning that they will be opportunely selected according to the features of the specific environment that will be painted. Considering the necessity of performing experimental campaigns, two robots have been set up inside as many laboratories depicted in Fig. 3. The first one is widely described in the next subsection 3.2, and it has been used to develop the scaled down painting end tool for the miniature approach. The other one, shown in the second part of Fig. 3, was not tested again, as it will be employed in the next step of our experiments, being more urgent to study the feasibility of a miniature painting end-tool at this first stage. Once that painting in small and full scale is shown to be possible, then further experiments concerning the two integrated technologies will be made. Moreover we will address the problem of high quality multicolor painting: this way we exploit all the capabilities that are put at disposal by the adoption of automated technologies. Multicolor means not only that architects can choose which color to attribute to walls, but that they could decide to reproduce drawings, whose resolution is determined in this paper: anyway, the smaller the end-tool the better the resolution.

Hence that approach is preferred at the miniature scale, because resolution is higher than at full scale, as it behaves in inverse proportion to the size of sprayed paint flux.

Conceiving the miniature laboratory

The main component of the 1:6 scaled-down laboratory is its miniature robot, which is depicted in Fig. 4. It is a 6 degree of freedom robot controlled by a 32 channel servo controller connected to the serial port of a PC, having a nominal reach of 0.4 m and a payload of 40 N. The robot is made up of a supporting rotating base fixed on a two degree of freedom hexapod for horizontal moves (not exploited for the experiments proposed in this contribution), an arm with three rotary motions whose end is equipped with a rotary wrist and a gripping hand (Fig. 4-a). The whole robot owns 6 servos (one for each degree of freedom excluding the hexapod). For this testing the gripping hand was substituted with the aerograph, that utilized the servo formerly used for gripping to rotate horizontally. Every servo is controlled by a small preassembled servo controller with some interesting features. It has high resolution (1uS) for accurate positioning and extremely smooth moves. The range is 0.75uS to 2.25uS for an angular range of about 170°. The motion control can be immediate response, speed controlled, timed motion, or a combination. A unique "Group Move" allows any combination of servos to begin and end motion at the same time, even if the servos have to move different distances. Every servo's position or movement can be queried to provide feedback to the host computer. Thanks to this arrangement it was possible to reproduce the movements that are used to paint plane walls with aerographs like in Figg. 4-b and 4- c. The robot is powered by one current and voltage regulator (range from 0 to 12 V and from 0 to 6 A).

The laboratory is surrounded by small plaster made surfaces to reproduce building walls mounted on three sides. Finally, one compressor was used to send high pressure air flows towards the aerograph installed on the end tool of the robot and to the color mixing device described in the next paragraph.

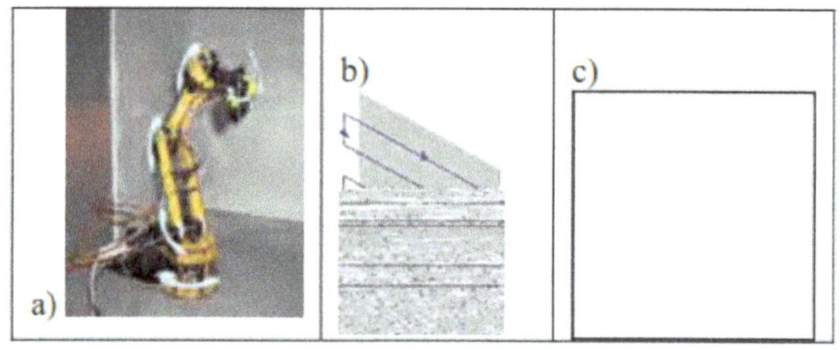

Figure 4: Scaled down robot (a) and trajectories used for wall painting (b and c).

Mixing device of the miniture laboratory

Every elementary painting equipment is made up of at least the following components: one spraying gun, a tank containing paint and a device to lead paint out of the tank and send it to the spray gun. The first step towards the development of the mixing device needs the choice of the best painting technology. There are three main approaches for painting: Airless; Aerography; Airmix.

The airless painting system works at very high pressure: paint is sent to the spray gun at about $1.2 \cdot 10^6$ Pa (to be created by a pump), whose precise value depends on its physical and state characteristics (density, viscosity, temperature, surface tension, solvent volatility, speed). In this case the jet stream shape is uniquely determined by the nozzle slot, which is generally chosen as vertically oriented. Instead the flow rate of paint is determined by the nozzle extension, which is generally limited from $3.1 \cdot 10^{-8}$ to $1.7 \cdot 10^{-6}$ m^2 (Perrin, H. and Zanassi, R., 1982). Paint flows through the nozzle at an average speed of about 140 m/s and it is held at about 0.25 m from the wall. Instead aerography produces a jet where air is mixed with paint, which is in turn sucked by air into the gun. It is different from the airless type, that directly atomizes paint through high pressures; in this case paint is sucked by high pressure air (about $4 \cdot 10^4$ Pa) flowing through the spray gun towards the nozzle. Sometimes the gun is fed by gravity or by low pressure, in order to help paint to pass through the gun, but the principle does not change, because in this second case high pressure air is always in charge of paint atomization. This technique is older than the previous one but is not suitable for quick drying paints and produces bigger drops and thicker films, hence lower quality products. In addition, its application is not simple because it generates a paint cloud around the nozzle and it could be dangerous for workers. The airmix system tradeoffs between the two former possibilities, and it is very useful for high precision works.

It is practiced through a nozzle that owns three holes: one in the centre working in airless mode; other two holes on the sides blowing high pressure air to mold the whole paint jet into the desired shape. Paint is blown between 3 and $6 \cdot 10^5$ Pa, and air is tuned according to the wished final result, whose pressure's magnitude is lower than for aerography (generally between 1 and $3 \cdot 10^4$ Pa). The success of this new system is however correlated to another requirement: it would revolutionize the concept of building painting just in case it will be able to draw high resolution images. In such a case the robot should be able to change the color of the sprayed painting while moving its arm, which means that it would reproduce also complex textures. In this case the system would overcome other similar applications, known just in the industrial field: they basically use very heavy paint distributors, which would be difficult to move. Moreover, they mix colors inside the distributors, which is far from the nozzle, requiring long time delays to switch from one color to another.

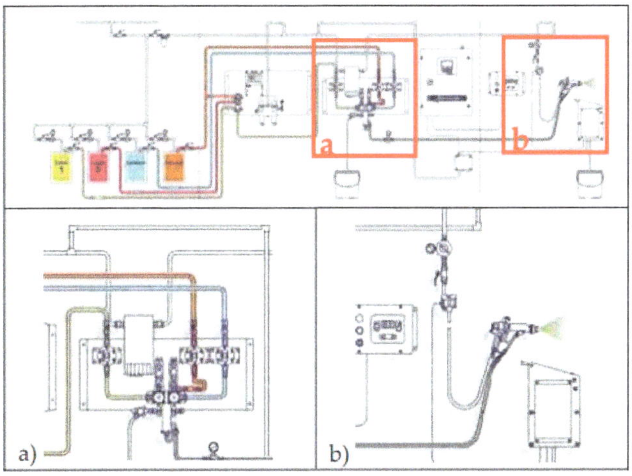

Figure 5: Example of mixing device for industrial applications and two details of the main figure, included in box a and box b.

Fig. 5 shows the concept on which it is based, where we can notice that it is too heavy to be moved throughout building sites and so large that a long delay between the time when it starts to change the color mix (through a change in pressures) and the time it comes out into a different color sprayed on the wall is required. For that reason our system is equipped with a high precision aeroghraph, that was adapted from a normal aerograph in such a way that it can be used for airmix spraying.

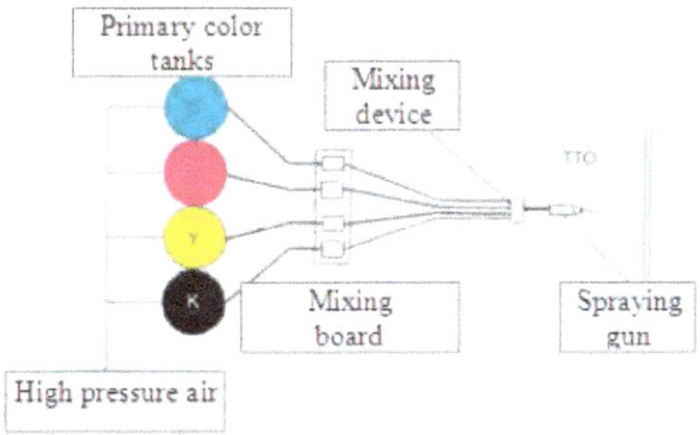

Figure 6: Sketch of the final multicolor mixing device.

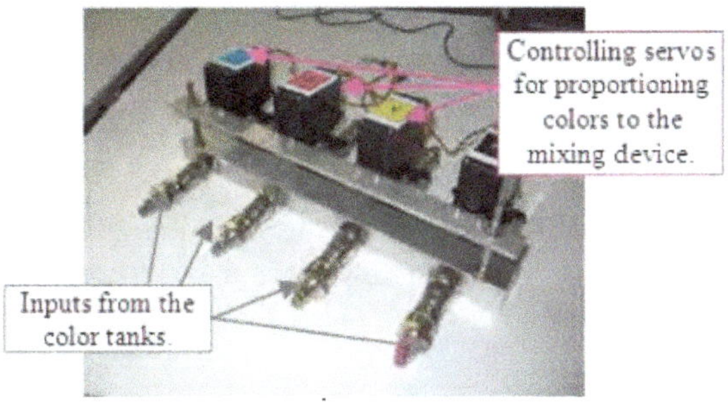

Figure 7: The mixing board.

Fig. 6 shows the functioning of the whole system: the four primary colors have been put inside high pressure tanks to be sent to the mixing board (Fig. 7), which turns up or down its valves to proportion the colors' relative amounts, previous to their blending into the mixing device close to the gun (Extension [1]).

All the system has been designed to work with four primary colors: cyan, magenta, yellow and black. Before showing the whole experimental set up it is necessary to understand how the mixing board in Fig. 7 works. The four color tanks push paint towards the mixing board at high pressure of $1.5{\cdot}10^5$ Pa, which is equipped with four ball valves. Every valve is connected to a servo (deflecting torque equals to 0.132 kg·m) which turns up or down the amount of paints flowing through it, establishing the percentage of primary colors to be mixed in the following device and hence deciding the hue of the final color. Every servo is driven by the same controller as for the robot, such that color proportions are coordinated with the robot's movements.

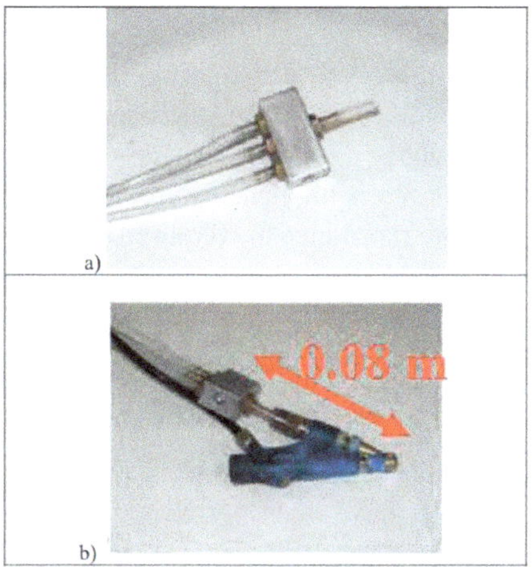

Figure 8: The mixing device (a) and its relative position to the spraying gun with the nozzle for airmix painting (b).

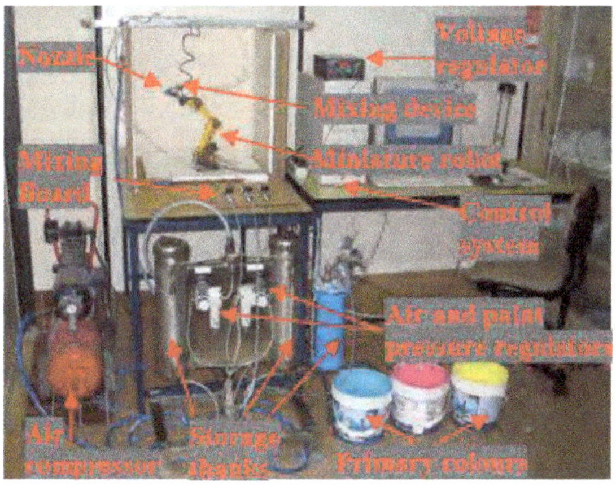

Figure 9: The whole multicolor miniature painting system.

From the mixing board the primary colors move towards the mixing device in Fig. 8-a, which is a modelled aluminum parallelepiped containing one transversal positioned steel spiral, helping the colors to be mixed into the final one by changing the flow movement into a turbulent one. This device is very close to the spraying gun (Fig. 8-b), that is the place where paint is mixed with air flowing at $3.5{\cdot}10^5$ Pa as soon as it goes out of the gun. To be noticed that the mixing device is positioned at a very short distance from the gun, equals to 0.08 m, in order to lower its inertia of color changing. Hence paint is partially atomized in the gun by the high pressure set in the nozzle, and that process is completed at the time when it is sprayed, that's to say just out of the gun before hitting the wall. In Fig. 9 it is possible to appreciate the whole experimental equipment, as installed in the scaleddown laboratory. High air pressure ($6\cdot10^4$ Pa) is sent by the air compressor towards the two pressure regulators: the first is used to regulate air pressure at $2.8\cdot10^4$ Pa before it is directed towards the nozzle; the second limits air pressure at $1.5\cdot10^4$ Pa before it enters the two storage thanks. This way paint comes out from the tanks at a final pressure of $0.8\cdot10^4$ Pa, that was considered to be good for this kind of paint and gun.

No pressure drops have been recorder along the other path followed by air. All the system is equipped with three safety valves. Then paint is sent to the mixing board, where the four servos drive their valves to the required opening, according to the inputs from the control system. The four paint colors flow towards the mixing device, that increases turbulent actions in order to spray a homogeneous painting, that is athomized at the gun-nozzle level, fixed on the robot arm, which is in turn powered by the voltage regulator. The control system gives the warranty that the robot's movements are coordinated with the valves' openings to reproduce the required color textures.

THE EXPERIMENTAL CAMPAIGN

Set up of the spraying parameters

Even if the system has been designed for four colors, however all the experiments have been performed with three colors: cyan, magenta and yellow. We know that in this case a narrower range of colors can be reproduced, but it is enough for our experimental purposes.

Preliminary experiments to set the opportune spraying parameters have been performed as detailed in (Naticchia B., Giretti A. and Carbonari A., 2006). There are two different sets of paramters: those ones which are fixed by standards and by the technology adopted, and those ones depending on the expected quality for the final product. The first group embraces type of paint (in this case waterbase paint), type of application (in this case airmix), spraying pressures (in this case $2.8\cdot10^4$ Pa for air and $0.8\cdot10^4$ Pa for paint), end-tool trajectories (please refer to Fig. 4), distance from the wall (in this case 0.08 m). Instead variable parameters are given by the distance step between every line of the trajectories and the couple speed-paint flow, which determines the paint thickness on the wall. A detailed study to set the most opportune values for these variables has been carried out in the way already detailed in the aforementioned paper. The quality of the final work has been evaluted using two indicators:

- saturation: indicates color intensity and purity ;
- thickness uniformity: evaluates thickness variation on the wall surface, that must be as constant as possible to avoid that visual perception changes with respect to the direction from which the surface is looked at.
- The first was measured using an image processing software, restituting how far from the totally 100% saturated condition is a surface after its coating. The second is important to avoid too big variations of paint thickness on the wall, that could cause a particular reflection of light, that would be perceived by human eyes as a vertical or horizontal stripped surface. The procedure for parameters optimiziation asks for the collection of one statistical sample (formed by at least 10 measures) whose characteristics vary according to the parameter to be set. In the previous experiments speed of arm motion was left to vary between 0.036 and 0.11 m/s. For our experiments we chose to work at an average speed of 0.073 m/s , as it trades off between a low speed strategy (that reduces the size of the pixel) and the necessity to move fast in order to prevent the formation of droplets (in case too thick paint is sprayed on walls). At that speed it was found out that the paint flow giving the best results is equal to $3.4\cdot10^{-7}$ m^3/s, as it sprinkles on the wall a paint layer thinner than $0.044\cdot10^{-3}$ kg/m^2, which has been shown to be the maximum allowable limit in (Naticchia B., Giretti A. and Carbonari A., 2006). The distance step between every line of the trajectories was set equals to 0.045 m, as suggested by the results published in the same paper. Once that these parameters have been set, other experiments to verify the accuracy and to table the features of the multicolor system have been perfomed, as expalined in the next paragraph.

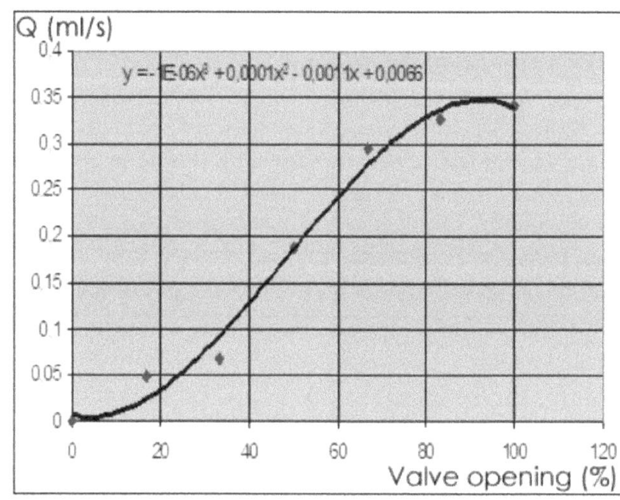

Figure 10: Characteristic curve of the ball valves.

Tests on the four colour spraying device

Given the preliminary experiments of the previous section, the multicolor system has been characterized and tested, in order to infer its features. Tests have been performed regarding:

- analysis of precision;
- level of resolution;
- accuracy in color reproduction.

The analysis of precision has the main aim of establishing the accuracy of the system when it is asked to paint a given shaded rectangular portion of wall. First, the behavior of the ball valve has been analyzed: its characteristic curve describing the paint flow rate (Q) with respect to its rotation around its axis (measured in percentage) from the closed position has been drawn.

The valve has been turned on through steps of 15 degrees, and the corresponding flow rate has been measured as pictured in Fig. 10. The relation coming out from Fig. 10 is of basic importance: thanks to that, the controller will be able to compute the paint flow sprayed by the gun, starting from the opening of each valve of the mixing board (which is equipped with four identical ball valves). Working the relation in the opposite way, it is possible to compute the percentage opening required for each ball valve if a pre-established paint flow is required. This information has been used to compute the relative opening of each valve when a known color mixing is required. Once we know the value of the total flow ($Q = 0.34 \cdot 10^{-3}$ l/s), it is necessary to work out single values for each primary color. The following relation must hold, meaning that the sum of all the paint flows cannot exceed the maximum:

$$1 = \frac{Q_C}{Q} + \frac{Q_M}{Q} + \frac{Q_Y}{Q} \qquad (1)$$

Where Q is the total paint flow rate, while Q_C, Q_M and Q_Y are the flow rates assigned to each primary color (cyan, magenta and yellow). From the theory of colors explained in the next paragraph it is possible to compute every relative flow rate and then the relation in Fig. 10 can be used to compute the required opening for each valve. At this stage it is possible to perform analysis of precision. It has been made by generating a sample of size 9, formed by nine shaded portions of walls of the kind in Fig. 11, changing from magenta to yellow and other 9 shaded from cyan to magenta. Through a software for imaging processing, the saturation of each sample on a 14 equally spaced column grid has been measured. In Fig. 11 we can appreciate that on each point of the grid, standard deviations measured for magenta on the selected points are low, and on the average equal to 8%. Even better results have been obtained for the other two primary colors, where cyan deviation from the mean are on the average equal to 5.5%. Table 1 sums up the first set of those results.

The time delay has been measured by filming the movements of the servos on the mixing board and the colors sprayed on the walls: the time difference between color switching of the ball valve and the time when it is observed on the wall is the unknown time delay. One of these tests is shown in Fig. 12. It has been found out that the switching time is equal to 3 s, corresponding to one pixel (0.13x0.20) m wide, which is the finest resolution feasible with this system.

Procedure for colour reproduction

One of the most important requirements for such a system is that it can accurately reproduce the colors of every pixel of drawings. For that purpose a general procedure for color reproduction has been set, that is based on the theory of colors in (Perrin, H. and Zanassi, R., 1982).

Table 1: Precision evaluation at each point of the grid superimposed to sprayed samples like in Fig. 11

Mean	86	84	83	79	71	68	63	58	54	47	41	36	35
s.d	2.5	3	2	2.9	4.4	3.3	4.4	4.1	6	7.5	5.5	5.1	7.3
% s.d.	2.9	3.5	2.4	3.7	6.1	4.9	7	7	11	16	13	14	21

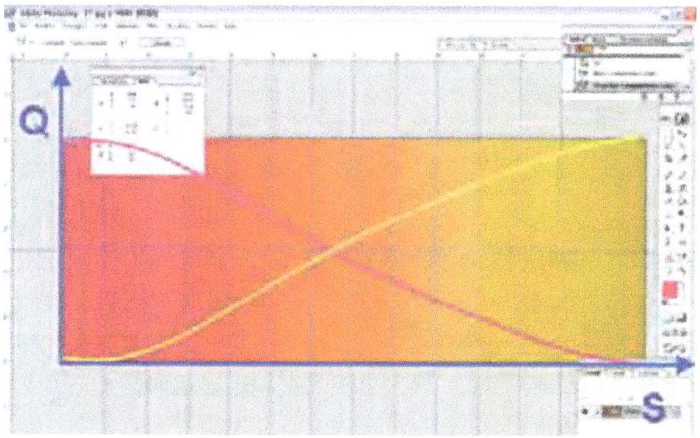

Figure 11: One of the samples for precision evaluation.

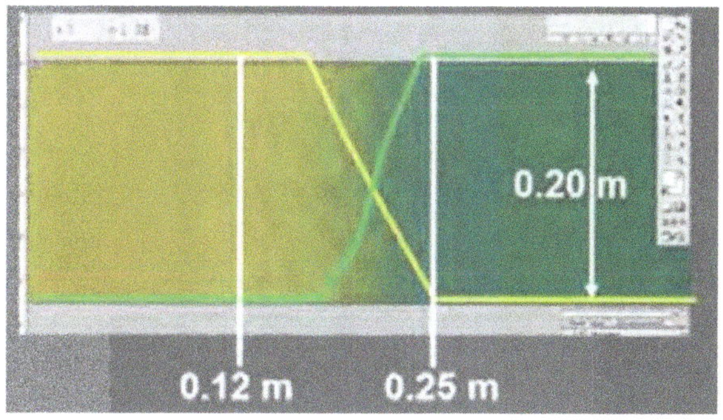

Figure 12: System resolution.

As we are dealing with a system working like printing machines, every color can be thought as composition of the three primary ones: cyan, magenta and yellow (generally also black is included to obtain a wider range in the visible spectrum, however in this experiment it was neglected). We will explain the general procedure while applying it to one real case. First of all one color belonging to the visible spectrum has been generated with the help of a software for image processing (Fig. 13- a), having the following normalized coordinates (C,M,Y) = (0.43, 0.06, 0.51). Every normalized coordinate is computed from the original CIE (Commission Internationale de l'Eclairage) coordinates (c,m,y) by defining for cyan:

$$C = \frac{c}{c + m + y} \qquad (2)$$

and the same holds for magenta and yellow. Then it must be considered that the primary colors used by our system are not the same primary colors considered by the image processing software. Hence three wall samples have been painted using these primary colors and obtaining the following normalized coordinates:

1st primary color (C*) = (0.67, 0.2, 0.13);
2nd primary color (M*) = (0.12, 0.80, 0.08);
3rd primary color (Y*) = (0.05, 0.00, 0.95). (3)

The relative amount of the system's primary colors (C^*, M^*, Y^*) to be used for the reproduction of the original color (C, M, Y) have been computed by solving the following system:

$$\begin{cases} C^* + \alpha_c M^* + \beta_c Y^* = C \\ \alpha_M C^* + M^* + \beta_M Y^* = M \\ \alpha_Y C^* + \beta_Y M^* + Y^* = Y \end{cases} \tag{4}$$

Where the coefficients ˆ and Ά allow for the non purity of our primary colors. For instance, if we know that the system must spray a certain amount of cyan (C), we know that it is present not only in its cyan (C*) but has also other contributions in its magenta and yellow (α_c=12% and β_c=5%), as suggested by eq. (3), from which all the other coefficients have been determined. The unknown variables have thus been determined to be (C*, M*, Y*) = (0.476, 0.025, 0.54). Then applying eq. (1) in the following form:

$$1 = C^* + M^* + Y^* = \frac{Q_C}{Q} + \frac{Q_M}{Q} + \frac{Q_Y}{Q} \tag{5}$$

and by the knowledge of the optimum flow rate Q, it was possible to compute the single flow rate values Q_C, Q_M and Q_Y; from these values and from the diagram in Fig. 10, valves' openings were determined and the final color was sprayed on a plaster wall sample (Fig. 13-b), as shown in Extension [2] and in Fig. 14.

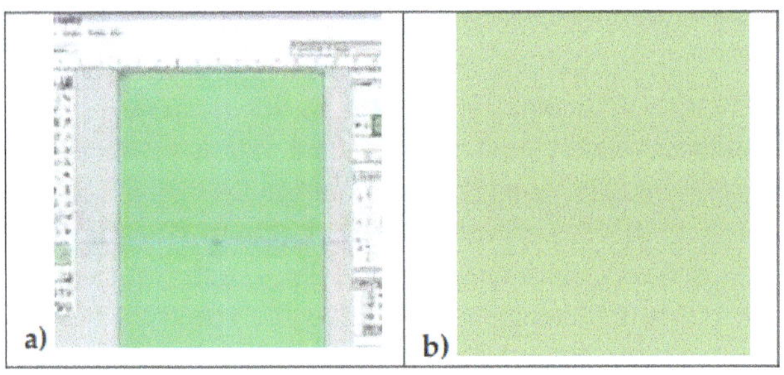

Figure 13: Original color (a) and its reproduction (b).

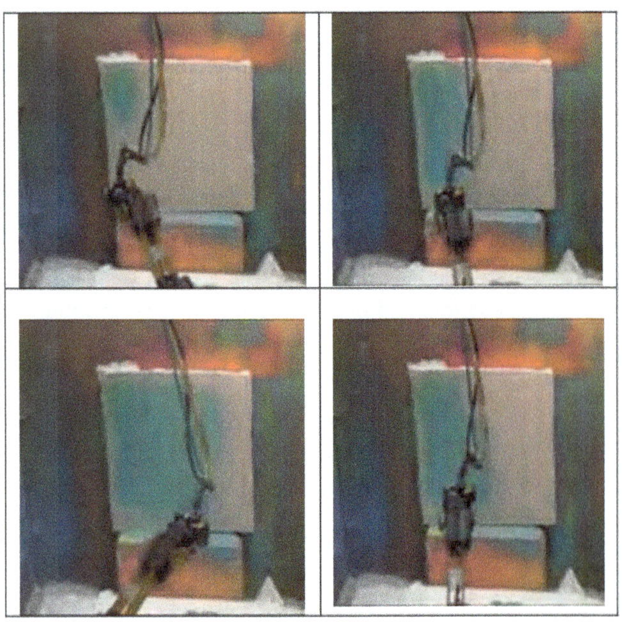

Figure 14: Four clockwise directed snapshots of the miniature robot's spray painting on a plaster wall.

Observing the two figures it is possible to appreciate that they look very similar. But this check has been performed also using a scanner acquisition system: it gave back the normalized values (0.53, 0.58, 0.31) for Fig. 13-b, which have been corrected by the knowledge of the shift introduced by that scanner (previously determined scanning pure primary colors from (Perrin, H. and Zanassi, R., 1982)), that is (+0.20, +0.06, -0.26). Hence the corrected value of the color sprayed by our system was (0.33, 0.1, 0.57), whose difference with the original one is (-0.1,+0.04, +0.06), that is the approximation introduced by the whole mechanical system.

Results

The first positive result pertains the ability of the small and lightweight robot to carry its end-tool, even when it sprays at quite high pressures, causing a strong recoil. Furthermore, analysis of precision gave very good results, as every shift is never higher than 10%. Also the finest pixel, (0.13x0.20) m wide, would allow users to perform quite detailed drawings, when painting very big walls. As far as reproduction of colors, we found that the shift introduced by the system is never higher than 10% (just in the case of cyan). Moreover we can notice that a very good coverage of the plaster wall has been obtained, even if just one layer of paint has been laid. We know that better results can be obtained by applying two layers of coating on walls, that will be experimented in later steps.

CONCLUSION AND FUTURE RESEARCH

Thanks to the mixing equipment developed throughout the research step described in this paper, we have shown that automated painting can be not only aimed at improving productivity, but also quality. In addition, two kinds of complementing ergonomics can be thought, one at reduced scale and the other one at human scale. Our future research will be aimed at discerning how it is possible to perform the most efficient choice among them, according to the working place and to the technology chosen. Moreover, it will be necessary to develop two technologies with very similar performances, in order to provide a whole system with interchangeable tools (reduced and human scale).

The small scale technology has been shown to have high precision, quite high resolution and very good accuracy in color reproduction. We know that even better results could be obtained by developing new painting products, more suitable for the narrow nozzles used for this small applications, by improving the precision of the mechanical system, of the nozzle and better choice of the primary colors.

When moving to full scale experiments, it will be necessary to reproduce the same precision level, which should be obtained by providing a very accurate control system, similar to the one developed for small scale experiments: adoption of one mixing board equipped with four valves, each of them connected to a servo which turns up or down the amount of paints flowing through it and establishing the percentage of primary colors to be mixed for the final color. A robot arm with high precision is required, too. Accuracy in color reproduction should be faced in a way similar to what was done in this contribution: an automated system to translate the normalized coordinates of the colors to be reproduced, into the movement speed of the robot end tool and valve openings of the mixing board. Much more work will be probably necessary to obtain high resolution. Because of the size of full scale robots, probably also the resolution of the human scale robot will be lower. Of course we will be not able to obtain pixels as small as (0.13x0.20) m, however we think to be able to exploit the features of the airmix system to mold its shape as needed: air pressure and air flow passing through the two holes on the sides of the airmix nozzle can be tuned according to shape and dimension wished for the final paint flow. This will determine the threshold under which the adoption of the small scale system will be preferred to the human scale approach. Another peculiarity of the small scale system is of course the ability to access some difficult places of buildings under construction, where human size robots could not be allowed.

ACKNOWLEDGMENTS

We have been indebted to prof. Mario De Grassi, full professor at the Polytechnic University of Marche – DACS Department of the Faculty of Engineering, for his helpful support and suggestions. We would like to thank the editor of this special issue for his kind invitation to publish our paper. Also the comments of the anonomous reviewers revealed to be critical for the improvement of the clarity and quality of the contents of this paper.

INDEX TO MULTIMEDIA EXTENSION

The multimedia extensions to this article can be found by following the hyperlinks from: www.ars-journal.com

[1] mixing board.wmp: video showing how the mixing board works.

[2] system validation.wmp : video showing the robot at work when spraying the final wall portion used for validation of the procedure for color reproduction.

REFERENCES

1. Bach, Fr. W., Rachkov, M., Seevers, J. and Hahn, M., (1995). High tractive power wall climbing robot, Automation in Construction, Vol. 4, No. 3, October 1995, pp. 213-224, Print ISSN: 0926-5805.
2. Forsberg, J., Aarenstrup, R., Wernersson, A., (1997). A construction Robot for Autonomous Plastering of Walls and Ceilings, Proceedings of the 14th International Symposium on Automation and Robotics in Construction (ISARC), June 8-11 1997, Pittsburgh, PA, USA.
3. Goldspray S.L., Manufacturing Company in Torrente Valencia - Spain, available on line at: www.goldspray.com/InglesWEB/Index2.htm, 2001.
4. Hougn, D.F., Benjaafar, S., Bonney, j. C., Budenske, J. R., Dvorak, M., Gini, M., French. H., Krantz, D. G., Li, P. Y., Malver, F., Nelson, B., Papanikolopoulos, N., Rybski, P. E., Stoeter, S. A., Voyles, R., Yesin, K. B., (2000). A miniature robotic system for reconnaissance and surveillance, Proceedings of the 2000 IEEE International Conference on Robotics and Automation, pp. 501-507, April 2000, San Francisco, CA.
5. Kahane, B. and Rosenfeld, Y., (2004). Balancing human and robot integration in building tasks, Computer Aided Civil and Infrastructure Engineering, Vol. 19, No. 6, November 2004, pp. 393-410, Print ISSN: 1093-9687, on-line ISSN: 1467-8667.
6. Naticchia B., Giretti A., Carbonari A., (2006). Set up of a robotized system for interior wall painting, Proceedings of the 23rd International Symposium on Automation and Robotics in Construction (ISARC), October 3-5, 2006, Tokyo, Japan.
7. Perrin, H., Zanassi, R., (1982) "Verniciatura a.spruzzo", vol. 5(2), La Rivista del Colore, Milano.
8. Stoeter, S. A., (2005). "Autonomous Stair-Climbing With Miniature Jumping Robots", IEEE Transactions on Systems, Men and Cybernetics-Part B : Cybernetics,Vol. 35, no. 2, April 2005, pp. 313-325, ISSN: 1083-4419.
9. Warszawsky, A. and Rosenfeld, Y., (1994). Robot for interior finishing works in buildings : feasibility analysis, ASCE Journal of Construction Engineering and Management, Vol. 120, No. 1, March/April 1994, pp. 132-151, Print ISSN: 0733-9364.
10. Yamazaki Y., Maeda J., (1998). The SMART system : an integrated application of automation and information tehcnology in production process, Journal of Computers in industry, Vol 35, No. 1, February 1998, pp. 87-99.
11. Yasuyoshi M. and Roozbeh K., (1993). Experiencing computer integrated construction, International Journal of Construction Engineering and Management, Vol. 119, No. 2, pp. 307-322, May/June 1993.
12. Yilma, B., Seif, M. A., (1999). Behavior-based artificial intelligence in miniature mobile robot, Mechatronics, Vol. 9, No. 2, March 1999, pp. 185-206, ISSN : 0957-4158.
13. Zhang, H., Zhang, J., Zong, G., (2006). Effective pneumatic scheme and control strategy of a climbing robot for class wall cleaning on high rise buildings, International Journal of Advanced Robotic Systems, Vol. 3, No. 2, June 2006, pp. 183-190, ISSN : 1729-8806.

CHAPTER 8

A WIRELESS MONITORING SYSTEM FOR CRACKS ON THE SURFACE OF REACTOR CONTAINMENT BUILDINGS

Jianguo Zhou, Yaming Xu and Tao Zhang

Keywords: Structural health monitoring, wireless sensor networks, containment crack monitoring, real-life deployment

INTRODUCTION

Civil infrastructure usually plays an important role in society's prosperity [1]. The safety of these structures during their normal operations cannot be neglected as any health problem as a result from them may cause severe life and property loss, which has happened many times in the past. Due to aging, loadings, environmental conditions and extreme events like earthquakes, structures may deteriorate in their performance slowly, or be damaged severely. Then, structural health monitoring (SHM) [2], which is defined as detecting changes to the material or geometric properties of structures, becomes essential to the safety of these structures. With proper tools and techniques, continuous, real-time and automated monitoring systems could evaluate reliability of a structure, identify potential harmful factors and provide assessment for maintenance decisions during its service life. Traditional wired monitoring systems have advantages in acquiring high-fidelity data and robustness, but they still have some drawbacks. Such systems require electronic cables to deliver data sampled by limit sensors to a central device, which usually could be time consuming and expensive for installation and maintenance. In addition, installing these cabled sensors on large civil structures could sometimes interfere with their everyday normal operations. Fortunately, with developments in Micro-Electro-Mechanical System (MEMS) technology, wireless communications and digital electronics, the advent of wireless sensor network (WSN) [3] technology brings new opportunities for structural health monitoring. Compared with wired systems, monitoring systems using WSNs could benefit from their obvious advantages like cable-free and plug and play deployment, which allows engineers to reduce time and cost for installation. Furthermore, these tiny sensor nodes make it possible for dense network deployments, thus allowing better understanding of the monitored structure.

Currently, many SHM systems based on WSNs have been implemented on various civil structures including bridges, tunnels, and tall buildings and so on. Utilizing 64 Micaz nodes with two different accuracy accelerometers that measure ambient vibrations, Kim and Pakzad *et al.* [4,5] implemented an up to 46-hop WSN monitoring system on the Golden Gate Bridge. Jang, Cho and Rice *et al.* [6,7] deployed and evaluated a wireless smart sensor networks monitoring system consisting of 70 Imote2 nodes with custom-designed SHM sensor boards on the Jindo Bridge. Feltrin and Flouri *et al.* [8,9,10] presented an SHM system based on WSNs that addressed challenges for long-term deployments and was used in deployments for monitoring a cable stayed bridge and a timber footbridge. Bischoff *et al.* [11] proposed an event-based monitoring application using eight Tmote Sky nodes with resistance strain gages on a steel railway bridge. Meanwhile, an integrated SHM system for highway bridges based on customized WSN platforms was developed by Hu *et al.* [12] As for tunnels, Bennett *et al.* [13] designed a WSN based underground monitoring system using Micaz nodes with crackmeters and inclinometers, and the SHM system was tested in the Prague Metro and the London Underground. Zonta *et al.* [14] utilized 16 3MATE! nodes with accelerometers, deformation gauges and thermometers to monitor the health status of a medieval

Citation: Jianguo Zhou, Yaming Xu and Tao Zhang, A Wireless Monitoring System for Cracks on the Surface of Reactor Containment Buildings, doi:10.3390/s16060883.

tower. Ou *et al.* [15] installed a WSN monitoring system on a tall building called Diwang Tower with eight sensor nodes and comparisons were made with the cable-based system.

In this paper, we implement a wireless monitoring system for nuclear power plants, especially for monitoring changes in crack width on the surface of reactor containment buildings during pressure tests. Customized WSN platforms are developed and the sensors used include crackmeters and thermometers. Issues like route discovery, time synchronization, data collection and duty-cycling schemes are considered in software design. After some simulation tests, full scale deployment is made at the Daya Bay Nuclear Power Station with 30 sensor nodes. The paper is organized as follows: Section 2 presents background and motivation. Section 3 describes hardware selection and assembly for sensor nodes. Software design issues like route discovery, time synchronization, data transfer protocol and wake-up mechanism are discussed in Section 4. Section 5 presents some simulation tests for evaluating the performance of the system. Section 6 describes the field test of the wireless monitoring system. Finally, Section 7 concludes the paper.

BACKGROUND AND MOTIVATION

The generation of electricity through nuclear energy has many advantages such as lowering greenhouse gas emission, being reliable, and being independent from natural aspects. However, the safety of nuclear power plants are most important, as any slight mistakes could lead to severe disasters. Chernobyl and Fukushima nuclear accidents are just two good examples. Reactor containment building (RCB), as the fourth and final defense layer for a nuclear reactor, is designed to keep radioactive substances from getting into the environment in operational states and in accident conditions like loss of coolant accidents (LOCAs), and protect the nuclear power plant against external natural and human induced events [16]. Then, usually RCBs have to undergo pressure tests to verify that they comply with the highest safety standards. Pressure tests are conducted during commissioning and in-service tests, which are also called initial and periodic tests. They comprise several measurement phases at different pressure levels to verify that the structural behavior of the RCB is consistent with the design and demonstrate that the leak rate of the RCB does not exceed the specified maximum leak rate. Commissioning tests for the RCB are carried out prior to the first criticality of the reactor and in-service tests are usually carried out every ten years. Pressure tests could last as long as two weeks and one very important task during tests is to inspect the surface of the RCB and measure the changes of crack width on the concrete structure in real-time.

Considering that cracks to be concerned about may appear anywhere of the RCB, on the dome or on the perimeter wall, it is impractical to carry out the monitoring task with wired systems. Currently, used battery powered devices that store data locally and gather them after pressure tests also have drawbacks. The monitoring data cannot be obtained by users in real-time for further analysis and users are not informed if some devices stop working during the test, thus leading to incomplete data. The advantages of WSNs and the successful cases of using them to monitor other structures motivate us to introduce this technology for real-time crack width changes monitoring on the surface of RCBs. As far as we know, this is the first time to use wireless sensor networks technology for RCB monitoring. We develop customized platforms and corresponding software. The deployment in Daya Bay Nuclear Power Station shows the effectiveness of the system.

HARDWARE SELECTION AND ASSEMBLY

Each wireless sensor node in a structural monitoring system is responsible for three primary functions: data acquisition, signal processing and wireless communication. Correspondingly, a sensor node usually consists of a microprocessor, radio transceiver, data storage, batteries and sensor boards. Up to now, many commercial wireless nodes have been used for structural monitoring. In addition, some researchers have also designed their own platforms to carry out monitoring tasks. Table 1 shows the features of typical commercial nodes [4,5,7,8,17,18,19,20] and some customized ones [12,21,22] used in structural monitoring. Based on the experience of past researchers, we design our own nodes with off-the-shelf components. For sensors, crackmeters and thermometers are selected according to monitoring contents. Our monitoring system typically comprises multiple sensor nodes, several cluster nodes and a sink node that is connected to a laptop.

Table 1: Typical nodes used in structural monitoring

Node	Microprocessor	Radio Chip	Data Storage (Bytes)	Batteries
Mica2	ATmega128L	CC1000	512 K	2 × AA
MicaZ	ATmega128L	CC2420	512 K	2 × AA
Imote2	XScalePXA271	CC2420	32 M	3 × AAA
Tmote-sky	MSP430F1611	CC2420	1 M	2 × AA
S-mote	MSP430F1611	CC2420	1 M	2 × AA lithium
WiSSM	ATmega128	Maxstream 9XCite	128 K	5 × AA

Sensor Selection

The most commonly used sensors for structural monitoring are accelerometers, with which the structural vibration response can be captured. In addition, two or three axes, high or low accuracy accelerometers are selected based on specific monitoring requirements. Our monitoring targets are width changes of the cracks on the surface of RCBs, then crackmeters are chosen, and they are based on linear variable differential transformers (LVDTs). LVDTs are displacement transducers and are widely used in industrial applications. They are made up of a central core, a primary coil and a secondary coil with two equal parts that are connected with opposite polarities. The primary coil is supplied with an alternating current, and the two parts of the secondary coil induce different voltages because of different mutual inductances when the core moves between the coils. The difference between the two voltages is the output voltage of the sensor. LVDTs have the advantage of high immunity to electrical noise, and as there is no mechanical contact between the core and coils, LVDTs could increase the expected life. The LVDT (Soway, Shenzhen, China) we choose can be seen in Figure 1a. It has a measurement range from −5 mm to +5 mm with a resolution of 1 μm. The output of the sensor is digital with RS485 interface and the work voltage is 12 Volt direct-current. It also has a built-in temperature compensation mechanism. The precision of the sensor is tested through comparison with the linear comparator, and the result shows that it meets the monitoring requirement. Besides crackmeters, we also use PT100 platinum resistance temperature detectors (Fenghe technology, Wuhan, China) to capture the temperature of the environment for data analysis purposes. The resistance of PT100 remains 100 Ω at 0 ∘C and changes with temperature. In addition, the temperature information can be obtained through signal converter according to the continuous function relationship between the resistance and temperature. The PT100 platinum resistance temperature detector is shown in Figure 1b.

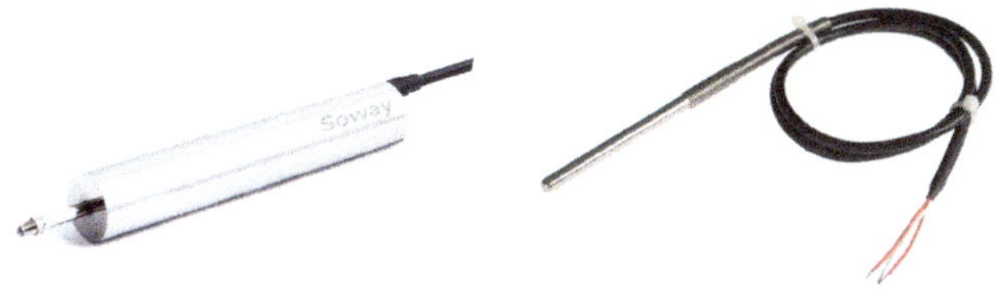

(a) (b)

Figure 1: Sensors used for crack width change monitoring. (**a**) LVDT crackmeter; (**b**) PT100 temperature detector.

Node Assembly

The components used by nodes mentioned in Table 1 give us a good reference for design customized platforms. In addition, the main components of our customized platforms are the C8051F930 microcontroller and the SI4432 radio chip, both of which are products of Silicon Labs (Austin, TX, US). The 8-bit microcontroller has 4 kB of RAM, 64 kB of on-chip flash memory and maximum 25 MHz clock speed. Its high-speed and low-power features suit the monitoring task well. The reason why the SI4432 is chosen is that low frequency bands are preferred in nuclear power plants. The 433 MHz radio-frequency transceiver offers extremely low receive sensitivity (−121 dBm) coupled with industry leading +20 dBm output power ensures extended range and improved link performance. The advanced radio features of the device reduces overall current consumption and allows the use of the low-cost C8051F930 microcontrollers. The SI4432's high level of integration also reduces cost while simplifying the nodes design. For nodes measuring width changes of cracks, the LVDT crackmeter and PT100 temperature detector are connected to them. These nodes send commands to sensors, receive feedback from them and then transmit the data wirelessly. In addition, a data storage module for backup in case of data loss in wireless transmission is also attached to the sensor node. It can store up to 5000 data samples, which is enough according to the required sampling period and the duration of pressure tests. A small LCD screen is used for display consideration. As the nodes may carry out monitoring tasks under harsh weather conditions like storm and snow, the hardware components need to be protected for normal operation. Then, a plastic box is used for housing and an external RS485 connector allows the LVDT crackmeter connection, data uploading, battery charging and node reprogramming without opening the enclosure. Figure 2 shows our customized sensor node and its logical structure. There are no sensors connected to cluster nodes in our monitoring system, and the function of cluster nodes is to relay the data transmitted from sensor nodes to the sink node. The only sink node in the system is connected to a laptop through a USB cable. It collects all monitoring data from cluster nodes, or directly from sensor nodes, and stores it in the laptop for further analysis.

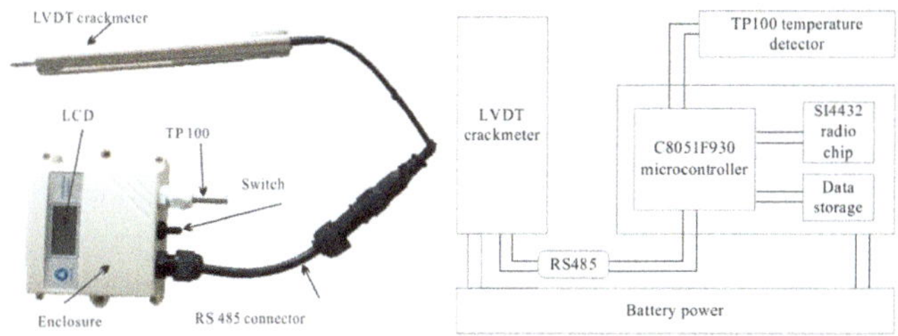

(a) (b)

Figure 2: The customized sensor node and its logical structure. (**a**) The sensor node; (**b**) The logical structure of the node.

Energy Consumption

The sensor nodes are usually battery-powered; therefore, the energy of them is limited, which means that it should be used properly. Energy consumption is one of the most important issues in wireless sensor networks. Every operation of the sensor node affects its life span and the lifetime of the whole monitoring network, respectively. In our system, the sensor nodes are powered by rechargeable lithium batteries with a total capacity of 4000 mAh. In order to make sure that the capacity is large enough to work through pressure test periods, the operation current under idle mode, sampling mode and transmitting mode is measured. We find each sampling state lasts about four seconds and transmitting one packet costs approximately two seconds. According to the typical 30 min sampling period, sensor nodes only sample and transmit twice in one hour. Then, the power consumption under each mode is calculated. Table 2 shows operation current and power consumption of the sensor node in different modes mentioned above. It can be seen from the table that the total power consumption of our sensor node is 2.245 mAh. In addition, the capacity of the batteries ensures a continuous monitoring service life up to 74 days if the initialization stage of the sensor node is not taken into account, which covers the two week pressure test completely.

Table 2: Operation current and power consumption in different modes

Operation Mode	Operation Current	Power Consumption
Idle	2 mA	2 mAh
Data sampling	80 mA	0.178 mAh
Data transmitting	60 mA	0.067 mAh

SOFTWARE IMPLEMENTATION

The sensor node in the monitoring system can do nothing without software. Software is the soul of the sensor node to execute sensing, computing, and communication. In addition, the operating system of sensor networks is the cornerstone of software. Taking the features of WSNs into account, many embedded operating systems special for sensor networks like TinyOS, SOS, MANTIS OS and Contiki have been developed by researchers. TinyOS [23,24] is an open source, lightweight operating system developed by UC Berkeley for low-power wireless sensor nodes. It provides a set of important services and abstractions, making it easier to build sensor network applications. Applications and the system itself are written in a C dialect, called nesC language. It has been run on many generic platforms and used in a wide variety of monitoring applications for WSNs [7,12,25]. SOS [26] is an operating system for mote-class WSNs developed by the NESL at University of California (Los Angeles, CA, USA). It uses a common kernel to implement basic services. In addition, the prominent characteristic of SOS is its dynamic reconfigurability, which allows dynamic addition, modification, and removal of network services. However, currently, SOS is no longer under active development. MANTIS OS [27] is an open source, multi-threaded operating system written in C for wireless sensor networking platforms. It supports many platforms and has been used in several applications. Contiki [28] is a tiny, highly portable operating system for small devices developed by a world-wide team of developers. It is a very promising operating system as it is easy to connect to the Internet. We have not currently introduced any operating systems into our monitoring system, and we just utilize C language to implement the software functions. In initialization stage, route discovery and time synchronization of nodes are considered, and we focus on transmitting protocol and duty cycling in monitoring stage. The graphic user interface (GUI) at the sink node side is also mentioned.

Initialization Stage

In the initialization stage, nodes need to prepare for data collection, and there are two main things they should complete during this stage: one is route discovery and the other is time synchronization. Route discovery is very important as it is the precondition of data collection. In our monitoring system, we implement a hierarchy routing protocol similar to LEACH protocol [29]. The difference is that there is no cluster heads selection phase because the sink nodes and cluster nodes in our system are existing cluster heads. First, the sink node is assigned a level 0 and broadcasts a route discovery message containing its identity and level. All of the cluster nodes and sensor nodes receiving this message set the sink node as their parent and assign themselves a level 1. Then, the cluster nodes in level 1 broadcast a new a route discovery message, and this process continues until all nodes in the network assign a level. Nodes that already have a parent will ignore later broadcast messages. In this way, a routing tree rooted at the sink node is established and all the sensor nodes send data directly to the sink node or through cluster nodes in the network. Time synchronization is one of the fundamental issues in WSNs. All nodes in the monitoring network sharing a common timescale makes the measured data meaningful. Some scheduling protocols like time division multiple access (TDMA) require time synchronization so that transmissions do not interfere. In addition, an accurate time is essential for efficient duty cycling and helps save energy. The required synchronization accuracy is not high in our monitoring system as the typical sampling period is 30 min. In addition, the algorithm we achieved in our system is a simplified flooding time synchronization protocol (FTSP) [30]. The synchronization process is conducted between neighbor levels of the spanning-tree established in route discovery phase. Each cluster head broadcasts a message time-stamped at the media access control (MAC) layer, and child nodes of that cluster head record the corresponding time from their respective local clocks at message reception. Only clocks offset between them are estimated, and, once synchronized, the network can keep the synchronization error in the second range during the whole monitoring period.

Monitoring Stage

The main task for sensor nodes in the monitoring stage is to measure changes of crack width and send them to cluster heads. As the widths of concerned cracks change slowly during pressure tests, a 30 min sampling period is enough for capturing the variation of cracks. In addition, duty cycling schemes are employed by sensor nodes. Every 30 min, sensor nodes wake up to get data from LVDT crackmeters and PT100 temperature detectors, store it and send it to cluster heads. Then, they go into idle mode and turn off their sensors and radios. In this way, sensor nodes can save a lot of energy, as they sleep most of the time. Although cluster heads do not need to monitor the cracks, they may need to always keep their radio on to receive packets from sensor nodes. Then cluster heads may consume much more power than sensor nodes, and they need to be powered by batteries with more capacity or use alternating current (AC) power when possible. Thanks to the time synchronization scheme, TDMA is used by sensor nodes for media access control to prevent transmission interference. Sensor nodes turn off their radio after sending messages out in order to save energy, which means usually they do not receive messages from cluster heads during the monitoring stage. Then, an unreliable data transfer protocol is used to adapt this feature. Sensor nodes just send measured data once to cluster heads, and there is no feedback from cluster heads about packet loss and thus no retransmission occurs. Data loss is inevitable in wireless transmission, and tests in Section 5 about data loss rate of our system under different sampling periods show that data loss rate under typical 30 min sampling period is tolerable. In addition, this means that the unreliable data transfer protocol is appropriate for our system. Unlike other high sampling rate applications, the amount of data in our system is low because of low sampling periods, and sensor nodes just send what they measure without any compression. The GUI that runs on the laptop receives all the monitoring data of the network with the help of the sink node, and it also provides functions for storing, inquiring and drawing data. The GUI is shown in Figure 3, which is implemented with C# language.

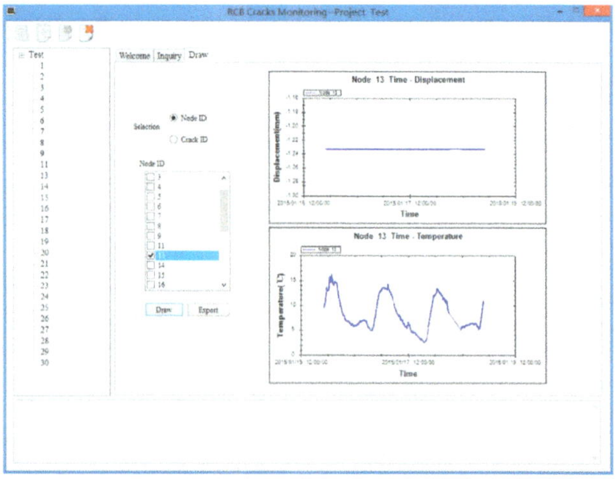

Figure 3: Graphical user interface (GUI) of the monitoring system.

WIRELESS TRANSMISSION TESTS

When the wireless monitoring system is implemented, transmission tests are carried out before full scale deployment to find out the transmission range of sensor nodes and the data loss rate of the system.

The purpose of transmission range test is to determine how far the radio we choose can transmit at most in the open air and guide the deployment of cluster nodes. Considering the structure of RCB and the place where the sink node has to be put, both horizontal and vertical transmission range tests are carried out. In horizontal tests, the sink node is connected to the laptop for receiving messages from sensor nodes, and one sensor node is placed near the sink node to send messages as references. Three other sensor nodes are placed at a horizontal distance of 30 m from the sink node to send messages, and then at the distance of 50 m, 70 m, 100 m, 120 m and 150 m, respectively. The transmission power levels of these sensor nodes are kept the same during the test, the sampling period is 1 min, and three sensor nodes are placed at each distance for 5 min. Test results show that the sink node can receive all messages when sensor nodes are placed at the distance of 30 m, 50 m, 70 m and 100 m, and it cannot receive all the messages at the distance of 120 m, while no messages are received at a distance of 150 m except those from the reference sensor node. From this test, we know that our sensor nodes ensure a horizontal transmission distance of 100 m. Most sensor nodes will be deployed at the dome area of the RCB, but the sink node needs to be placed on the ground, which is at a vertical distance of 50 m from the dome due to space and power supply constraints. Then, a vertical transmission range test is conducted on a 20-story building. Similar to the horizontal test, the sink node and reference sensor node are placed on the ground, three sensor nodes are placed at level 5 with line of sight, then at level 10 and 15. The sampling period and duration at each level are the same as the horizontal test. Results show that even when sensor nodes are placed at level 15, a vertical distance of about 50 m, the sink node on the ground can receive all the messages from them, which means that it is feasible to put the sink node on the ground for convenience.

As mentioned in Section 4, an unreliable data transfer protocol is used for saving energy consideration, and this may cause data loss during wireless transmission due to interference and other environment factors. Then, a data loss rate test is carried out to evaluate the reliability of the wireless monitoring system. Thirty sensor nodes are casually deployed on the top floor of our office building and several LVDT sensors are installed on the surface of wall cracks. The sampling period is set at 1 min, then 15 and 30 min, respectively. All sensor nodes get and send data to the sink node under different sampling periods, and each test lasts 24 h for each sampling period. In addition, data loss rate is calculated based on the sampling period and the messages received at the sink node. Figure 4 shows the average data receiving rate under different sampling periods. It can be seen from the figure that about 83% of data is received at the sink node under the 1 min sampling period during the 24 h test. In addition, the data loss rate under 15 and 30 min sampling periods is relatively low compared with that of the 1 min sampling period. Considering that the typical sampling period for our system is 30 min, the data loss rate is tolerable and the unreliable data transfer protocol meets the demand of our system and saves energy at the same time.

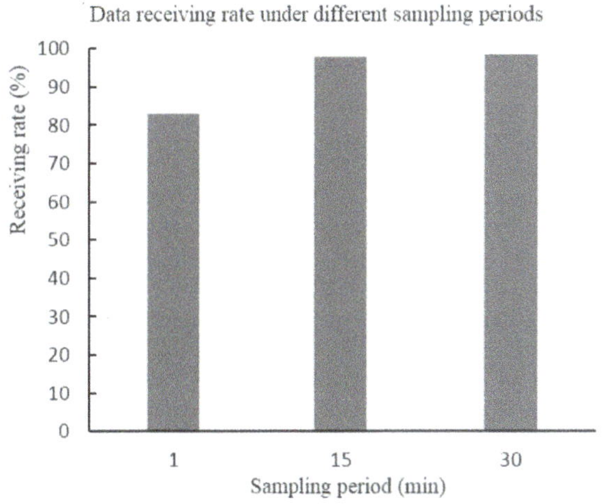

Figure 4: Average data receiving rate under different sampling periods.

REAL-LIFE DEPLOYMENT

Daya Bay Nuclear Power Station is located in Guangdong province, which is the first commercial nuclear power station in mainland China. As one of the earliest and largest joint venture projects launched under China's Open Door Policy, it began to be built in 1987 and was completed in 1993, and French and British technology were used during the construction. The plant has two pressurized water reactor generating units with a total electrical capacity of 1968 MW. It produces around 14 billion kWh of electricity per year, and the primary consumers of this nuclear power station are Hong Kong and Guangdong province, consuming 70% and 30%, respectively. Daya Bay Nuclear Power Station is under strict management and operation to maintain safety. Our full scale deployment is carried out during its every ten years' in-service pressure test.

The pressure test is conducted to check the integrity and leak rate of the concrete containment building that houses the nuclear reactor. In addition, one important task during the test is to inspect the surface of the RCB and measure the changes of crack width on the concrete structure. Before the pressure test, a manual inspection is conducted to find out cracks that need to be monitored on the surface of the RCB. Then, the locations of concerned cracks are reported and the deployment of sensor nodes is based on the report. Figure 5 shows the geometry of concerned cracks and they may appear on the dome or on the perimeter wall of the RCB.

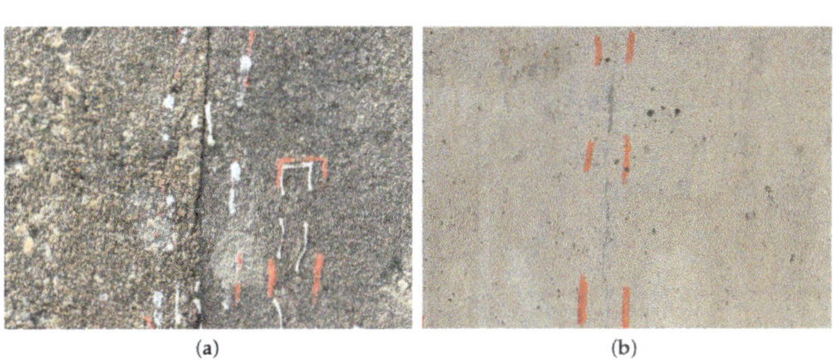

Figure 5: The geometry of cracks on the surface of RCB. (**a**) Crack shape one; (**b**) Crack shape two.

Sensor nodes with LVDT crackmeters are installed on the surface of concerned cracks. The details of installation are illustrated in Figure 6. First, the LVDT base is fixed on one side of the crack with plastic clamps, making the measurement direction of the sensor vertical to the crack's direction. Then, the retractable part of the LVDT is fixed on the other side of the crack using a mounting bracket according to the range of the sensor. As we mainly focus on the changes of cracks' width during the pressure test, the initial width of cracks are not measured.

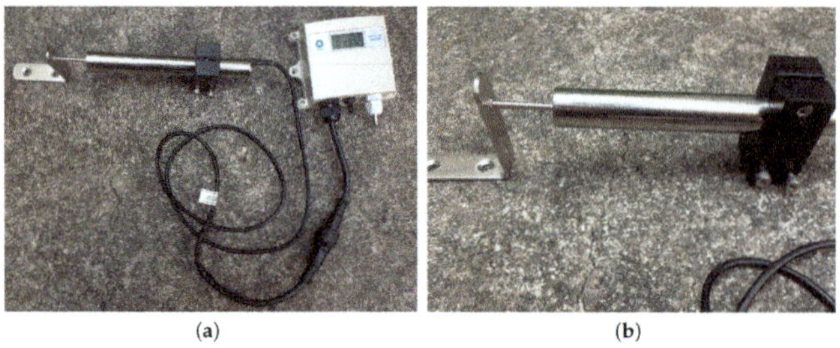

Figure 6: The installation of LVDT crackmeters. (**a**) The sensor node; (**b**) Installation details diagram.

The quantity and function of nodes used in the full scale deployment is shown in Table 3. Thirty sensor nodes are installed on the surface of concerned cracks according to the results of manual inspection. Considering the transmission range of sensor nodes and the structure of the containment building, four cluster nodes are used for data relay purpose in the wireless monitoring system. They are evenly deployed on the top and around the dome area. The only sink node responsible for data collection is connected to the laptop and placed on the ground for continuous power supply. Figure 7shows the schematic layout of the wireless monitoring system on the RCB.

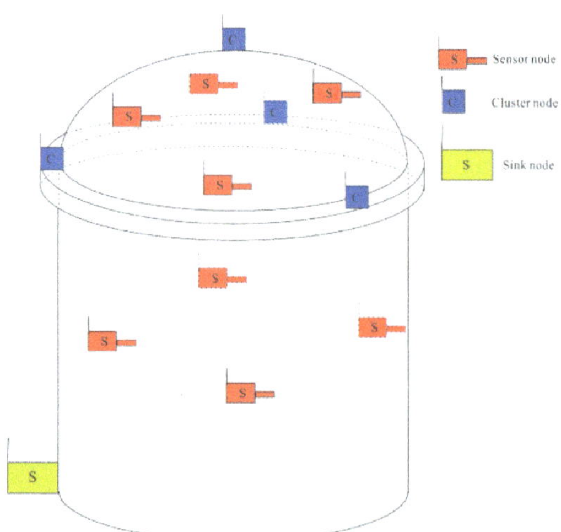

Figure 7: Schematic layout of the wireless monitoring system in the RCB.

Table 3: Quantity and function of nodes used in the full scale deployment

Node Types	Quantity	Function
Sensor node	30	Data sampling and transmission
Cluster node	4	Data relay
Sink node	1	Data collection

When the deployment of the wireless monitoring system is completed, the pressure test starts. The reactor is shut down and compressed air is pumped into the containment building, the building is pressurized to predetermined pressure under several measurement phases at different pressure levels, which can be seen in Figure 8. At first, the pressure stays at 0 bar (gage) for 9 h, then it goes up to a certain pressure and stays there for some time. In addition, the maximum pressure is 4.2 bar (gage), which lasts for 24 h. After that, the pressure decreases to 0 bar (gage) gradually. The whole pressure test process lasts about seven days.

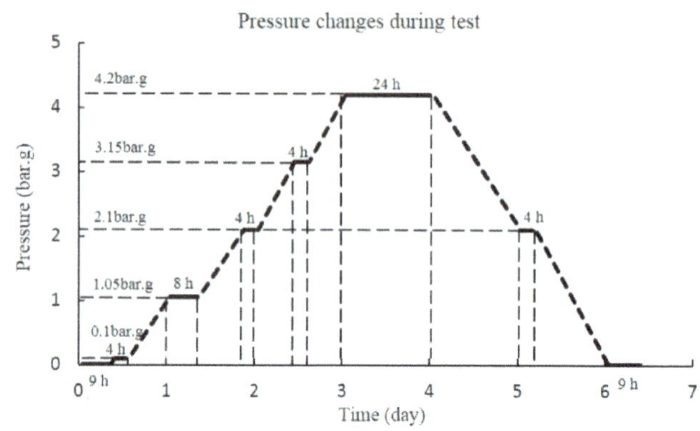

Figure 8: Pressure changes during the pressure test.

Just before the pressure test, the wireless monitoring system runs route discovery and time synchronization algorithms to prepare for data collection. Then, the monitoring stage starts along with the pressure test, and sensor nodes record the changes of cracks' width due to thermal and pressure impacts and send them to cluster heads every 30 min. Cluster nodes help sensor nodes relay data to the sink node. The sink node collects, stores and displays all the data in real-time. The monitoring stage lasts seven days until the pressure test stops. The sink node can acquire data from all of the 30 sensor nodes, and the average data receiving rate is similar to the transmission test that we have done before, which verifies the efficiency of the wireless monitoring system. In addition, only about 10% of the energy is consumed on average, which is consistent with our previous theoretical calculations.

CONCLUSIONS

The advent of wireless sensor networks brings both opportunities and challenges for structural health monitoring. We enjoy the convenience of wireless transmission but suffer limited resources at the same time. In this paper, we present a real-time monitoring system for cracks on the surface of reactor containment buildings during pressure tests based on wireless sensor networks. Sensors like crackmeters and temperature detectors are chosen and customized platforms are designed and implemented to meet the monitoring requirements. Meanwhile, software protocols like route discovery, time synchronization and data transfer are developed to take the limited resources into account. A graphical user interface is programmed to receive, store and display data. The monitoring system is evaluated with several experimental tests before real-life deployment. In addition, the full scale deployment of the system at Daya Bay Nuclear Power Station with 30 sensor nodes is quite successful. Despite the successful deployment, there are still some improvements that can be made for our system. Currently only crackmeters and temperature detectors are used in our nodes, and more sensors may be added for multiple purposes in the future. The topological structure of the monitoring network is established at the initialization stage and is static; thus, it is not robust to failures, especially for cluster head errors. Therefore, dynamic topology will be a better choice for the system. The system is mainly designed for low sampling frequency tasks as tailored time synchronization protocols are developed. However, it may need to be redesigned to suit some high sampling frequency monitoring projects. Similarly, the unreliable data transfer protocol only works well with low sampling frequency, reliable protocols or data recovery mechanisms that are required for high sampling frequency tasks. We believe our system can be easily adapted to many monitoring tasks if these improvements are made.

ACKNOWLEDGMENTS

This work has been supported by the National Natural Science Foundation of China under grant No. 41474005, and Key Laboratory of Advanced Engineering Surveying of National Administration of Surveying, Mapping and Geoinformation under grant No. TJES1202.

AUTHOR CONTRIBUTIONS

The corresponding author Jianguo Zhou, who is responsible for the overall work, proposed the research idea, conducted the experiments and prepared the manuscript. The second author Yaming Xu provided general guidance during the entire research period. The third author Tao Zhang helped in design and implementation of hardware and software.

REFERENCES

1. Brownjohn, J. Structural health monitoring of civil infrastructure. *Philos. Trans. R. Soc. Lond. A: Math. Phys. Eng. Sci.*2007, *365*, 589–622.
2. Chang, P.; Flatau, A.; Liu, S. Review paper: Health monitoring of civil infrastructure. *Struct. Health Monit.* 2003, *2*, 257–267.
3. Akyildiz, I.; Su, W.; Sankarasubramaniam, Y.; Cayirci, E. Wireless sensor networks: A survey. *Comput. Netw.* 2002, *38*, 393–422.
4. Kim, S.; Pakzad, S.; Culler, D.; Demmel, J.; Fenves, G.; Glaser, S.; Turon, M. Health Monitoring of Civil Infrastructures Using Wireless Sensor Networks. In Proceedings of the 6th International Conference on Information Processing in Sensor Networks (IPSN), Cambridge, MA, USA, 25–27 April 2007; pp. 254–263.
5. Pakzad, S.; Fenves, G.; Kim, S.; Culler, D. Design and Implementation of Scalable Wireless Sensor Network for Structural Monitoring. *J. Infrastruct. Syst.* 2008, *14*, 89–101.
6. Jang, S.; Jo, H.; Cho, S.; Mechitov, K.; Rice, J.; Sim, S.; Jung, H.; Yun, C.; Spencer, B.; Agha, G. Structural health monitoring of a cable-stayed bridge using smart sensor technology: Deployment and evaluation. *Smart Struct. Syst.*2010, *6*, 439–459.
7. Rice, J.; Mechitov, K.; Sim, S.; Nagayama, T.; Jang, S.; Kim, R.; Spencer, B.; Agha, G.; Fujino, Y. Flexible smart sensor framework for autonomous structural health monitoring. *Smart Struct. Syst.* 2010, *6*, 423–438.
8. Feltrin, G.; Meyer, J.; Bischoff, R.; Motavalli, M. Long-term monitoring of cable stays with a wireless sensor network. *Struct. Infrastruct. Eng.* 2010, *6*, 535–548.
9. Flouri, K.; Saukh, O.; Sauter, R.; Jalsan, K.; Bischoff, R.; Meyer, J.; Feltrin, G. A versatile software architecture for civil structure monitoring with wireless sensor networks. *Smart Struct. Syst.* 2012, *10*, 209–228.
10. Feltrin, G.; Jalsan, K.; Flouri, K. Vibration monitoring of a footbridge with a wireless sensor network. *J. Vib. Control*2013, *19*, 2285–2300.
11. Bischoff, R.; Meyer, J.; Enochsson, O.; Feltrin, G.; Elfgren, L. Event-based strain monitoring on a railway bridge with a wireless sensor network. In Proceedings of the 4th International Conference on Structural Health Monitoring of Intelligent Infrastructure (SHMII-4), Zurich, Switzerland, 22–24 July 2009; pp. 74–82.
12. Hu, X.; Wang, B.; Ji, H. A Wireless Sensor Network-Based Structural Health Monitoring System for Highway Bridges. *Comput. Aided Civil Infrastruct. Eng.* 2013, *28*, 193–209.
13. Bennett, P.; Soga, K.; Wassell, I.; Fidler, P.; Abe, K.; Kobayashi, Y.; Vanicek, M. Wireless sensor networks for underground railway applications: Case studies in Prague and London. *Smart Struct. Syst.* 2010, *6*, 619–639.
14. Zonta, D.; Wu, H.; Pozzi, M.; Zanon, P.; Ceriotti, M.; Mottola, L.; Picco, G.; Murphy, A.; Guna, S.; Corra, M. Wireless sensor networks for permanent health monitoring of historic buildings. *Smart Struct. Syst.* 2010, *6*, 595–618.
15. Ou, J.; Li, H.; Xiao, Y.; Li, Q. Health dynamic measurement of tall building using wireless sensor network. In Proceedings of the Smart Structures and Materials 2005, San Diego, CA, USA, 7–10 March 2005; pp. 205–216.
16. Amyot, P. Construction of the containment building for the Gentilly nuclear power reactor. *Nucl. Eng. Des.* 1969, *9*, 479–487.
17. Xu, N.; Rangwala, S.; Chintalapudi, K.; Ganesan, D.; Broad, A.; Govindan, R.; Estrin, D. A wireless sensor network for structural monitoring. In Proceedings of the 2nd iNternational Conference on Embedded Networked Sensor Systems (SenSys), Baltimore, MD, USA, 3–5 November 2004; pp. 13–24.
18. Chintalapudi, K.; Fu, T.; Jeongyeup, P.; Kothari, N.; Rangwala, S.; Caffrey, J.; Govindan, R.; Johnson, E.; Masri, S. Monitoring civil structures with a wireless sensor network. *IEEE Internet Comput.* 2006, *10*, 26–34.
19. Nagayama, T.; Spencer, B.; Rice, J. Autonomous decentralized structural health monitoring using smart sensors. *Struct. Control Health Monit.* 2009, *16*, 842–859.
20. Whelan, M.; Janoyan, K. Design of a robust, high-rate wireless sensor network for static and dynamic structural monitoring. *J. Intell. Mater. Syst. Struct.* 2009, *20*, 849–863.

21. Lynch, J.; Wang, Y.; Loh, K.; Yi, J.; Yun, C. Performance monitoring of the Geumdang Bridge using a dense network of high-resolution wireless sensors. *Smart Mater. Struct.* 2006, *15*, 1561–1575.

22. Lynch, J. An overview of wireless structural health monitoring for civil structures. *Philos. Trans. R. Soc. A: Math. Phys. Eng. Sci.* 2007, *365*, 345–372.

23. Gay, D.; Levis, P.; Culler, D. Software design patterns for TinyOS. *ACM Trans. Embed. Comput. Syst.* 2007, *6*, 40–49.

24. Archer, W.; Levis, P.; Regehr, J. Interface Contracts for TinyOS. In Proceedings of the 6th International Conference on Information Processing in Sensor Networks (IPSN), Cambridge, MA, USA, 25–27 April 2007; pp. 158–165.

25. Linderman, L.; Mechitov, K.; Spencer, B. TinyOS-based real-time wireless data acquisition framework for structural health monitoring and control. *Struct. Control Health Monit.* 2013, *20*, 1007–1020.

26. Han, C.; Kumar, R.; Shea, R.; Kohler, E.; Srivastava, M. A dynamic operating system for sensor nodes. In Proceedings of the 3rd International Conference on Mobile Systems, Applications, and Services, Seattle, WC, USA, 6–8 June 2005; pp. 163–176.

27. Bhatti, S.; Carlson, J.; Dai, H.; Deng, J.; Rose, J.; Sheth, A.; Shucker, B.; Gruenwald, C.; Torgerson, A.; Han, R. MANTIS OS: An Embedded Multithreaded Operating System for Wireless Micro Sensor Platforms. *Mob. Netw. Appl.* 2005, *10*, 563–579.

28. Dunkels, A.; Gronvall, B.; Voigt, T. Contiki—A lightweight and flexible operating system for tiny networked sensors. In Proceedings of the 29th Annual IEEE International Conference on Local Computer Networks, Tampa, FL, USA, 16–18 November 2004; pp. 455–462.

29. Heinzelman, W.; Chandrakasan, A.; Balakrishnan, H. Energy-efficient communication protocol for wireless microsensor networks. In Proceedings of the 33rd Annual Hawaii International Conference on System Sciences, Maui, HI, USA, 4–7 January 2000.

30. Maroti, M.; Kusy, B.; Simon, G.; Ledeczi, A. The flooding time synchronization protocol. In Proceedings of the 2nd international conference on Embedded networked sensor systems(SenSys), Baltimore, MD, USA, 3–5 November 2004; pp. 39–49.

CHAPTER 9 GEOSPATIAL INFORMATION CATEGORIES MAPPING IN A CROSS-LINGUAL ENVIRONMENT

Xi Kuai, Lin Li, Heng Luo, Shen Hang, Zhijun Zhang and Yu Liu

Keywords: Geographic information systems, semantic interoperability, cross-lingual, lightweight ontology, topographic map

INTRODUCTION

The vision of a "Digital Earth" articulated by US Vice President Al Gore [1,2,3] has contributed significantly to the growth in global geospatial information (GI) on physical and social environments. However, how to query, retrieve, and manipulate those data from heterogeneous sources has challenged the GI community [2,3,4,5]. Thus, an approach to integrating GI data from various heterogeneous sources has found increased importance [6].

A data integration process is not as simple as joining several systems because any effort at information sharing runs into the problem of semantic heterogeneity [7]. Semantic heterogeneity occurs when enabling interoperability across geographic information systems (GIS) [8,9,10,11] because GIS are often designed to address data from highly distributed, multidisciplinary, and cross-lingual data sources with different application demands [12]. Clarifying the semantics of data is therefore a crucial step toward successful data integration [13]. To achieve this, domain ontologies are built as a mediator to exchange information in such a way that the precise meaning of the data (*i.e.*, semantics) is readily retrievable beyond simple keyword matching via knowledge representation languages and reasoning [7,13,14,15]. Thus, ontology engineering has been regarded as an effective means of providing seamless connection between component GIS at the semantic level [8,12,16].

While the GI community widely acknowledges the utility of ontology technologies, two main problems need to be solved for GI ontology engineering and sharing are as follows: (1) traditional ontology research and technologies focusing on terminology and schema cannot answer the question surrounding how to engineer GI ontologies and integrate them with GIS or Spatial Data Infrastructures (SDI) [6]; and (2) mechanisms still need to be explored for GI ontology mapping in cross-lingual environments to facilitate semantic integration between GI ontologies described in different natural languages [17,18,19,20].

The reason for the first problem is that GI features and categories are a product of spatial cognition and social convention; thus, the ontology engineering works in GI domains are different from others, in which the location, topology, mereology, and other spatial relations play a major role in the identification and representation of GI semantics [14]. For example, from a feature-driven ontology perspective, the geographic categories "river" and "bank" should be specified into different classes, and normally, the spatial relation "adjacent-to" between these two categories is missing. Moreover, geographic and non-geographic entities are ontologically distinct in a number of ways [21]. To enhance the semantic expressiveness and overcome the issue of semantic heterogeneity during the GI ontology engineering process, the spatial-related characteristics of GI categories must be considered to enrich the spatial-related semantics of the given ontology.

Although a majority of current GI ontologies have been developed in English with English vocabularies, the amount of multilingual content on the Semantic Web and thus the number

Citation: Xi Kuai , Lin Li, Heng Luo, Shen Hang, Zhijun Zhang and Yu Liu, Geospatial Information Categories Mapping in a Cross-lingual Environment, doi:10.3390/ijgi5060090.

of vocabularies/ontologies in multiple languages continue to grow [22]. Thus, methods for matching vocabularies across languages have become increasingly more important for promoting the accessibility of the data in multiple languages by end users [23]. As a motivating scenario, if a user wants to query the water level data along the Mekong River (The seventh longest river in the world, covering six different countries—Cambodia, Laos, Myanmar, Thailand, Vietnam and China—and the official languages of each country are different), there are several data providers offering the related GI data via their national GIS in their native natural languages. This situation has generated a substantial challenge to integrating highly heterogeneous GI data across natural language barriers.

The purpose of this study is to establish a formal ontology model for cross-lingual geospatial information ontology mapping. Starting from two GI classification standards with different natural languages—Chinese and English(for the sake of simplicity and clarity, this study was restricted to the "surface water" categories from these two standards)—a set of semantic primitives are extracted from the free-text definition of the categories in the standards by applying Natural Language Processing (NLP) techniques. Then, an ontology-driven approach is used, and the formal ontology model is established to formally represent these semantic primitives using semantic statements, in which the spatial-related properties and relations are considered as crucial statements for the representation and identification of the semantics of the GI categories. To overcome the natural language barrier, the statements in Chinese are translated into English by using machine translation tools, and the mapping relationships between statements are determined within an English context, which then serve as the basis for the similarity calculation between categories in different GI ontologies. Finally, a similarity calculation algorithm is designed to distance the semantic similarity between GI categories in different ontologies, and the final mapping relationships between pairs of categories are determined based on calculated similarity values. The contributions of the proposed approach include (1) the construction of the spatial-related semantic properties and relations to serve the requirement of the presentation and identification of the spatial characteristics of the GI categories and (2) the algorithms of GI ontology mapping in a cross-lingual environment based on formally represented and comparable semantic statements.

The remainder of this paper is organized as follows. Section 2 presents the related works in the literature. Then, the main procedure of our methodology is presented in Section 3. Next, a case study demonstrating the application of our method is shown in Section 4. Finally, conclusions are drawn, and future works are noted.

RELATED WORKS

Semantic Interpretation

Knowledge acquisition (KA) is a broad field that encompasses the processes of extracting, creating, and structuring knowledge from heterogeneous resources [24]. Semantic interpretation (SI) for KA is defined as the composition of two sub-processes: the extraction of semantic primitives from the free-text definition in ontologies and semantic enrichment based on the extracted semantic primitives.

The research on semantic primitive extraction builds on a large body of works within the fields of Natural Language Processing (NLP) [25]. NLP and text mining are research fields aimed at exploiting rich knowledge resources with the goal of understanding, extracting and retrieving semantic information from unstructured written text. Knowledge resources that have been used for these purposes include the entire range of terminologies, including lexicons, controlled vocabularies, thesauri, and ontologies [26,27]. Although numerous methods and algorithms have been developed recently (such as symbolic, statistical, and hybrid approaches) [26], a fully automated algorithm for semantic extraction using NLP techniques seems unachievable, and a manual process as an assistant is normally inevitable.

For semantic enrichment, authors in [28,29] proposed a systematic methodology to explore and identify semantic information provided by categories in geographic ontologies, in which the semantic representations of categories are enriched with a set of semantic properties and relations to reveal similarities and heterogeneities between these categories. Authors in [30] presented an axiomatic formalization of a theory of top-level relations (parthood relations, sub-universal relations, and cross-categorical relations) between three categories of geospatial-related entities, namely, individuals, universals, and collections. In addition, they demonstrated how a more exact understanding helps to overcome the semantic heterogeneous problems in the information integration process. In [13], the semantics of a concept in GI ontologies were presented using an extendable and structural definition framework composed of a number of RDF triple statements, and a comparison algorithm was designed to determine the semantic relationships of concepts between different domains. The primary objective of these studies was to extract and represent the semantics of concepts/entities based on structural common vocabularies, which make the semantics of the concepts/entities comparable. However, the structural common vocabularies in these works are determined by domain experts manually; thus, the objectivity and automation of the algorithms (avoiding *ad hoc* manual procedures and subjective experts' knowledge) remain quite limited.

Ontology Mapping

Mapping relationship discovery for ontologies has attracted considerable attention in recent years. Various approaches based on processes to find similarities between different but related ontologies have emerged [31]. With respect to the literature specifically oriented toward geospatial information (GI) ontologies, authors in [32] performed an analysis of the different models of semantic similarity measurement and evaluated these models with respect to the particular requirements of geospatial data. Authors in [7,33] systematically surveyed several of the most recent and often-referenced works on integrating GI and GI ontology mapping by applying comparison criteria, such as logical inference, mapping approaches, degree of automation, and geospatial relativity. In addition, a general conclusion is proposed that, for the ontology mapping task, the use of formal ontologies and, consequently, the use of reasoners should be mandatory.

In recent years, Volunteered Geographic Information (VGI) has been proposed for GI ontology mapping in a web environment. Authors in [34,35] devised a mechanism for computing the semantic similarity of the Open Street Map (OSM) geographic classes using volunteered lexical definitions to alleviate the semantic gap between different VGI producers. Another set of studies focused on introducing an artificial neural network approach to simulate the human perception and measure the semantic similarity between spatial entities for the purpose of improving the automaticity of the ontology mapping process [36,37].

All these proposals, combining the use of different models of semantic similarity measurement, have emerged to provide solutions to existing GI ontology mapping problems in English environments. However, the semantic web and ontology engineering have experienced significant advancements in standards and techniques, and increasingly more domain ontologies and localization content in the semantic web are described using native natural languages [23]. There is a pressing need for cross-lingual ontology mapping mechanisms in the GI community that are designed to reconcile semantics of different ontologies in multilingual environments and to improve the accessibility of various GI ontologies across language barriers [38].

Author in [39] groups the existing cross lingual ontology mapping(CLOM) algorithms into following categories: manual processing [40,41,42], corpus-based approach [43], linguistic enrichment [44], indirect alignment composition [45], and translation-based approach [39,46]. Compared to these CLOM approaches, translation-based CLOM is currently a very popular approach that is exercised by several researchers [47,48,49,50], which is enabled by translations achieved through the use of machine translation (MT) tools, bilingual/multilingual thesauri, dictionaries *etc.* Typically, these approaches rely only on string-based lexical comparisons of entity names and descriptions [51,52,53,54], while the comparisons between semantic interpretation, e.g., model-theoretic semantics of entities are missing.

METHODOLOGIES

The main procedure for our methodologies is divided into two sub-processes, as shown in Figure 1. In the semantic interpretation process, two GI formal ontologies, namely, O_A and O_B, are established from the free-text definition of the corresponding classification standards with different natural languages. In the ontology mapping process, all the category names and semantic statements in O_A are translated from L_A into L_B, and the mapping relationships between category names and semantic statements are determined within the same language context, which then serve as the basis for the similarity calculation between categories in different GI ontologies. Finally, a similarity calculation algorithm is designed, and the final mapping results between pairs of categories in different classification standards are determined.

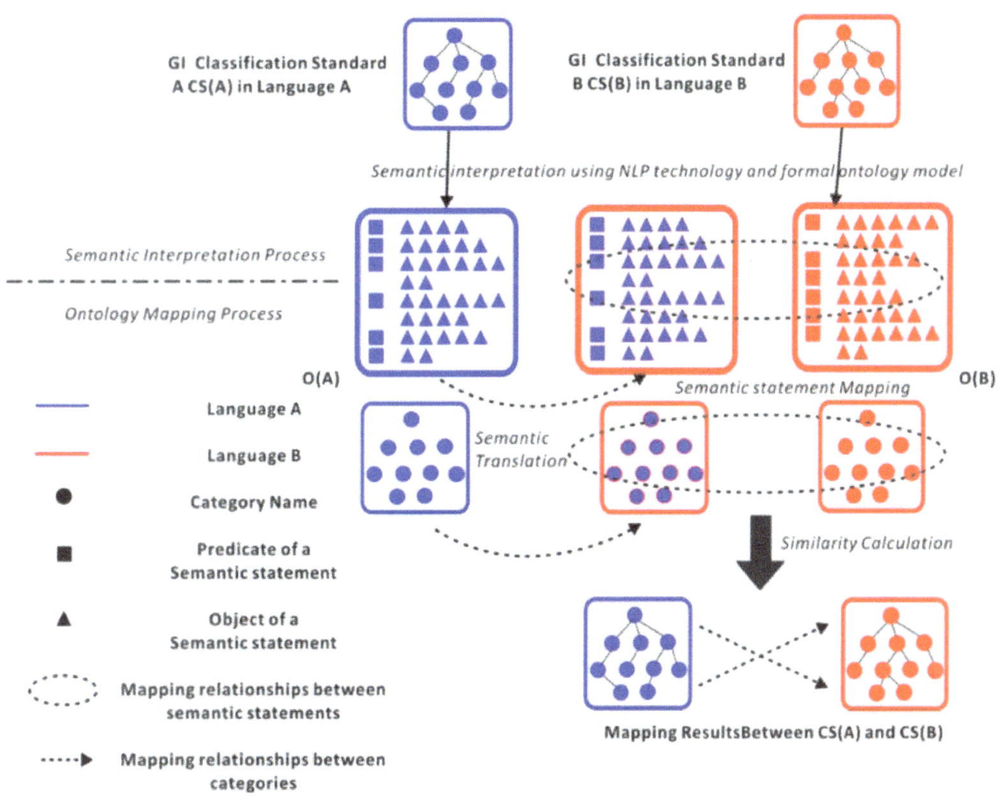

Figure 1: Main procedure for our methodologies.

Semantic Interpretation

Semantic Primitive Extraction

In geospatial information repositories, free-text definitions are often the primary and only available objective descriptions of categories. Semantic primitives are syntactic and lexical patterns in the free-text definition and can be extracted using NLP tools

[55]. The fields of studies on NLP have developed methods and algorithms for information retrieval and extraction from free-text knowledge resources. The methodology adopted here for analyzing definitions and extracting semantic primitives was introduced by [56]. In this research, the lexical patterns of nominal phrases and verb phrases are considered as semantic primitives. An example is illustrated in Figure 2, and the main steps of the process are as follows:

1. One category definition in free-text format is chosen as the input natural language material;
2. Word segmentation is performed to split the whole sentence into individual words;
3. Words are categorized and tagged into their parts-of-speech tag sets (see Table 1 and Table 2) and labeled accordingly;
4. The nominal phrases and verb phrases are chunked, and the sentence structure is analyzed to extract lexical patterns as the semantic primitives.

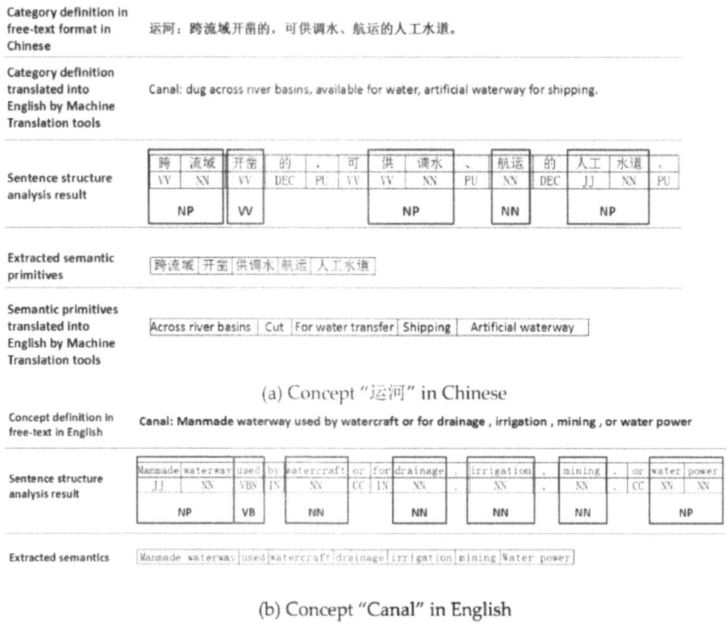

Figure 2: Extract the semantic primitives from the free-text definitions by applying NLP tools in Chinese and English.

Table 1. Summary of the Penn Treebank Part-of-Speech Tag sets in English

Part of Speech	Abbr	Part of Speech	Abbr	Part of Speech	Abbr
Adjective	JJ	Exclamation	UH	Possessive wh-pronoun	WP$
Adjective comparative	JJR	Existential	EX	Predeterminer	PDT
Adjective superlative	JJS	Foreign word	FW	Proper noun plural	NNPS
Adverb	RB	Gerund	VBG	Proper noun	NNP
Adverb comparative	RBR	List item marker	LS	Symbol	SYM
Adverb superlative	RBS	Modal verb	MD	to	TO
Article	DT	Participle past	VBN	Verb base form	VB
Cardinal number	CD	Particle	RP	Verb present tense	VBP
Common noun plural	NNS	Past tense verb	VBD	Verb 3rd person singular	VBZ
Common noun singular or mass	NN	Personal pronoun	PRP	Wh-determiner	WDT
Conjunction coordinating	CC	Possessive ending	POS	Wh-pronoun	WP
Conjunction subordinating	IN	Possessive pronoun	PRP$	Wh-adverb	WRB

Table 2: Summary of the Penn Treebank Part-of-Speech Tag sets in Chinese

Part of Speech	Abbr	Part of Speech	Abbr	Part of Speech	Abbr
adverb	AD	determiner	DT	proper noun	NR
aspect marker	AS	for words "dengdeng"("等等")	ETC	temporal noun	NT
in ba-construction	BA	foreign words	FW	ordinal number	OD
coordinating conjunction	CC	interjection	IJ	onomatopoeia	ON
cardinal number	CD	other noun-modifier	JJ	preposition excl. "bei"("被") and "ba"("把")	P
subordinating conjunction	CS	"bei"("被") in long bei-const	LB	pronoun	PN
"de"("的")in a relative-clause	DEC	localizer	LC	punctuation	PU
Associative "de"	DEG	measure word	M	"bei"("被") in short bei-const	SB
"de"("得") inV-deconst. and V-de-R	DER	other particle	MSP	sentence-final particle	SP
"di"("地") before VP	DEV	common noun	NN	predicative adjective	VA
"shi"("是")	VC	"you"("有") as the main verb	VE	other verb	VV

Construction of the Formal Ontology Model

From Wikipedia an "ontology in information science" is a formal naming and definition of the types, properties, and interrelationships of the concepts that really or fundamentally exist for a particular domain of discourse. It is thus a practical application of philosophical ontology, with taxonomy. In addition, a domain ontology (or domain-specific ontology) represents concepts that belong to a general domain. Thus, for a formal representation [57,58], the domain ontology (denoted by O_{Domain}), and concepts in the domain could be summarized by Equations (1)–(3).

$$O_{Domain} = \{S(C_{Domain}), S(R_C), S(H_C), S(P_C)\} \quad (1)$$

$$C_{Domain} = \{T_C, D_C\} \quad (2)$$

$$D_C = \{R_C, H_C, P_C\}, R_C \in S(R_C), H_C \in S(H_C), P_C \in S(P_C) \quad (3)$$

In Equation (1), $S(C_{Domain})$ represents the set of concepts in a domain, and the semantics of each concept in the domain are categorized into different groups, namely, $S(H_C)$, $S(R_C)$, and $S(P_C)$; $S(H_C)$ represents the set of the hierarchical relations about the taxonomic information in O_{Domain}, $S(R_C)$ represents the set of other interrelations between these concepts, and $S(P_C)$ represents the set of the semantic properties belong to the concepts in this domain.

In Equation (2), the semantics of a concept in the domain are considered as the composition of terminology of this concept (denoted by T_C) and structural definition of this concept (denoted by D_C). Unlike the free-text format of definition, D_C commonly consists of the semantic properties of the concept (P_C), the hierarchical relation (H_C) and other interrelations (R_C) between this concept and other concepts in the domain. Thus, from Equations (2) and (3), a certain concept in the domain, C_{Domain} can be deduced as a function of T_C, R_C, H_C, and P_C in Equation (4)

$$C_{Domain} = \{T_C, R_C, H_C, P_C\}, R_C \in S(R_C), H_C \in S(H_C), P_C \in S(P_C) \quad (4)$$

in which R_C, H_C, P_C are used to represent the semantics of this concept, and belong to $S(R_C)$, $S(H_C)$, $S(P_C)$, respectively.

Considering the situation in the GI domain, we use the word "category" instead of "concept". Because the semantic characteristics of the GI category are highly correlated in space and time [59], the spatial- and temporal-related semantic properties and relations should be included in the model as crucial vocabularies for the representation and identification of the semantics of the GI categories. Thus, the GI ontology O_{GI} and the semantics of a certain category C_{GI} in O_{GI} can be represented as Equations (5) and (6):

$$O_{GI} = \{S(T_C), S(R_S), S(R_T), S(R_C), S(H_C), S(P_S), S(P_T), S(P_C)\} \quad (5)$$

$$C_{GI} = \{(T_C = V_{T_C}) \cap (R_S = V_{R_S}) \cap (R_T = V_{R_T}) \cap (R_C = V_{R_C}) \cap (H_C = V_{H_C}) \cap (P_S = V_{P_S}) \cap (P_T = V_{P_T}) \cap (P_C = V_{P_C})\} \quad (6)$$

In Equation (5), $S(T_C)$ represents the set of the category names in O_{GI}; $S(R_S)$, $S(R_T)$ represent the set of the spatial-related and temporal-related semantic relations between categories; $S(H_C)$represents the set of hierarchical relations; $S(P_S)$, $S(P_T)$ represent the set of the spatial-related and temporal-related semantic properties belong to the categories in O_{GI}; and $S(R_C)$, $S(P_C)$ represent the set of other semantic properties and relations in O_{GI}. And in Equation (6), V_x represents the values of certain semantic properties/relations of C_{GI}; T_C, R_S, R_T, R_C, H_C, P_S, P_T, P_C are used to represent the semantics of C_{GI}, and belong to $S(T_C)$, $S(R_S)$, $S(R_T)$, $S(R_C)$, $S(H_C)$, $S(P_S)$, $S(P_T)$, $S(P_C)$, respectively.

In order to solve the problems of geographic representation, authors in [60] distinguished three main theoretical tools that are required for the purposes of developing an overall formal theory of spatial representation, namely, mereology, location, and topology, these theoretical tools are selected as the basis for defining spatial-related semantics in our formal ontology model. In addition, geographic entities in reality is essentially dynamic, authors in [61] pointed out that a good ontology must be capable of accounting for spatial reality both synchronically (as it exists at a time) and diachronically(as it unfolds through time), thus the "time point" and "time period" properties should be used to describe dynamic characteristics of the geographic entities in our model. Moreover, in order to specify semantic relations and properties used in geographic definitions, authors in [28] analyzed several geographic ontologies and identified patterns which were systematically used to express specific semantic relations and properties, including hierarchical relations, part-whole relations and neighborhood relations, and semantic properties such as purpose, nature, material, size, and so on. Based on previous researches and our formal ontology model in Equation (5), the semantic property and relation types in our model are subdivided and shown in Figure 3.

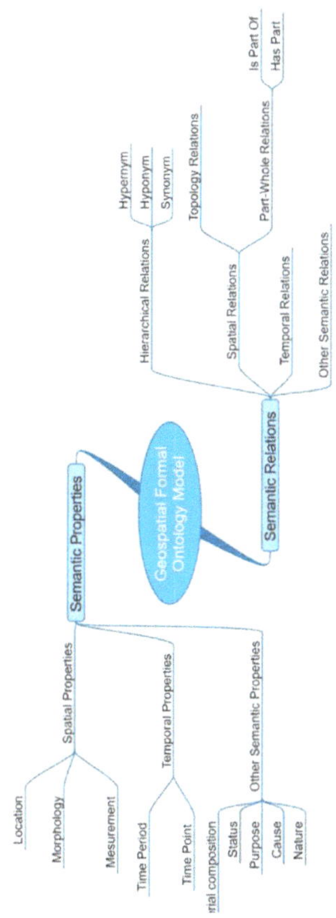

Figure 3: Semantic property and relation groups in the Geospatial Formal Ontology Model.

Transformation from Semantic Primitives to Formal Ontology Model

In order to make the semantic primitives structural and comparable, domain experts are responsible for analyzing these semantic primitives and transforming them into different groups of semantic properties/relations in our geospatial formal ontology model. The famous triple statement *Subject-Predicate-Object* and the web ontology language (OWL) are selected as the basis for presenting the semantic properties/relations and their values in a machine-readable manner. The *Subject*represents a C_{GI} in O_{GI}; the *Predicate* is a certain semantic property or semantic relation type illustrated in Figure 3, in which all of the semantic relations are presented by object property and the *Object* in these semantic relations is another C_{GI} in O_{GI}or an "owl:class" object type, while most of the semantic properties are presented by object property too, and a few of them are presented by datatype property in OWL syntax, and the *Object* in these semantic properties is a "rdfs:literal" datatype. The following rules are adopted to handle the formalization process:

(1) The GI category can be represented by a number of semantic relations/properties; however, the number of semantic relations/properties involved should be minimized to avoid redundancy.

(2) Not every GI category must cover all semantic relations/properties in the model. The situation whereby two different categories use the same set of semantic relations/properties to represent their semantics cannot be guaranteed.

(3) The semantic information of a certain category in our model is the combination of different semantic relations-properties and their values. This combination should represent all the semantic information of the category and be able to distinguish the different geospatial categories within and beyond domain ontologies to avoid ambiguity.

(4) The hypernym, hyponym, and synonym relations should be included in the hierarchical relation group. If category A is a hyponym of category B, A must inherent all the semantic properties/relations of B to retain semantic consistency.

According to the above-mentioned rules, the semantic primitives can be specified into these properties/relations types as structure statements for identification and representation of the GI categories. For example, the free-text definition of the "canal" category in English is "manmade waterway used by watercraft or for drainage, irrigation, mining, or water power". In addition, the semantic primitives of the "canal" category are extracted by applying NLP tools to the set of phrases including "manmade waterway", "used", "watercraft", "drainage", "irrigation", "mining", and "water power". Then, transforming these semantic primitives into the proposed formal ontology model, the semantics of the category "canal" can be represented as a set of several semantic statements as follows:

$$C_{Canal} = \{T_C = \text{"canal"} \cap H_C = \text{"Hypernym : waterway"} \cap P_C = \text{"(Purpose : watercraft)} \cup \text{(Purpose : drainage)} \cup \text{(Purpose : irrigation)} \cup \text{(Purpose : mining)} \cup \text{(Purpose : waterpower)"} \cap P_C = \text{"Nature : Manmade"}\} \quad (7)$$

In addition, the representation in OWL format is illustrated in Figure 4.

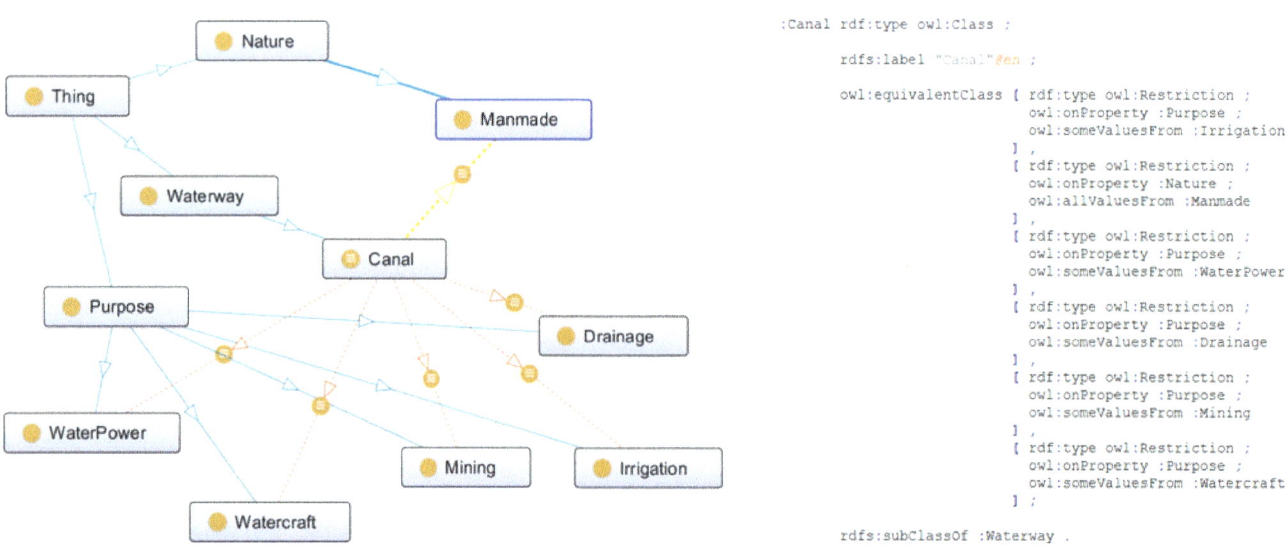

(a) (b)

Figure 4: Representation of the category "canal" in OWL format: (**a**) The OntoGraf view in Protégé and (**b**) the semantic statement presentation in turtle file format.

Ontology Mapping Algorithms

Semantics Translation

Assume that we have formal ontologies O_A, O_B presented in different natural languages, namely, language A (L_A) and B (L_B), respectively. According to the geospatial formal ontology model introduced in Section 3.1.2, the semantics of ontologies O_A and O_B consist of category name sets $S\ (CN_A)$ and $S\ (CN_B)$ and semantic statement sets $S\ (SS_A)$ and $S\ (SS_B)$, labeled in different natural languages, in which the semantic statement consists of semantic property/relation types (as illustrated in Figure 3 in Section 3.1.2) and their corresponding values. In order to cross the natural language barrier between O_A and O_B, algorithm 1 illustrates the process of semantics translation between L_A and L_B:

1: **Input:** Formal ontologies $O_A(S(CN_A), S(SS_A))$ in L_A
2: **Output:** Translation candidate result set of the semantics in O_A, $O'_A\ (S(TC(CN_A)), S(TC(SS_A\text{-}object)))$ in
3: L_B.
4: **Symbols:**
5: $S(TC(CN_A))$—Translation candidate result set of $S(CN_A)$ in L_B.
6: $S(TC(SS_A))$—Translation candidate result set of $S(SS_A)$ in L_B.
7: ss_A*-object*—The *Object* part of the semantic statement ss_A.
8: 1:**for each category name** cn_A in $S(CN_A)$, translate cn_A in L_A into cn_A' in L_B by using different Machine Translation (MT) web services (Google Translator API at" http://translate.google.cn/", Bing Translator API at" http://www.bing.com/translator/?ref=SALL&mkt=zh-CN", and Baidu Translator API at "http://fanyi.baidu.com/?aldtype=16047#zh/en/"), collect all of the translation results about cn_A, into the translation candidate results $TC(cn_A)$, and store all of the category name translation candidate results into the translation candidate set $S(TC(CN_A))$;
9: 2:**for each semantic statement** ss_A in $S(SS_A)$, according to the OWL triple statement syntax, it can be subdivided into three part, *Subject*, *Predicate*, and *Object*, translate ss_A*-object* in L_A into ss_A'*-object* in L_B by using different Machine Translation (MT) web services, collect all of the translation results about ss_A*-object*, into the translation candidate results $TC(ss_A\text{-}object)$, and store all of the semantic statements translation candidate results into the translation candidate set $S(TC(SS_A\text{-}object))$.

10: Take the "运河" category in Chinese as an example, the semantic primitives of the "运河" category are extracted by applying NLP tools to the set of phrases including "跨流域", "开凿", "供调水", "航运", "人工水道". Then, transforming these semantic primitives into the proposed formal ontology model, the semantics of the category "运河"can be represented as a set of several semantic statements as follows:
11:

$$C_{\text{运河}} = \Big\{T_C =''\ \text{运河}'' \cap H_C =''\ Hypernym : \text{水道}'' \cap P_C =''\ (Purpose : \text{调水}) \cup \\ (Purpose : \text{航运})'' \cap P_C =''\ Nature : \text{人工}'' \cap R_S =''\ Topology : \text{跨流域}''\Big\} \tag{8}$$

12: And the semantics translation result of Cin English is as follows:

$$\begin{aligned} C_{\text{运河}} = \{T_C = \text{“(Canal)”} \cap H_C = \text{“}Hypernym : \text{(Waterway, Aqueduct)”} \cap \\ P_C = \text{“}(Purpose : \text{(Water transfer, Diversion)}) \cup \\ (Purpose : \text{(Shipping)})\text{”} \cap P_C = \text{“}Nature : \text{(Manual,Artificial)”} \cap \\ R_S = \text{“}Topology : \text{(Inter-basin, Across river basins)”}\} \end{aligned} \tag{9}$$

Algorithm 1: Semantics Translation.

Semantic Statement Mapping

To determine the mapping relationships between categories in different GI ontologies, the mapping relationships at the semantic statement level should be determined first because the semantic statement presents the most detailed semantic characteristics of the compared categories. Once their relationships are determined, the similarity between categories can be determined quantitatively. Algorithm 2 shows the comparison process for category names and semantic statements between O_A and O_B. In addition, all the mapping results $M(O_A, O_B)$ are stored as the basis for the similarity calculation between the concepts in different GI ontologies.

Input: O'_A $(S(TC(CN_A)), S(TC(SS_A\text{-}object)))$ in L_B, Formal ontologies $O_B(S(CN_B), S(SS_B))$ in L_B
Output: Mapping result set $M(O_A, O_B)$ about category names and semantic statements between
O_A and O_B.
Symbols:
$T(ss)$—semantic property/relation types for a certain semantic statement ss.
$M(O_A', O_B)$—mapping relationships about category names and semantic statements between O_A'
and O_B.
1: **for each category name** cn_A in $S(CN_A)$, find the translation candidate results of cn_A, $TC(cn_A)$,
  2: **for each translation candidate** $tc(cn_A)$ in $TC(cn_A)$, search $S(CN_B)$ in O_B, find the matched
  category name cn_B in $S(CN_B)$ by applying Equation(10),
    3: If there is a translation candidate $tc(cn_A)$ has the mapping relationship "exact match"
    with cn_B, store the mapping result $m(cn_A, cn_B$, 'exact match') in $M(O_A, O_B)$;
    4: else If there is a translation candidate $tc(cn_A)$ has the mapping relationship
    "close match" with cn_B, store the mapping result $m(cn_A, cn_B$, 'close match') in $M(O_A, O_B)$;
5: **for each semantic statement** ss_A in $S(SS_A)$, find the translation candidate results of ss_A-*object*,
$TC(ss_A$-*object*),
  6: **for each translation candidate** $tc(ss_A$-*object*) in $TC(ss_A$-*object*), search $S(SS_B$-*object*) in O_B,
  find the matched semantic statement *Object*, ss_B-*object* in $S(SS_B)$ by applying Equation(10),
    7: If there is a translation candidate $tc(ss_A$-*object*) has the mapping relationship "exact
    match" with ss_B-*object*, and $T(ss_A)$ equals $T(ss_B)$, store the mapping result $m(ss_A, ss_B$, 'exact
    match') in $M(O_A, O_B)$;
    8: else If there is a translation candidate $tc(ss_A$-*object*) has the mapping relationship "close
    match" with ss_B-*object*, and $T(ss_A)$ equals $T(ss_B)$, store the mapping result $m(ss_A, ss_B$, 'close
    match') in $M(O_A, O_B)$.

$$m(A,B) = \begin{cases} exactly\ match, & A\ is\ the\ same\ word\ or\ synonym\ of\ B \\ close\ match, & A\ is\ the\ near\ synonym\ of\ B \\ not\ match, & otherwise \end{cases} \tag{10}$$

Algorithm 2: Semantic Statement Mapping.

Similarity Calculation

Given two categories, C_a and C_b in the formal ontologies O_A and O_B, respectively, based on the $M(O_A, O_B)$, the semantic similarity between C_a and C_b can be calculated using algorithm 3.

Input: Categories $C_a(CN_a, SS_a)$ in O_A, $C_b(CN_b, SS_b)$ in O_B and mapping relationship set $M(O_A, O_B)$
about category names and semantic statements between O_A and O_B.
Output: Semantic similarity value between C_a and C_b, $Sim(a, b)$.
Symbols:
$Cot(SS_a)$—the number of semantic statements in SS_a.
$Cot(SS_b)$—the number of semantic statements in SS_b.
$m(CN_a, CN_b)$—mapping relationship between CN_a and CN_b.
$m(SS_a(i), SS_b(j))$—mapping relationship between $SS_a(i)$ in C_a and $SS_b(j)$ in C_b.
$Pt(SS_{ab})$—the sum of the match point value between SS_a and SS_b.
$Pt(CN_{ab})$—the match point value between CN_a and CN_b.
1: **for each semantic statement** $SS_a(i)$ in SS_a, find the matched semantic statement $SS_b(j)$ in SS_b
based on the mapping relationship set $M(O_A, O_B)$;
  If $m(SS_a(i), SS_b(j))$ = "exact match", then the match point value between $SS_a(i)$ and $SS_b(j)$ is
assigned 1;
  Else if $m(SS_a(i), SS_b(j))$ = "close match", then the match point value between $SS_a(i)$ and $SS_b(j)$ is
  assigned 0.5;
2: Record the sum of the match point values between SS_a and SS_b as $Pt(SS_{ab})$ and the number of
matched statements between SS_a and SS_b as $Cot(SS_{ab})$;
3: find the mapping relationship between CN_a and CN_b based on $M(O_A, O_B)$,
  If m(CN_a, CN_b) = "exact match", then the match point value between CN_a and CN_b is assigned 1;
  Else if m(CN_a, CN_b) = "close match", then the match point value between CN_a and CN_b is
  assigned 0.5;
4: Record the match point value between CN_a and CN_b as $Pt(CN_{ab})$;
5: the similarity of categories C_a and C_b can be calculated using the following equation:

$$Sim(a,b) = \begin{cases} \frac{1}{2} * \frac{Pt(SS_{ab})}{Cot(SS_a)} + \frac{1}{2} * \frac{Pt(SS_{ab})}{Cot(SS_b)}, & if\ m(CN_a, CN_b) = "not\ match" \\ \frac{1}{3} * \frac{Pt(SS_{ab})}{Cot(SS_a)} + \frac{1}{3} * \frac{Pt(SS_{ab})}{Cot(SS_b)} + \frac{Pt(CN_{ab})}{3}, & if\ m(CN_a, CN_b) = "exact\ match"/"close\ match" \end{cases} \tag{11}$$

In addition, the mapping relationships between category pairs C_a and C_b, namely, $MR(a, b)$, can be determined using the following equation:

$$MR(a,b) = \begin{cases} exact\ match, & if\ Sim(a,b) = 1 \\ close\ match, & if\ 0.5 <= Sim(a,b) < 1 \\ related, & if\ \ 0 < Sim(a,b) < 0.5 \\ not\ match, & if\ \ Sim(a,b) = 0 \end{cases} \quad (12)$$

Algorithm 3: Similarity Calculation.

A CASE STUDY

Study Material

To illustrate the methodologies, two different classification standards in two corresponding natural languages have been selected for use in the mapping process. CS_C is developed based on the national topographic map standards in China (Standards of "Cartographic symbols for national fundamental scale maps" and "Specifications for feature classification and codes of fundamental geographic information"). CS_A is developed by the U.S. Geological Survey in America (http://cegis.usgs.gov/ttl/USTopographic.ttl). Both standards are digital literature materials; the category names and their free-text definitions are provided as source information for our experiment. In addition, for the sake of simplicity and clarity, our study was restricted to the "surface water" categories from these two classification standards. Table 3 briefly lists the characteristics of these two selected dataset, with detailed explanations as follows:

(1) oth standards have their own classification system to address the categories of "surface water". The categories in CS_C are organized using a four-level hierarchy with six major categories. By contrast, the categories in CS_A are organized by a four-level hierarchy with 81 major categories, which means that the hierarchical structure of CS_A does not closely match that of CS_C.

(2) The free-text definitions in both standards are used as category definitions.

(3) The number of categories in CS_C is 74, and the number of concepts in CS_A is 92; thus, the CS_A covers more category types than does CS_C.

(4) The natural language in CS_C is Chinese, whereas the natural language in CS_A is English, which means that there is a natural language barrier between these two GI classification standards.

Table 3: Characteristics of CS_C and CS_A

Characteristic	CS_C	CS_A
Number of categories	74	92
Classification system	Taxonomy (without overlap)	Taxonomy (without overlap)
Levels of hierarchy	4	4
Number of major categories	6	81
Definition	Free-text, unstructured	Free-text, unstructured
Attribute	Id, Category name	Category name, Source of the definition
Language	Chinese	English

Results

The well-defined category definitions in both CS_C and CS_A serve as the basis for our study. The Web Ontology Language (OWL) API is integrated to facilitate the implementation of the proposed algorithm in Eclipse with the JAVA language, and the experiment results are as follows.

Semantic Statement Mappings

The semantic primitives are extracted using the Stanford Natural Language Processing Tools (http://nlp.stanford.edu/software/) and are transformed into the formal ontologies O_C and O_A with the set of category names and semantic statements by domain experts and encoded by the OWL via Protégé. Using the semantic statement mapping algorithm introduced in Section 3.2.2, the number of mapping relationships between the statements in O_C and O_A is recorded, and the mapping results for different semantic property/relation types are shown in Table 4.

Table 4: Condition of the mapping statements between O_C and O_A

		Number of Semantic Statements in O_C	Number of Semantic Statements in O_A	Number of Mapping Statements	Mapping Rate
Property Types		80	111	44	29.93%
Spatial Properties	Location	7	8	1	7.14%
	Morphology	4	23	3	12.50%
	Measurement	2	1	0	0.00%
Temporal Properties	Time Period	2	4	3	75%
	Time Point	3	1	1	33%
Other Semantic Properties	Material Composition	15	22	11	42.31%
	Nature	2	2	2	100.00%
	Status	19	22	8	24.24%
	Cause	2	6	1	14.29%
	Purpose	24	22	14	43.75%
Relation Types		62	70	28	26.92%
Hierarchical Relations		25	24	9	22.50%
Spatial Relations	Topology Relations	27	29	14	33.33%
	Part-Whole Relations	9	17	5	23.81%
Temporal Relations		0	0	0	0.00%
Other Related Relations		1	0	0	0.00%
Total		142	181	72	28.69%

The total number of semantic statements in O_C is 142, and the total number of such statements in O_A is 181. In addition, the total mapping rate of the semantic statements between O_C and O_A is 28.69%. The details of the mapping relationships between semantic statements in each type can be found in Appendix.

For the semantic statement about the semantic property types, the most matched type is "purpose". This is because the semantic property type of "purpose" is used to represent the manmade category, which includes "ditch", "canal", and "dam", and the free-text definitions in both the Chinese and American classification standards for these types of categories are very similar. The semantic information about purpose and functionality are considered as the crucial characteristics of the categories. It is easy to understand that the semantic property type "nature" has the highest mapping rate, namely, 100%, because there are only two values for this type of semantic statement, namely, "natural" and "manmade", in both $O_C$and O_A. Considering the semantic property type "location", there are seven semantic statements in O_C, and eight in O_A, but the mapping rate of this type is extremely low(only one semantic statement is mapped with mapping rate 7.14%). That's because the semantic property type "location" is used to describe the region environment where certain geographic category is at, and a lot of the categories in O_A are bay-related or glacier-related, such as "glacier", "ice cap" and "iceberg tongue" with semantic property value of "location", "mountainous area", "regions of perennial frost", and "coast", respectively, and there are no such categories in O_C. For the semantic statement about the relation types, the most matched type is "spatial relation", which is also the type with the highest mapping rate, indicating that the spatial-related relations play a major role in the identification and representation of GI semantics.

Similarity Calculation and Category Mappings

The similarities between concepts are calculated using the semantic statement mapping relationships and Algorithm 3 proposed in Section 3.2.3. Three typical examples of the mapping results between categories are chosen for further discussion. Table 5 shows the names and free-text definitions of the compared category pairs. In addition, the corresponding semantic statements, calculated similarity values and final mapping relationships between these category pairs are presented in Table 6.

Table 5: Names and free-text definitions of the compared concept pairs

Concept Pairs	Concepts	Names	Free-Text Definitions
Pair 1	Concept 1 in O_C	溢洪道	水库的泄洪水道，用以排泄水库预定蓄水高度以上的洪水。
	Translation of Concept 1 in O_C	Spillway	Reservoir spillway channel to drain reservoir reservation head above the flood.
	Concept 1 in O_A	Spillway	A passage for surplus water to run over or around a dam.
Pair 2	Concept 2 in O_C	干河床（干涸河）	降水或融雪后短暂时间内有水的河床或河流改道后遗留的河道。
	Translation of Concept 2 in O_C	Arroyo (dry river)	Precipitation or snowmelt water within a short time after the river or river diversions left after the river.
	Concept 2 in O_A	Wash	The usually dry portion of a stream bed that contains water only during or after a local rainstorm or heavy snowmelt.
Pair 3	Concept 3 in O_C	水系	江、河、湖、海、井、泉、水库、池塘、沟渠等自然和人工水体及连通体系的总称。
	Translation of Concept 3 in O_C	Water System	River, river, lake, sea, wells, springs and reservoirs, ponds, ditches, and other natural and artificial water bodies and the connected system in general.
	Concept 3 in O_A	Surface Water	The water portion of the Earth's surface, including the surface of sea and inland waters

Table 6: Example of categories definitions and similarity calculation

Concepts	Semantic Statements	Translation of Semantic Statement in O_C	Mapping Relationships between Statement	Similarity Values	Mapping Results
Concept 1 in O_C	(Hypernym: 水道)⊓ (Is-Part-Of:水库) ⊓ (Purpose: 排泄洪水)	(Hypernym: Waterways) ⊓ (Is-Part-Of:Reservoir) ⊓ (Purpose:Drain flood)	"Spillway" Exact match "Spillway"(Concept Name) "Hypernym:Waterways" Close match "Hypernym:Passage" "Is-Part-Of:Reservoir"Close match "Is-Part-Of:Dam" "Purpose:Drain flood" Exact match "Purpose:Surplus Water"	0.78	Close Match
Concept 1 in O_A	(Hypernym: Passage) ⊓ (Is-Part-Of: Dam) ⊓ (Purpose: Surplus Water)				
Concept 2 in O_C	(Hypernym:河床)⊓(Material:水)⊓(Status: 干涸)⊓(Temporality:降雪或融雪后)	(Hypernym: riverbed) ⊓(Material: water) ⊓(Status: dry) ⊓(Temporality: After the rainfall or snowmelt)	"Hypernym:riverbed" Exact match "Hypernym:Streambed" "Material:water" Exact match "Material:Water" "Status:dry"Exact match "Status:Dry" "Temporality:After the rainfall or snowmelt" Exact match "Temporality:during or after a local rainstorm or heavy snowmelt"	1.0	Exact Match
Concept 2 in O_A	(Hypernym: Streambed) ⊓(Material: Water) ⊓(Status: Dry) ⊓(Temporality: during or after a local rainstorm or heavy snowmelt)				
Concept 3 in O_C	(Hyponym:江)⊓ (Hyponym:河) ⊓ (Hyponym:湖)⊓ (Hyponym:海)⊓ (Hyponym:井) ⊓ (Hyponym:泉) ⊓ (Hyponym:水库) ⊓ (Hyponym:池塘)⊓ (Hyponym: 沟渠) ⊓ (Hyponym:水体) ⊓(Nature:自然 ⊔ Nature: 人工) ⊓ (Material: 水)	(Hyponym: river) ⊓ (Hyponym: river) ⊓ (Hyponym: lake) ⊓ (Hyponym: sea) ⊓ (Hyponym: well) ⊓ (Hyponym: spring) ⊓ (Hyponym: reservoir) ⊓ (Hyponym: pond) ⊓ (Hyponym: ditch) ⊓ (Hyponym: body of water) ⊓ (Nature: natural ⊔ Nature: artificial) ⊓ (Material: water)	"Hyponym: river" Exact match "Hyponym:River" "Hyponym: river" Exact match "Hyponym:Stream" "Hyponym:lake" Exact match "Hyponym:Lake" "Hyponym:sea" Exact match "Hyponym:Sea" "Hyponym:spring" Exact match "Hyponym:Spring" "Hyponym:reservoir" Exact match "Hyponym:Reservoir" "Hyponym:pond" Exact match "Pond" "Hyponym:ditch" Exact match "Ditch" "Hyponym:body of water" Exact match "Hyponym:Water body" "Nature:natural" Exact match "Nature:Natural" "Nature:artificial" Exact match "Nature:Manmade" "Material:water" Exact match "Material:Water"	0.92	Close Match
Concept 3 in O_A	(Material: Water) ⊓ (Hyponym: Sea) ⊓ (Hyponym: Inland Water) ⊓ (Is-Part-Of: Earth's surface) ⊓ [(Hyponym: River) ⊓ (Hyponym: Stream)⊓ (Hyponym: Lake) ⊓ (Hyponym: Spring)⊓ (Hyponym: Reservoir) ⊓ (Hyponym: Pond)⊓ (Hyponym: Ditch) ⊓ (Hyponym: Water body)⊓ (Nature: Natural)⊓ (Nature: Manmade)] (The semantic statements in "(…)" were not the semantic information extracted from the free-text definition and were inferred based on the semantic statements in other concepts, which have a hierarchical relation with the concept. They were added to the concept by the domain expert manually.)				

Example 1: Concept pair of "spillway" in O_C ***and "spillway" in O_A***

These two concepts are comparable because the mapping relationship between their concept names is "exact match". Because their concept names and four semantic statements are matched (detailed mapping relationships are illustrated in Table 6, line 1 and 2), the second condition in Equation (9) is used to calculate the final similarity between "spillway" in $O_C$and "spillway" in O_A. The similarity value between these two concepts is calculated as 0.78; thus, the mapping relationship between these two concepts is "close match". This example demonstrates the simplest case for the calculation of the semantic similarity between concepts.

Example 2: Concept pair of "arroyo (dry river)" in O_C ***and "wash" in O_A***

In this example, the mapping relationship between the concept name of "arroyo (dry river)" and "wash" cannot be determined based on the mapping algorithm in Section 3.2.1. However, the similarity value between these two concepts is higher than the value in example (1). This is because all the semantic statements used to represent the semantic meaning of these two concepts are correspondingly matched (detailed mapping relationships are illustrated in Table 6, line 3 and 4), and all the mapping relationships between them are "exact match". The first condition in Equation 9 is used to calculate the final similarity between the concepts "arroyo (dry river)" in O_C and "wash" in O_A. The similarity value between these two concepts is calculated as 1.0; thus, the mapping relationship between these two concepts is "exact match". This example demonstrates a common situation in the cross-lingual environment in that two concepts have the same semantic meaning while their names are definitely different. Moreover, the utility of applying our methodologies to the complex application of cross-lingual GI ontology integration has been proven.

Example 3: Concept pair of concept 3 "Water System" in O_C ***and concept 3 "Surface water" in O_A***

At first glance, the semantic statements between the concept "water system" and "surface water" are not matched very well, and the concept names of these two concepts cannot be matched either.

This is because these two concepts are both the top concept in their own taxonomies, and these two concepts are abstract concepts in that they do not represent real-world objects with detailed characteristic entities, for example, rivers, lakes, and oceans. Thus, the definitions of this category in different languages may be very different, even when they are conveying the same meaning. Therefore, the solution for the semantic meaning representation of this type concept is not the same as the solution used in Examples (1) and (2). The sematic meaning of the hyponym-related concepts should be considered to infer the integrated semantic meaning of this abstract concept. After the implicit semantic statements have been inferred out (detailed mapping relationships are illustrated in Table 6, line 5 and 6), the first condition in Equation (9) is used to calculate the final similarity between the concepts "water system" in O_C and "surface water" in O_A. The similarity value between these two concepts is calculated as 0.92; thus, the mapping relationship between these two concepts is "close match".

CONCLUSIONS AND FUTURE WORK

The presented research focuses on the determination of semantic mapping relationships between categories in different GI ontologies with natural language barriers. The proposed formal ontology model in this study is used to represent and identify the semantic characteristics of the GI categories with OWL-based semantic statements transformed from free-text definitions of two GI classification standards. A new similarity calculation algorithm based on this formal ontology model is presented to distance the semantic similarities and identify the mapping relationships between categories.

In particular, we work with two classification standards of topographic maps in Chinese and American English. The conducted experiment indicates that the proposed approach successfully determines the mapping relationships between categories in different GI ontologies and facilitates ontology integration in a cross-lingual environment. Due to the usages of the multilingual supported NLP tools in our experiment, it is easy to replicate our model to determine the mapping relationships between other GI ontologies, which may be described using other native natural languages, in addition to Chinese. However, this model has only been applied to geospatial information (GI) integration at the category level, and research on GI integration at the data level has not been fulfilled. That will form the basis for future study. In addition, publishing the mapping information in a cross-lingual context as linked data in a semantic web environment should also be considered.

ACKNOWLEDGMENTS

This research is supported by the National Administration of Surveying, Mapping and Geoinformation, China, under the Special Fund for Surveying, Mapping and Geographical Information Scientific Research in the Public Interest (No. 201412014), and Specialized Research Fund for the Doctoral Program of Higher Education (No. 20120141110048).

AUTHOR CONTRIBUTIONS

This research was mainly performed and prepared by Xi Kuai and Lin Li. Xi Kuai and Lin Li contributed with ideas, conceived and designed the study. Xi Kuai wrote the paper. Heng Luo, Hang Shen and Yu Liu contributed the tools, and analyze the results of the experiment. Zhijun Zhang reviewed and edited the manuscript.

APPENDIX: DETAIL OF THE MAPPING STATEMENTS BETWEEN O_C AND O_A

Table A1: Detail of the Mapping Statements between O_C and O_A

Property Types	Semantic Statements in O_C in Chinese	Translation of Semantic Statements in O_C in English	Semantic Statements in O_A in English	Mapping Relations	Semantic Statements in O_C in Chinese	Translation of Semantic Statements in O_C in English	Semantic Statements in O_A in English	Mapping Relations
Material	水	water	water	Exact match	石	stone	stones	Exact match
	水蒸气	water vapor	vapors	Exact match	木桩	wooden stake	wood	Close match
	泥	mud	mud	Exact match	草地	grassland	grassy	Close match
	砖	brick	brick	Exact match	砾石	gravel	gravel	Exact match
	沙	sand	sand	Exact match	礁石	reef	reef	Exact match
	水泥	cement	concrete	Exact match				
Nature	自然的	natural	natural	Exact match				
	人造的	manmade	manmade	Exact match				
Status	流动	flow	flowing	Exact match	倾泻	pour	moving outward an downslope	Close match
	独立	stand along	free standing	Exact match	高潮时被水体淹没，低潮时露出	submerged at high tide the water, exposed at low tide	alternately covered and left bare by the tide	Exact match
	有水潮浸	tide water immersion	washed by waves or tides	Close match	涌出	emission	issue from the ground	Close match
	干涸	dried up	dry	Exact match	洪水泛滥	flood	subject to flooding	Exact match
Temporality	长期	long-term	permanent	Close match	降水或融雪后短时间内	within a short time after rainfall or snowmelt	during or after a local rainstorm or heavy snowmelt	Exact match
	终年	all year round	permanent	Exact match	季节性	seasonal	occasionally	Close match
Location	沙地	sandy	desert	Close match				
Purpose	引水	water diversion	run water	Exact match	减缓水流流速	slow water flow rate	restrain current or tide	Close match
	输水	water delivery	conveying water	Exact match	保护港口	protection of harbor	protect harbor	Exact match
	贮水	water storage	contain water	Exact match	护岸	bank protection	sustain an embankment	Exact match
	将水位升高或降低，使船能在不同高低水位的水道间通行	To raise or lower the water level, at different high and low water level so the ship channel traffic	raise and lower vessels as they pass from one level to another.	Exact match	抬高水位	raising of water level	raise the level of water	Exact match
	控制流量	control flow	control the flow of water	Exact match	通行船只	passage vessel	route for watercraft	Exact match
	灌溉	irrigation	irrigation	Exact match	拦截河流	blocked rivers	Across the course of a stream	Close match
	调节水流方向	adjusting to the flow direction	direct current or tide	Exact match	扬水	pump up water-	Pump	Exact match
Morphology	陡坡	steep slope	a vertical or near vertical descent	Close match	坝式	dam type	dam	Exact match
	虹吸式	siphon	siphon	Exact match				
Cause	堆积	accumulation	accumulate	Exact match				
				Relation Types				
Hierarchical Relation	源头	source	source	Exact match	设施	facilities	facility	Exact match
	河床	riverbed	channel bottom	Exact match	构筑物	structure	construction	Exact match
	区域	regional	region	Exact match	通道	channel	path	Exact match
	地带	zone	zone	Exact match	水道	waterways	waterway	Exact match
	设备	device	device	Exact match				
Spatial Relation	地面上	on the ground	on the surface of the land	Close match	水体平均大潮高潮面与水体最低低潮面之间	mean high water springs of water and water between the lowest low water	Between high water and low water marks	Exact match
	水体内	in body of water	in water	Exact match	沿河流	along the river	alongside a stream	Exact match
	海域内	within the sea	in the sea	Exact match	水陆间	between land and water	contact between a body of water and the land	Exact match
	水下	underwater	below the surface of water	Exact match	洼地内	in the depressions	surrounded by land	Close match
	跨流域	across river basins	across the course of a stream	Exact match	陆地上	on the land	Covered with the earth	Close match
	跨道路	cross roads	crossing road or trail	Exact match	海岸线与干出线之间	between the coastline and the dry line	Between high water and low water lines	Exact match
	海岸边	coastal	adjacent to the shore	Exact match	海岸边	the coast	offshore	Close match
Is-part-of	网状水系	network drainage	network of interlacing channels	Exact match	水库	reservoir	dam	Close match
	网状水系	network drainage	a drainage network	Exact match	河渠	canal	a river system	Close match
	闸室	chamber	lock chamber	Exact match				

REFERENCES

1. Gore, A. The digital earth: Understanding our planet in the 21st century. *Photogramm. Eng. Remote Sens.* 1999, *65*.
2. Craglia, M.; Goodchild, M.F.; Annoni, A.; Camara, G.; Gould, M.; Kuhn, W.; Mark, D.; Masser, I.; Maguire, D.; Liang, S.; *et al.* Next-generation digital earth: A position paper from the vespucci initiative for the advancement of geographic information science. *Int. J. Spat. Data Infrastruct. Res.* 2008, *3*, 146–167.
3. Craglia, M.; de Bie, K.; Jackson, D.; Pesaresi, M.; Remetey-Fülöpp, G.; Wang, C.; Annoni, A.; Bian, L.; Campbell, F.; Ehlers, M.; *et al.* Digital Earth 2020: Towards the vision for the next decade. *Int. J. Digit. Earth* 2012, *5*, 4–21.
4. Yue, P.; Di, L.; Yang, W.; Yu, G.; Zhao, P. Semantics-based automatic composition of geospatial Web service chains. *Comput. Geosci.* 2007, *33*, 649–665.
5. Janowicz, K.; Hitzler, P. The digital earth as knowledge engine. *Semant. Web* 2012, *3*, 213–221.
6. Janowicz, K. Observation-driven geo-ontology engineering. *Trans. GIS* 2012, *16*, 351–374.
7. Buccella, A.; Cechich, A.; Gendarmi, D.; Lanubile, F.; Semeraro, G.; Colagrossi, A. Building a global normalized ontology for integrating geographic data sources. *Comput. Geosci.* 2011, *37*, 893–916.
8. Bishr, Y. Overcoming the semantic and other barriers to GIS interoperability. *Int. J. Geogr. Inf. Sci.* 1998, *12*, 299–314.
9. Lemmens, R.L. Semantic Interoperability of Distributed Geoservices. Ph.D. Thesis, Delft University of Technology, Delft, The Netherlands, 2006.
10. Fallahi, G.R.; Frank, A.U.; Mesgari, M.S.; Rajabifard, A. An ontological structure for semantic interoperability of GIS and environmental modeling. *Int. J. Appl. Earth Obs. Geoinf.* 2008, *10*, 342–357.
11. Ma, X.; Wu, C.; Carranza, E.J.M.; Schetselaar, E.M.; van der Meer, F.D.; Liu, G.; Wange, X.; Zhang, X. Development of a controlled vocabulary for semantic interoperability of mineral exploration geodata for mining projects. *Comput. Geosci.* 2010, *36*, 1512–1522.
12. Kuhn, W. Geospatial semantics: Why, of what, and how? *J. Data Semant. III* 2005, *3534*, 1–24.
13. Hong, J.-H.; Kuo, C.-L. A semi-automatic lightweight ontology bridging for the semantic integration of cross-domain geospatial information. *Int. J. Geogr. Inf. Sci.* 2015, *29*, 1–25.
14. Fonseca, F.T.; Egenhofer, M.J.; Davis, C.A., Jr.; Borges, K.A.V. Ontologies and knowledge sharing in urban GIS. *Comput. Environ. Urban Syst.* 2000, *24*, 251–272.
15. Pundt, H.; Bishr, Y. Domain ontologies for data sharing–An example from environmental monitoring using field GIS. *Comput. Geosci.* 2002, *28*, 95–102.

16. Yang, C.; Raskin, R.; Goodchild, M.; Gahegan, M. Geospatial Cyberinfrastructure: Past, present and future. *Comput. Environ. Urban Syst.* 2010, *34*, 264–277.
17. Stoimenov, L.; Stanimirovic, A.; Djordjevic-Kajan, S. Discovering mappings between ontologies in semantic integration process. In Proceedings of the 9th AGILE Conference on Geographic Information Science, Visegrád, Hungary, 20–22 April 2006.
18. Janowicz, K.; Raubal, M.; Kuhn, W. The semantics of similarity in geographic information retrieval. *J. Spat. Inf. Sci.*2011, *2*, 29–57.
19. Schwering, A.; Raubal, M. Spatial relations for semantic similarity measurement. In *Perspectives in Conceptual Modeling*; Springer-Verlag: Heidelberg, Germany, 2005; pp. 259–269.
20. Hakimpour, F. *Using Ontologies to Resolve Semantic Heterogeneity for Integrating Spatial Database Schemata*; Zurich University: Zurich, Switzerland, 2003.
21. Mark, D.M.; Skupin, A.; Smith, B. Features, objects, and other things: Ontological distinctions in the geographic domain. In *Spatial Information Theory*; Springer: New York, NY, USA, 2001; pp. 489–502.
22. Stadler, C.; Jens, L.; Konrad, H.; Sören, A. Linkedgeodata: A core for a web of spatial open data. *Semantic Web* 2012, *3*, 333–354.
23. Trojahn, C.; Fu, B.; Zamazal, O.; Ritze, D. *State-of-the-Art in Multilingual and Cross-Lingual Ontology Matching*; Springer: Heidelberg, Germany, 2014.
24. Liu, K.; Hogan, W.R.; Crowley, R.S. Natural Language Processing methods and systems for biomedical ontology learning. *J. Biomed. Inform.* 2011, *44*, 163–179.
25. Buitelaar, P.; Cimiano, P.; Magnini, B. Ontology learning from text: Methods, evaluation and applications. *Comput. Linguist.* 2006, *32*, 569–572.
26. Bird, S.; Klein, E.; Loper, E. *Natural Language Processing with Python*; O'Reilly Vlg. GmbH & Co.: Sebastopol, CA, USA, 2009.
27. Jurafsky, D.; Martin, J.H. *Speech and Language Processing: An Introduction to Natural Language Processing, Computational Linguistics, and Speech Recognition*; Prentice Hall: Upper Saddle River, NJ, USA, 2000.
28. Kavouras, M.; Kokla, M.; Tomai, E. Comparing categories among geographic ontologies. *Comput. Geosci.* 2005, *31*, 145–154.
29. Kavouras, M.; Kokla, M. *Theories of Geographic Concepts: Ontological Approaches to Semantic Integration*; CRC Press: Boca Raton, FL, USA, 2008.
30. Bittner, T.; Donnelly, M.; Smith, B. A spatio-temporal ontology for geographic information integration. *Int. J. Geogr. Inf. Sci.* 2009, *23*, 765–798.
31. Zheng, J.G.; Fu, L.Y.; Ma, X.G.; Fox, P. SEM+: Tool for discovering concept mapping in Earth science related domain. *Earth Sci. Inform.* 2015, *8*, 1–8.
32. Schwering, A. Approaches to semantic similarity measurement for geo-spatial data: A survey. *Trans. GIS* 2008, *12*, 5–29.
33. Buccella, A.; Cechich, A.; Fillottrani, P. Ontology-driven geographic information integration: A survey of current approaches. *Comput. Geosci.* 2009, *35*, 710–723.
34. Ballatore, A.; Bertolotto, M.; Wilson, D.C. Geographic knowledge extraction and semantic similarity in OpenStreetMap. *Knowl. Inf. Syst.* 2013, *37*, 61–81.
35. Ballatore, A.; Wilson, D.C.; Bertolotto, M. Computing the semantic similarity of geographic terms using volunteered lexical definitions. *Int. J. Geogr. Inf. Sci.* 2013, *27*, 2099–2118.
36. Li, W.; Raskin, R.; Goodchild, M.F. Semantic similarity measurement based on knowledge mining: An artificial neural net approach. *Int. J. Geogr. Inf. Sci.* 2012, *26*, 1415–1435.
37. Xu, Y.; Xie, Z.; Chen, Z. Research on semantics of entity space similarity measure based on artificial neural networks. In Proceedings of the 23rd International Conference on Geoinformatics, Wuhan, China, 19–21 June 2015.
38. Laurini, R. Geographic ontologies, gazetteers and multilingualism. *Future Internet* 2015, *7*, 1–23.
39. Fu, B.; Brennan, R.; O'Sullivan, D. A configurable translation-based cross-lingual ontology mapping system to adjust mapping outcomes. *Web Semant. Sci. Serv. Agents World Wide Web* 2012, *15*, 15–36.
40. Sini, A.; Sini, M. Mapping AGROVOC and the Chinese Agricultural Thesaurus: Definitions, tools, procedures. *New Rev. Hypermedia Multimed.* 2006, *12*, 51–62.
41. Wang, S.; Isaac, A.; Schopman, B.; Schlobach, S.; van der Meij, L. Matching multi-lingual subject vocabularies. In *Research & Advanced Technology for Digital Libraries*; Springer: Berlin, Germany, 2009; pp. 125–137.
42. Meilicke, C.; García-Castrod, R.; Freitas, F.; van Hage, W.R.; Montiel-Ponsoda, E.; de Azevedo, R.R.; Stuckenschmidt, H.; Šváb-Zamazal, O.; Svátek, V.; Tamilin, A.; *et al.* MultiFarm: A benchmark for multilingual ontology matching. *J. Web Semant.* 2012, *15*, 62–68.
43. Ngai, G.; Carpuat, M.; Fung, P. Identifying concepts across languages: A First step towards a corpus-based approach to automatic ontology alignment. In Proceedings of the 19th international conference on Computational linguistics, Stroudsburg, PA, USA, August 2002.
44. Pazienza, M.T.; Stellato, A. Linguistically motivated ontology mapping for the semantic web. In Proceedings of the 2nd Italian Semantic Web Workshop, Trento, Italy, 14–16 December 2005; pp. 14–16.
45. Jung, J.J.; Håkansson, A.; Hartung, R. Indirect alignment between multilingual ontologies. In *Agent and Multi-Agent Systems: Technologies and Applications*; Springer: Berlin, Germany, 2009; Volume 5559, pp. 233–241.
46. Trojahn, C.; Quaresma, P.; Vieira, R. A Framework for multilingual ontology mapping. In Proceedings of the International Conference on Language Resources and Evaluation, Marrakech, Morocco, 28–30 May 2008; pp. 1034–1037.
47. Wang, S.; Englebienne, G.; Schlobach, S. Learning

concept mappings from instance similarity. In *The Semantic Web—ISWC 2008*; Springer: Berlin, Germany, 2008; pp. 339–355.

48. Zheng, Q.; Shao, C.; Li, J.; Wang, Z.; Hu, L. RiMOM2013 results for OAEI 2013. In Proceedings of the 8th International Conference on Workshop on Ontology Matching, Sydney, Australia, 21 October 2013.
49. Zhang, X.; Zhong, Q.; Shi, F.; Li, J.; Tang, J. RiMOM results for OAEI 2009. In Proceedings of the 4th International Conference on Workshop on Ontology Matching, Washington, DC, USA, 25 October 2009.
50. Wang, Z.; Zhang, X.; Hou, L.; Zhao, Y.; Li, J.; Qi, Y.; Tang, J. RiMOM results for OAEI 2010. In Proceedings of the 5th International Conference on Ontology Matching, Shanghai, China, 7 November 2010.
51. Euzenat, J.; Shvaiko, P. *Ontology Matching*; Springer: Berlin, Germany, 2007.
52. Ehrig, M.; Sure, Y. Ontology mapping–An integrated approach. In *The Semantic Web: Research and Applications*; Springer: Berlin, Heidelberg, Germany, 2004; pp. 76–91.
53. Kalfoglou, Y.; Schorlemmer, M. Ontology mapping: The state of the art. *Knowl. Eng. Rev.* 2003, *18*, 1–31.
54. Doan, A.H.; Madhavan, J.; Domingos, P.; Halevy, A. Ontology matching: A machine learning approach. In *International Handbooks on Information Systems*; Springer: Berlin, Germany, 2004; pp. 397–416.
55. Kantor, P. Foundations of statistical natural language processing. *Nat. Lang. Eng.* 1999, *26*, 91–92.
56. MacCartney, B. The Stanford Natural Language Processing Group. Available online: http://nlp.stanford.edu/ (accessed on 18 December 2015).
57. Guarino, N. Formal ontology, conceptual analysis and knowledge representation. *Int. J. Hum. Comput. Stud.* 1995, *43*, 625–640.
58. Herre, H. General Formal Ontology (GFO): A foundational ontology for conceptual modelling. In *Theory & Applications of Ontology Computer Applications*; Springer Netherlands: Dordrecht, The Netherlands, 2010; pp. 297–345.
59. Frank, A.U. Ontology for spatio-temporal databases. In *Spatio-Temporal Databases*; Springer: Berlin, Germany, 2003; pp. 9–77.
60. Casati, R.; Smith, B.; Varzi, A.C. Ontological tools for geographic representation. In *Formal Ontology in Information Systems*; IOS Press: Amsterdam, The Netherlands, 1998; pp. 77–85.
61. Grenon, P.; Smith, B. SNAP and SPAN: Towards dynamic spatial ontology. *Spat. Cognit. Comput.* 2004, *4*, 69–104.

CHAPTER 10 PASSIVE WIRELESS HERMETIC ENVIRONMENT MONITORING SYSTEM FOR SPRAY PAINTING WORKSHOP

Lifeng Wang , Jingjing Ma, Yan Huang, Dan Tang and Qing-An Huang

Keywords: Passive wireless, environment monitoring, inductive coupling, spray painting workshop

INTRODUCTION

The sensor technology has penetrated into industry and daily life in all aspects [1]. However, traditional sensors are not easy, or even unable, to measure some harsh environments, such as sealed environments, rotating parts, or very high temperature environments. The reason for this is that traditional sensors are active powered and/or wire connected. Aiming at these shortcomings of traditional sensors, passive wireless sensors have been invented and applied to these special environments [2,3,4]. The most widely used mechanism is inductive coupling [5].

The first passive wireless sensor was reported by Collins in 1967 [6], which measures intraocular pressure with a pair of flat coils. However, for a long time after this period, passive wireless sensors had not been given much attention. Until 1996, a totally integrated Inductor–capacitor (LC) type passive wireless intraocular pressure sensor was proposed by Puers et al. [7]. After 2000, with the development of the micromachining process, passive wireless sensing technology began to develop quickly. Besides non-contact pressure detection, passive wireless sensing technology has been used to measure many other parameters [8,9,10,11]. A passive wireless humidity sensor which could be used to monitor the moisture in sealed package was proposed by Zhang at al. [12]. Using ferroelectric ceramic as the temperature sensing material, a passive wireless temperature sensor is reported in [13], which can work in environments with temperatures up to 700 °C. In some cases, several parameters need to be monitored at the same time. Using several resonators with different resonant frequencies or one resonator with multi-resonant frequencies, passive wireless multi-parameter sensing could be achieved [14,15,16]. Besides, adding back-scatter modulation together with channel multiplexing into the inductive coupling system, a passive wireless multi-parameter sensing system could be easily realized [17,18].

In this study, passive wireless multi-parameter sensing technology was used to monitor the environmental parameters of a spray painting workshop. In order to ensure the quality of spraying and the safety of the working environment, the environmental parameters—including the temperature, humidity, and pressure—of a spray painting workshop need to be monitored in real time. For example, temperature affects the rate of solvent evaporation. However, the air in the spray painting workshop contains a large amount of flammable paint mist and organic gas, so wire-powered devices are forbidden; even the lamps used for lighting are all installed outside the glass wall of the spray painting workshop. Hence, the passive wireless sensing technology is very suitable for this situation, which has the advantages of operating without a power supply and remote sensing capability. In Section 2, the overall block diagram and the working principle of the system is introduced. Then, the circuit design of each part is given in detail. Because of the harsh environment of the spray painting room, the package of the monitoring system is quite important. Section 3 describes the package of the monitoring system. The measurement and demonstration of the system are shown in Section 4. Finally, some conclusions are given in Section 5.

Citation: Lifeng Wang, Jingjing Ma, Yan Huang, Dan Tang and Qing-An Huang, Passive Wireless Hermetic Environment Monitoring System for Spray Painting Workshop, doi:10.3390/s16081207.

CIRCUITS DESIGN

Overall Block Diagram

The passive wireless hermetic environment monitoring system for the spray painting workshop is composed of a transponder and a reader; the overall block diagram of the system is shown in Figure 1. The power and the data transmission between the transponder and the reader are realized by way of inductive coupling. Hence, the transponder can be used to monitor hermetic environments without wires or batteries.

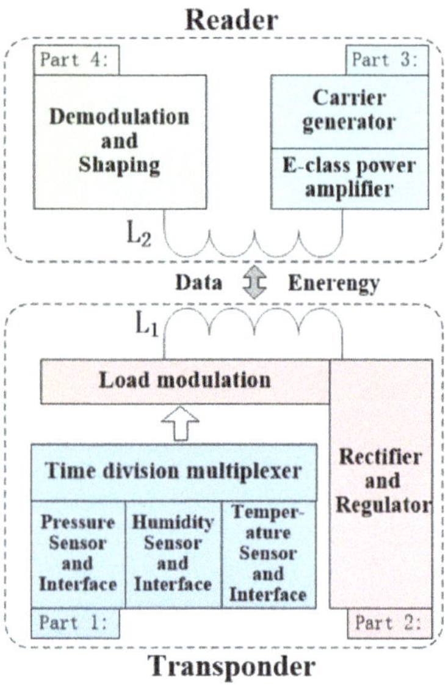

Figure 1: Overall block diagram of the passive wireless hermetic environment monitoring system. The system is composed of a reader and a transponder. The power and the data transmission between the transponder and the reader are realized by way of inductive coupling.

The working process of the system is as follows. Firstly, the carrier generator together with the E-class power amplifier in the reader generates a carrier signal. Then, the carrier signal in the reader is coupled to the transponder though their inductors. The coupled carrier signal in the transponder is then rectified and regulated as the voltage source for itself. With sufficient power supply, the temperature, humidity, and the pressure sensors together with their interfaces in the transponder can each generate a digital frequency signal, whose value is decided based on the environment. In order to realize the monitoring of multiple environmental parameters, the time division multiplexer is used to sequentially transmit three digital frequency signals. Then, the output signals of the time division multiplexer are modulated to the carrier signal by the load modulation circuit in the transponder, and the modulated carrier signal is, at the same time, coupled back to the reader. Finally, the digital frequency signals generated by the sensors and their interfaces are recovered from the modulated carrier signal by the demodulation and shaping circuits. The detail designs of the transponder and the reader are given below.

Design of the Transponder

Part 1: Sensor Interface Circuits and Time Division Multiplexer

The function of the sensor interface circuit is to convert the capacitive sensor signal to a digital frequency signal. A lot of circuits can convert capacitive variations to frequency signals. Here, we used the multi-harmonic oscillator, because of its low temperature drift characterization.

Figure 2 shows the schematic of the multi-harmonic oscillator. In Figure 2, the capacitor C_s represents a capacitive sensor. During the circuit design process, we use variable capacitors to represent capacitive sensors. The resistor R_f and the op amp constitute an integral circuit, which plays the role of frequency selection. The resistance R_1 and R_2 constitute the positive feedback circuit, and the diode D_1 and D_2 have the function of bidirectional limiting of voltage. The output frequency f of the multi-harmonic oscillator can be calculated by [19]:

$$f = \frac{1}{2R_f C_s ln(1 + \frac{2R_2}{R_1})} \quad (1)$$

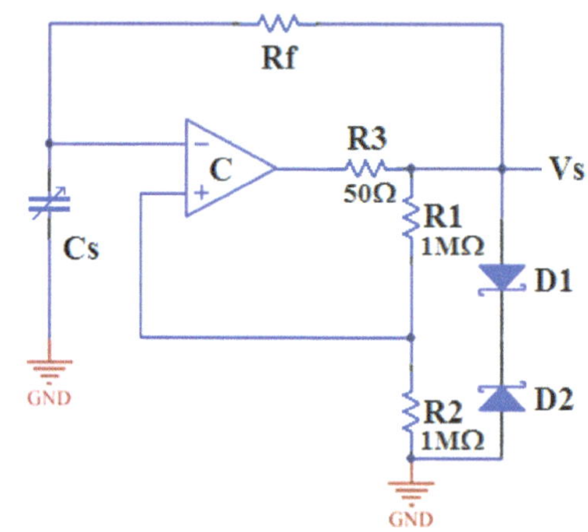

Figure 2: Schematic of the sensor interface circuit. By this multi harmonic oscillator, the capacitor C_s could be converted to frequency signal V_s.

There are three kinds of environmental sensors integrated in the transponder, but the monitoring system has only one data transmission channel. Therefore, the time division multiplexer is used to realize the time division multiplexing transmission. The time division multiplexer consists of the oscillator, the frequency divider and the multiplexer. Figure 3shows the output signals of the time division multiplexer. It can be seen that the sync signal and the signals of three sensors are sequentially outputted, and the transmission time for each signal was designed to be 64 ms.

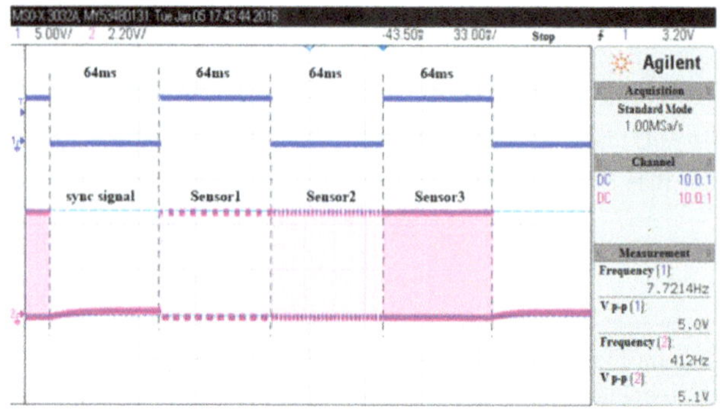

Figure 3: Output signals of the time division multiplexing circuit. The sync signal and the signals of three sensors are sequentially outputted. The transmission time for each signal is designed to be 64 ms.

Part 2: Rectifier, Regulator, and Load Modulation Circuit

The rectifier together with the regulator converts the received ac carrier signal to a stable dc voltage, which is used as the power supply for the transponder itself. The load modulation circuit modulates the low frequency data signals from the time division multiplexer to the high frequency carrier signal. The schematic of the rectifier, regulator, and load modulation circuit is shown in Figure 4. L_1 is the planar inductor that receives the carrier signal from the reader. The diode bridge plays the role of rectifier. The diode D_3, together with the capacitor C_2, filters the signal after rectification. Finally, a regulator chip U_1, TPS71550 [20], is used to make the dc voltage more smooth and stable.

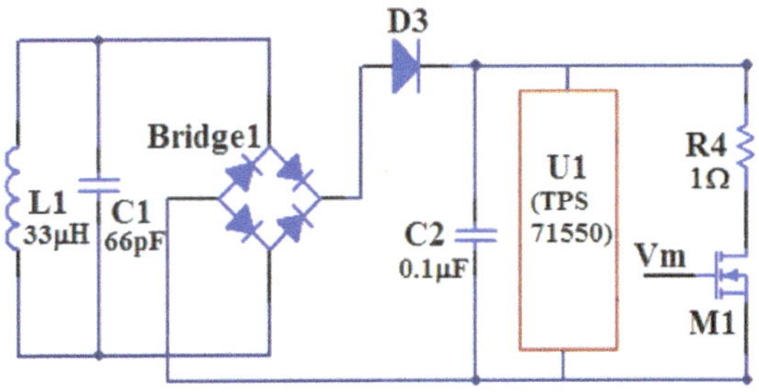

Figure 4: Schematic of the rectifier, regulator, and load modulation circuit. L_1 is the planar inductor that receives the carrier signal from the reader. The diode bridge plays the role of rectifier. The diode D_3, together with the capacitor C_2, filters the signal after rectification. The regulator chip U_1 (TPS71550) is used to make the dc voltage more smooth and stable. The NMOS transistor M1 as well as the resistor R4 forms the load modulation circuit.

The load modulation circuit consists of the NMOS transistor M_1 and the resistor R_4. The mechanism of the load modulation can be simply explained as follows. The low frequency signal V_m changes the on/off state of M_1, so the equivalent load impedance in the reader that coupled from the transponder synchronously changes, and the change of the load impedance of the reader finally leads to the amplitude variation of the carrier signal. The result is that the input low frequency signal V_m is modulated to the high frequency carrier signal. Hence, the load modulation we used here actually belongs to the AM modulation. For better recovery of the sensor data, the frequency of the sensor signal must be at least one order of magnitude lower than the carrier signal [21]. Besides, a large modulation depth is also very helpful, which can be acquired from the large impedance variation of the load modulation circuit and the high Q factors of the transponder inductor and the reader inductor.

Figure 5 shows the signal waves of the load modulation circuit. In Figure 5a, the bottom wave is the unmodulated sensor signals from the time division multiplexer, the top wave is the modulated carrier signal at the transponder inductor. In Figure 5b, the bottom wave is the modulated signals at the transponder inductor, and the top wave is the modulated carrier signal at the reader inductor. It can be seen that the sync signal and the three sensors' signals are successfully modulated to the carrier signal. The frequency of the carrier signal in our system is 5.1 MHz.

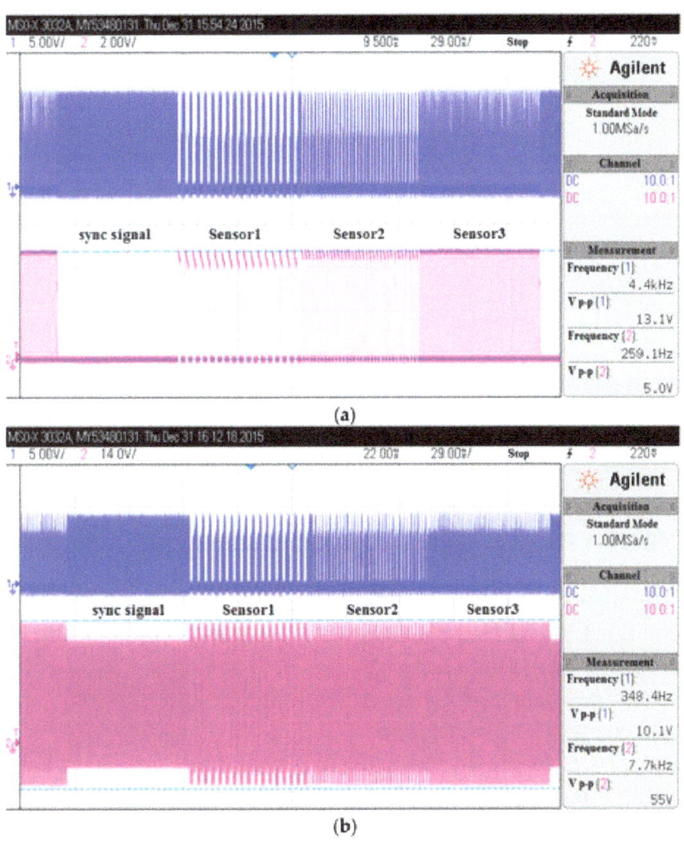

Figure 5: The signal waves of the load modulation circuit. (**a**) The bottom wave is the unmodulated sensor signals from the time division multiplexer, the top wave is the modulated carrier signal at the transponder inductor; (**b**) The bottom wave is the modulated signals at the transponder inductor, and the top wave is the modulated carrier signal at the reader inductor.

Design of the Reader

Part 3: Carrier Generator and E-Class Power Amplifier

In our passive wireless sensing system, the carrier signal not only transmits sensor data but also provides power for the transponder. Therefore, a carrier signal with large amplitude is needed. Firstly, an oscillator chip LTC6900 [22] is used to generate a 50% duty cycle square wave. The output frequency of LTC6900 is adjustable from 1 kHz to 20 MHz, which is high enough for our system. However, the amplitude of the output wave is less than 5 V, which is too weak to power the transponder. Hence, the output of the oscillator chip needs to be amplified. The E-class power amplifier, which has very high efficiency, is used here to amplify the output wave of the oscillator chip. Figure 6 shows the schematic of the E-class power amplifier. V_c is the output wave from the oscillator chip. If the capacitance of C_3 and (in parallel) the MOS transistor M_2 and the parameters of the LCR branch (L_2, C_4, and R_5) are regulated just right, the LCR branch will be on its resonant state and the carrier amplitude on inductor L_2 will reach the maximum value. When the LCR branch is on its resonant state, the amplitude of the output carrier signal will be decided by the dc power supply of the E-class amplifier. The output amplitude of the carrier signal reaches 63 V when the dc power supply is set to 12 V.

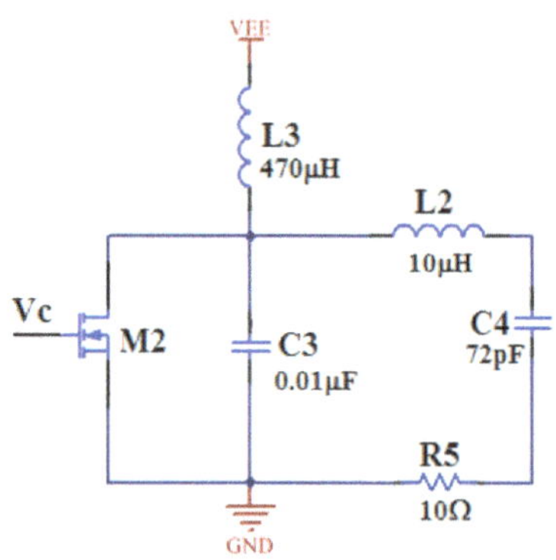

Figure 6: Schematic of the E-class power amplifier. When the LCR branch (L_2, C_4, and R_5) is on its resonant state, the amplitude of the output carrier signal will reach the maximum value.

Part 4: Demodulation and Shaping Circuits

After the reader inductor receives the modulated signal from the transponder inductor, demodulation and shaping circuits are needed to recover the sensor data. The demodulation circuit is realized by using a diode together with a capacitor and two resistors. The schematic of the demodulation circuit is shown in Figure 7. The diode has the property of unidirectional conductivity, so the capacitor C_5 will be charged if the input signal V_i is positive and will discharge if V_i is negative. The charging and the discharging processes are carried out repeatedly and alternatively, which realize the envelope detection of the input V_i. Appropriate value selection of the resistors and the capacitor can minimize the distortion of the demodulated signal.

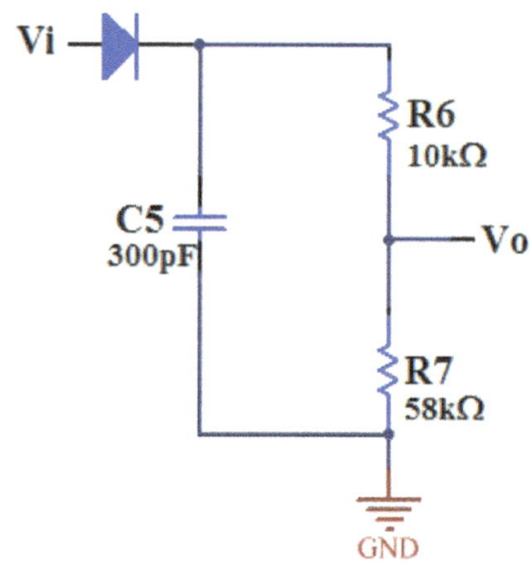

Figure 7: Schematic of the demodulation circuit. The envelope detection of the input V_i is realized by the diode, which has the property of unidirectional conductivity.

Figure 8a shows the input modulated signal and the output demodulated signal of the demodulation circuit: the signal wave below is the modulated signal and the signal wave above is the demodulated signal. It can be seen that the demodulated signal has a relatively small amplitude, and, at the same time, a lot of glitches. So the shaping circuit, which contains the voltage follower and the hysteresis comparator, is necessary. Figure 8b shows the signals of the shaping circuit: the signal wave above is the output of the voltage follower and the signal wave below is the final output signal from the hysteresis comparator. It can be seen that through the inductive coupling based backscatter modulation, diode based demodulation, and the shaping process, the sensor data generated by the transponder is finally outputted from the reader.

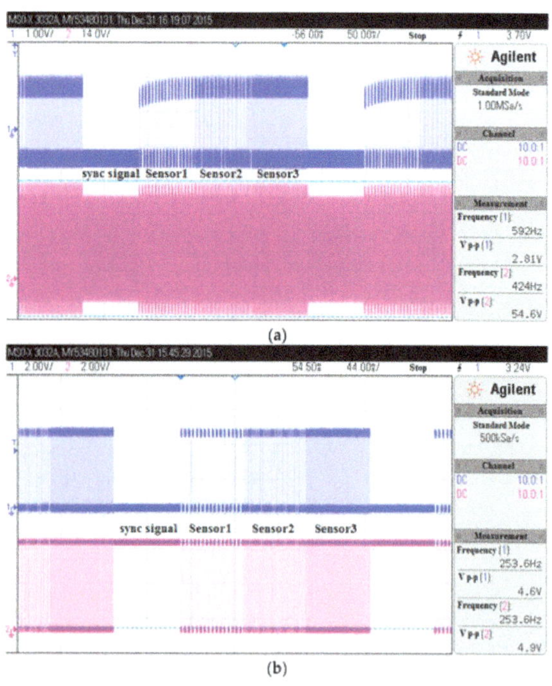

Figure 8: (**a**) The input modulated signal and the output demodulated signal of the demodulation circuit: the signal wave below is the modulated signal, and the signal wave above is the demodulated signal; (**b**) The signals of the shaping circuit: the signal wave above is the output of the voltage follower, and the signal wave below is the final output signal from the hysteresis comparator.

Figure 9 shows the photograph of the fabricated reader and the transponder. The area of the reader and transponder are both 5 cm × 8.8 cm, and the sizes of the inductors of the reader and the transponder are both 4 cm × 4 cm. Three environment monitoring sensors, including the capacitive temperature sensor, the capacitive humidity sensor, and the capacitive pressure sensor, are mounted on the transponder. Both the temperature and humidity sensors are the capacitive structures filled with corresponding sensitive dielectric materials. More details can be found in our previous works [15,23]. The capacitive pressure sensor is a commercial product, SCB10H from VTI [24], in which the pressure sensing mechanism is based on the bending of a silicon diagram.

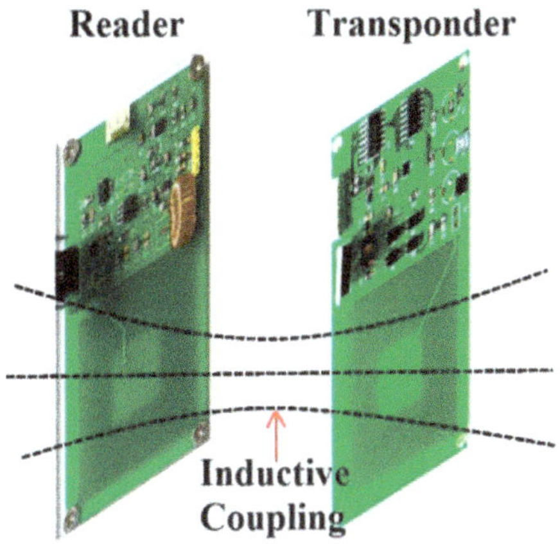

Figure 9: Photograph of the reader and the transponder. The sizes of the inductors of the reader and the transponder are both 4 cm × 4 cm. The inductive coupling between these two inductors realizes the passive wireless sensing.

PACKAGE OF THE TRANSPONDER

In a spray painting workshop, two kinds of contaminants are produced during the painting process: the paint mist and the organic gas. Among them, the paint mist is the main factor that leads to the sensor failure. The diameter of most paint mists are less than 10 μm, and they are very easy to adhere to objects. If no package is mounted, the paint mists will deposit on sensor chips and quickly lead to sensor failure.

However, the hermetically sealed package is not suitable for the transponder, which includes the humidity sensor and the pressure sensor. In the circumstance of the spray painting workshop, the package of the transponder should not only be permeable to air but also be resistant to the deposition of the paint mist. Figure 10 shows the schematic of the package of the transponder. Firstly, the transponder is mounted on the baseboard. Then, the transponder is covered with the film fixing cap. The filter film on the film fixing cap is used for resisting paint mist. Finally, the film fixing cap is covered with the porous cap, which is used for blocking dust and large particles.

Therefore, the filter film plays the most important role in the package of the transponder. Based on the environment of the spray painting workshop, we designed a composite film that is composed of a porous aluminum anodic oxide (AAO) base film and a polymeric polytetrafluoroethylene (PTFE) film. The AAO base film filters chemical macromolecules or impurity particles in the air, and provides flow channels for air.

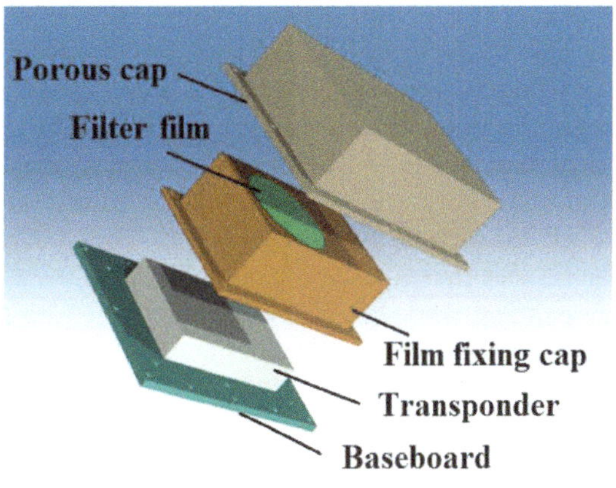

Figure 10: Schematic of the package of the transponder. Firstly, the transponder is mounted on the baseboard. Then, the transponder is covered with the film fixing cap. The filter film on the film fixing cap is used for resisting paint mist. Finally, the film fixing cap is covered with the porous cap, which is used for blocking dust and large particles.

The polymeric PTFE film is used to obtain a hydrophobic surface, in other words, to prevent paint mist adhesion. The fabrication of the AAO/PTFE composite film is not very complex. Firstly, the surface oxide film of the prepared aluminum wafer is removed by alkali solution. Secondly, the anodic oxidation process is performed for the first time and then the oxide film is also removed. Thirdly, the anodic oxidation process is performed for the second time, followed by the hole enlarging treatment and heat treatment. Fourthly, using the spin coating process, a hydrophobic and microporous PTFE membrane is coated on the AAO film. Finally, the PTFE film is hardened in vacuum constant temperature drying oven. The SEM photographs of the surface and the section view of the AAO/PTFE composite film are shown in Figure 11. In Figure 11a, the average diameter of the holes is about 200 nm. Figure 11b shows that the AAO film is covered with the PTFE layer, and the PTFE material already seeps into the holes of the AAO film. The thicknesses of the AAO film and the PTFE layer are about 10 μm and 1.5 μm, respectively.

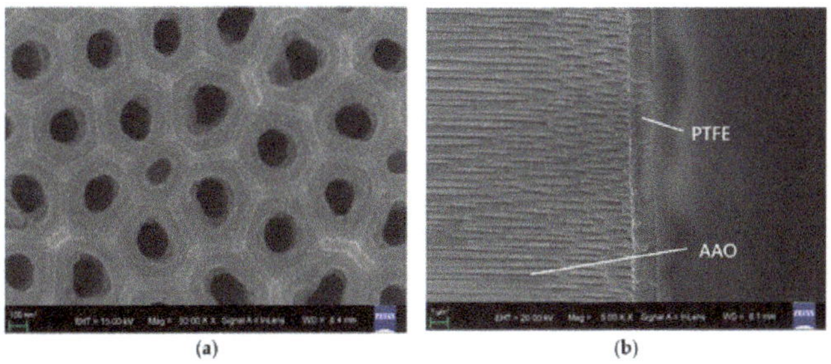

Figure 11: SEM photographs of the aluminum anodic oxide/ polytetrafluoroethylene (AAO/PTFE) composite film (**a**) the surface view; (**b**) the section view.

A contrast experiment was carried out to compare the performance of the fabricated composite filter film with two other kinds of filter films, PTFE film and polyvinylidene fluoride (PVDF) film. The PTFE and PVDF films are both large-scale commercial polymeric materials. They have good air permeability, good corrosion resistance, and low surface tension. Other films that have similar properties as above, such as polypropylene films [25] and Nafion films [26], may be possible candidates. Here, we mainly investigated the change of water vapor flux and the antipollution properties of each filter film after being exposed for a long time to pollution in the spray painting workshop. Figure 12 gives the transient responses of the humidity sensors before and after 600 h paint mist contamination; the sensors were covered by the AAO/PTFE film, the PTFE film, and the PVDF film, respectively. The photographs of the polluted films are also shown in Figure 12. It can be seen that, after a long time period of working in the spray painting workshop, the composite filter film AAO/PTFE has the minimum pollution and the smallest change of water vapor flux.

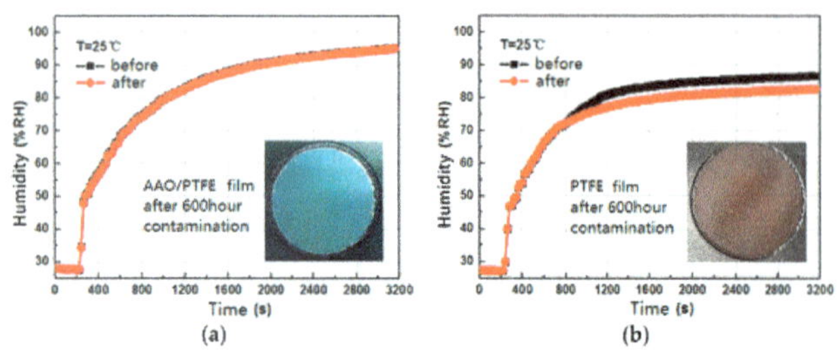

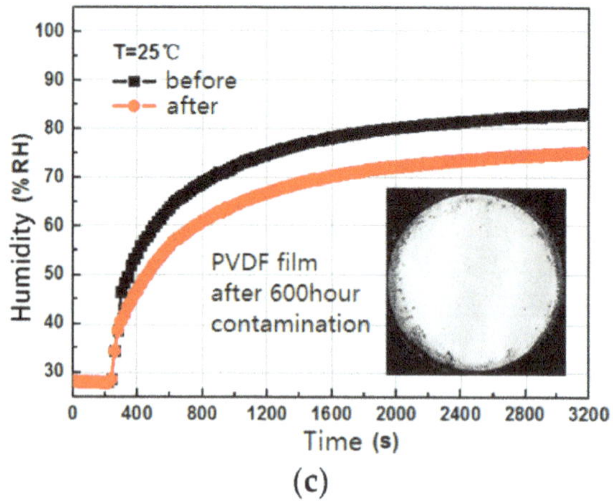

Figure 12: Transient responses of the humidity sensors covered by different filter films before and after 600 h paint mist contamination, (**a**) AAO/PTFE film; (**b**) PTFE film; and (**c**) polyvinylidene fluoride (PVDF) film.

MEASUREMENT AND DEMONSTRATION

By using the OMEGA205 C340 temperature and humidity chamber (made by Vötsch Industrietechnik), the temperature and humidity characterizations of the passive wireless monitoring system were measured. In Figure 13a, the variation of the output frequency is about 11 kHz when the temperature changes from 0 °C to 60 °C. Hence, the average temperature sensitivity of the monitoring system is 180 Hz/°C. Figure 13b shows that the output frequency changes about 8 kHz when the humidity varies from 15 %RH to 95 %RH, and the humidity sensitivity can be calculated as 100 Hz/%RH.

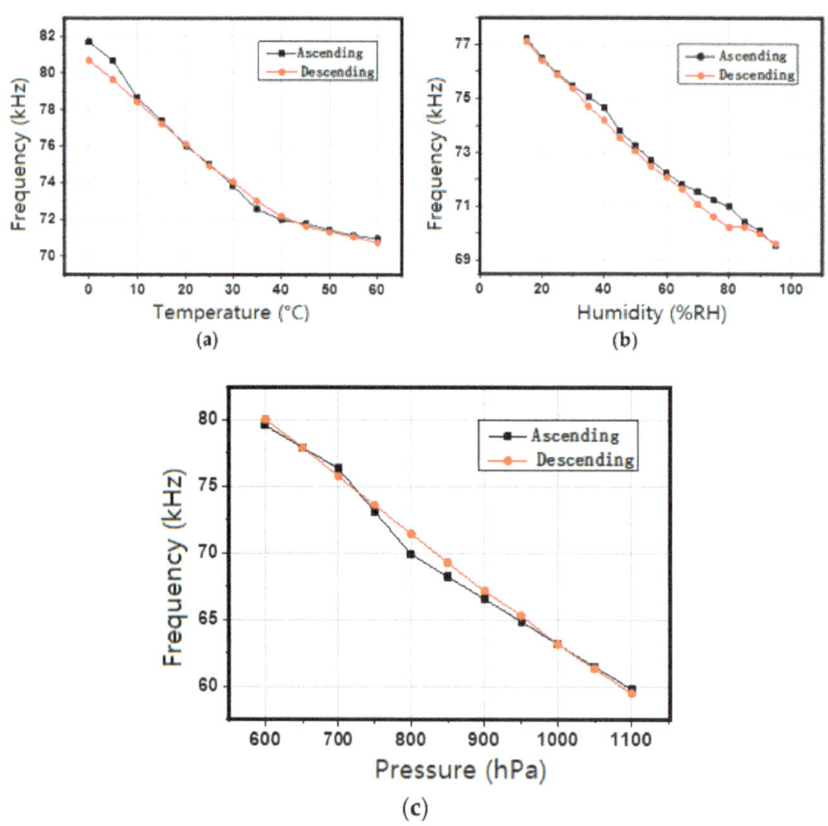

Figure 13: Output frequency versus (**a**) temperature; (**b**) humidity; (**c**) pressure.

The pressure characterization of the passive wireless monitoring system was measured by the GY-600 pressure chamber (made by XiangHe, GuoRuiZhi Test Equipment Co., Ltd., Beijing, China). The measured relationship between the pressure and the output frequency is shown in Figure 13c. The variation of the output frequency is about 21 kHz when the pressure changes from 600 hPa to 1100 hPa. So the pressure sensitivity of the monitoring system is 42 Hz/hPa. The passive wireless transmission distance of the monitoring system was also measured. The test platform as well as the measurement result is shown in Figure

14. When the distance between the transponder and the reader reaches 4 cm, the output signal of the reader is still working properly. Under fixed inductor size and input power, adjusting the resonant frequency of the transponder and the reader to the carrier signal frequency could be an efficient way to increase the telemetry distance. Because the wall of the spray painting workshop is made of glass, the effect of glass in the distance measurement needs to be considered. However, the measurement result shows that glass has no influence on the passive wireless sensing distance.

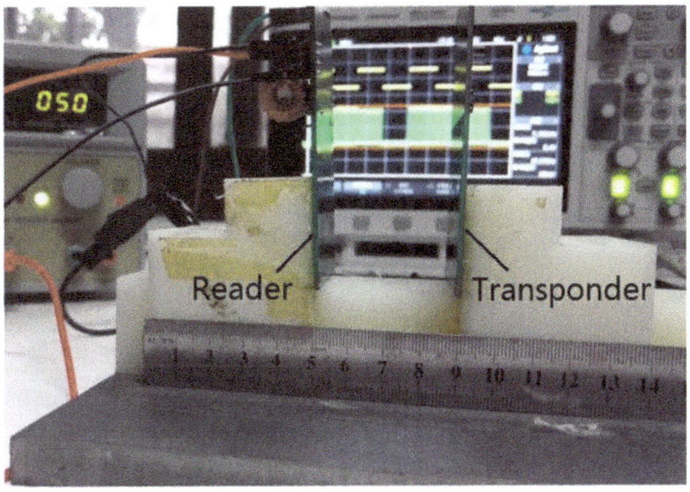

Figure 14: Measurement of passive wireless transmission distance. This ruler shows the passive wireless transmission distance between the transponder and the reader reaches 4 cm.

After the measurement, the passive wireless hermetic environment monitoring system is installed on the glass wall of the spray painting workshop for demonstration. Figure 15a is the photograph of the transponder and reader installed on the glass wall of the spray painting workshop. The one with porous cap inside the glass wall is the transponder, which monitors the environmental parameters of the spray painting workshop, and the one directly facing the transponder outside the glass wall is the reader, which receives data from the transponder and at the same time provides power for it.

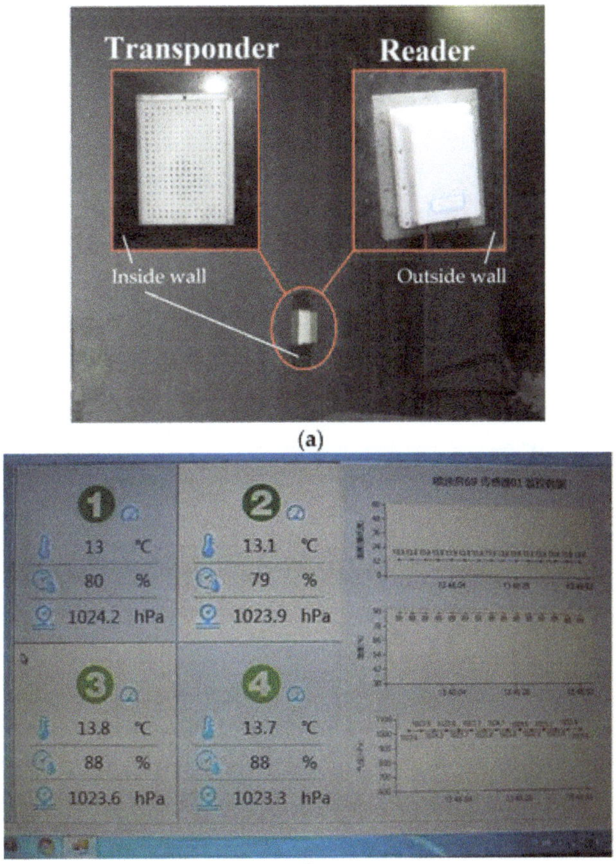

(a)

(b)

Figure 15: Demonstration of the monitoring systems in spray painting workshop (**a**) Installation of the transponder and reader on the glass wall of the spray painting workshop; (**b**) A screenshot of the monitoring software showing both real-time data and history records.

Figure 15b is a screenshot of the monitoring software, which could display both real-time data and historical records of the passive wireless monitoring system. Output results of four monitoring systems are shown on the screen at the same time. Among them, No. 1 and No. 2 monitoring systems are installed in one room, and No. 3 and No. 4 are in another room.

CONCLUSIONS

A passive wireless hermetic environment monitoring system was designed, fabricated and demonstrated for the spray painting workshop. The proposed system is composed of a transponder and a reader, and circuit design of each part is given in detail. Because the spray painting workshop contains a large amount of paint mist, the package of the transponder was designed carefully. Three different filter films for the package were compared. The experimental results show that the composite filter film AAO/PTFE has the minimum pollution and the smallest change of water vapor flux. After fabrication, the temperature, humidity and pressure characterizations and the remote sensing distance of the monitoring system were measured. Finally, the passive wireless hermetic environment monitoring system was installed on the glass wall of the spray painting workshop and was successfully demonstrated.

ACKNOWLEDGMENTS

This work is supported by the National Natural Science Foundation of China (61401084, 61136006) and the National High Technology Research and Development Program of China (2015AA042602).

AUTHOR CONTRIBUTIONS

Lifeng Wang and Qing-An Huang conceived and designed the monitoring system; Yan Huang and Dan Tang designed the circuits and did the measurements; Jingjing Ma designed and fabricated the package; Lifeng Wang, Jingjing Ma and Yan Huang did the demonstration; Lifeng Wang wrote the paper.

REFERENCES

1. Bogue, R. Recent developments in MEMS sensors: A review of applications, markets and technologies. *Sens. Rev.* 2013, *33*, 300–304.
2. Wang, Y.; Jia, Y.; Chen, Q.; Wang, Y. A Passive Wireless Temperature Sensor for Harsh Environment Applications. *Sensors* 2008, *8*, 7982–7995.
3. Wilson, W.C.; Atkinson, G.M. Passive wireless sensor applications for NASA's extreme aeronautical environments. *IEEE Sens. J.* 2014, *14*, 3745–3753.
4. Tan, Q.; Kang, H.; Xiong, J.; Qin, L.; Zhang, W.; Li, C.; Ding, L.; Zhang, X.; Yang, M. A Wireless Passive Pressure Microsensor Fabricated in HTCC MEMS Technology for Harsh Environments. *Sensors* 2013, *13*, 9896–9908.
5. Tumanski, S. Induction coil sensors—A review. *Meas. Sci. Technol.* 2007, *18*, R31–R46.
6. Collins, C.C. Miniature passive pressure transensor for implanting in the eye. *IEEE Trans. Biomed. Eng.* 1967, *14*, 74–83.
7. Schuylenbergh, K.V.; Puers, R. Passive telemetry by harmonics detection. In Proceedings of the 18th Annual International Conference of the IEEE Engineering in Medicine and Biology Society, 1996. Bridging Disciplines for Biomedicine, Amsterdam, The Netherlands, 31 October–3 November 1996; pp. 299–300.
8. Bhadra, S.; Blunt, W.; Dynowski, C.; McDonald, M.; Thomson, D.J.; Freund, M.S.; Cicek, N.; Bridges, G.E. Fluid embeddable coupled coil sensor for wireless pH monitoring in a bioreactor. *IEEE Trans. Instrum. Meas.* 2014, *63*, 1337–1346.
9. Zhang, C.; Guo, L.; Wang, L.F.; Huang, J.Q.; Huang, Q.A. Passive wireless integrated humidity sensor based on dual-layer spiral inductors. *Electron. Lett.* 2014, *50*, 1287–1289.
10. Ma, M.; Liu, Z.; Shan, W.; Li, Y.; Kalantar-zadeh, K.; Wlodarski, W. Passive wireless gas sensors based on the LTCC technique. In Proceedings of the 2015 IEEE MTT-S International Microwave Workshop Series on Advanced Materials and Processes for RF and THz Applications (IMWS-AMP), Suzhou, China, 1–3 July 2015; pp. 1–3.
11. Aschenbrenner, B.; Zagar, B.G. Analysis and validation of a planar high-frequency contactless absolute inductive position sensor. *IEEE Trans. Instrum. Meas.* 2015, *64*, 768–775.
12. Zhang, C.; Wang, L.F.; Huang, J.Q.; Huang, Q.A. An LC-type passive wireless humidity sensor system with portable telemetry unit. *J. Microelectromech. Syst.* 2015, *24*, 575–581.
13. Tan, Q.; Luo, T.; Xiong, J.; Kang, H.; Ji, X.; Zhang, Y.; Yang, M.; Wang, X.; Xue, C.; Liu, J.; et al. A harsh environment-oriented wireless passive temperature sensor realized by LTCC technology. *Sensors* 2014, *14*, 4154–4166.
14. Malocha, D.C.; Gallagher, M.; Fisher, B.; Humphries, J.; Gallagher, D.; Kozlovski, N. A Passive Wireless Multi-Sensor SAW Technology Device and System Perspectives. *Sensors* 2013, *13*, 5897–5922.
15. Dong, L.; Wang, L.F.; Huang, Q.A. Implementation of Multiparameter Monitoring by an LC-Type Passive Wireless Sensor Through Specific Winding Stacked Inductors. *IEEE Int. Things J.* 2015, *2*, 168–174.
16. DeRouin, A.; Ong, K.G. Multi-parameter sensing with a single magnetoelastic sensor by applying loads on the null locations of multiple resonant modes. *Smart Mater. Struct.* 2015, *25*, 035044.
17. DeHennis, A.D.; Wise, K.D. A wireless microsystem for the remote sensing of pressure, temperature, and relative humidity. *J. Microelectromech. Syst.* 2005, *14*, 12–22.
18. Chaimanonart, N.; Suster, M.A.; Young, D.J. Two-Channel passive data telemetry with remote RF powering for High-Performance wireless and batteryless strain sensing microsystem applications. *IEEE Sens. J.* 2009, *10*, 1375–1382.
19. Kang, H.G.; Chen, D.Q.; Zhang, L. *Fundamentals of Electronic Technology, Analog Part*, 6th ed.; Higher Education Press: Beijing, China, 2013; pp. 462–464. (In Chinese)

20. TPS715xx Datasheet, Texas Instruments Incorporated. Available online: http://www.ti.com (accessed on 1 November 2014).

21. Shanmugam, K.S. *Digital and Analog Communication Systems*; John Wiley & Sons Inc.: New York, NY, USA, 1979.

22. LTC6900 Datasheet, Linear Technology Corporation. Available online: http://www.linear.com (accessed on 1 November 2014).

23. Ren, Q.Y.; Wang, L.F.; Huang, J.Q.; Zhang, C.; Huang, Q.A. A Novel Capacitive Temperature Sensor for a Lab-on-a-chip System. In Proceedings of the IEEE Sensors, Valencia, Spain, 2–5 November 2014; pp. 436–439.

24. SCB10H Series Pressure Elements Datasheet, VTI Technologies. Available online: http://www.vti.fi (accessed on 1 September 2014).

25. Busico, V.; Cipullo, R. Microstructure of polypropylene. *Prog. Polym. Sci.* 2001, *26*, 443–533.

26. Mauritz, K.A.; Moore, R.B. State of understanding of Nafion. *Chem. Rev.* 2004, *104*, 4535–4585.

CHAPTER 11

TEMPORAL-SPATIAL EVOLUTION ANALYSIS OF LAKE SIZE DISTRIBUTION USING LANDSAT IMAGERY DATA

Lin Li , Hui Xia, Zheng Li and Zhijun Zhang

Keywords: AWEI, Landsat, the Yangtze River Basin, lake evolution

INTRODUCTION

Lakes not only contain abundant freshwater, agricultural and aquatic products, minerals, and other resources but also play an important role in natural cycles, climate regulation, and ecological balance [1,2]. Under the influence of LUCC (Land-Use and Land-Cover Change) and other forms of environmental change, surface water, as one of the most important resources on Earth, has undergone constant temporal and spatial evolution. For many years, the impacts of surface water evolution on ecology, society, health and the economy have been topics of a large number of academic studies [3,4,5,6,7,8,9,10]. For example, the areas of 22 relatively large lakes in the southeast region of Qiangtang Plateau have been expanding during 30 years of increasing ice melt, snowmelt, and rainfall, and decreasing evaporation [11].

Remote sensing technology provides a more convenient platform for real-time monitoring and management of surface water resources, as it offers a broad scope of observation and rapid data collection [12,13,14,15,16,17]. For example, MODIS image data were used to monitor and analyze the changes in the inundated area of Poyang Lake over 11 years [18], and a near real-time water surface detection method based on MODIS data was developed and tested to be applied to other existing or scheduled sensors with similar bands [19]. Researchers analyzed different periods of volume change and built underwater terrain models of the Baiyangdian wetland using multiple Landsat and HJ-1A/B images, and CCD images combined with water levels from the hydrological station in Hebei province, China [20]. Time-series analysis of remote sensing data for mapping temporary surface water bodies was also applied in large scales [21]. NOAA AVHRR data, with their characteristic short cycles, have been used for statistics and analysis of flood areas in northeast India, providing advanced technological support to remote-sensing monitoring, evaluation, and decision-making for disaster prevention [22]. The four lakes in the middle and lower Yangtze River Basin are important freshwater reservoirs and serve both society and ecology so that monitoring them is a valuable feat. The lakes are downstream of the Three Gorge Dam and analysis of their temporal-spatial evolution can be important to water management in the region.

In this study, Landsat TM/ETM data are applied to monitor the water area changes of four lakes in the middle and lower reaches of the Yangtze River basin: the Dongting, Poyang, Chaohu, and Taihu Lakes. We applied a previously-validated lake extraction algorithm called AWEI (Automated Water Extraction Index) to Landsat satellite imagery to measure the area of the four lakes [23], then estimated variability and trends in lake area, and evaluated the correlation with precipitation data. The spatial changes such as displacement of lake centroids were also analyzed, because this can reflect the driving force of the lake evolution process [24,25]. Overall, the objectives of this study are to contribute to the goal of feasibly and conveniently mapping lake areas for long-term lake monitoring, as well as providing a way to analyze variation in lakes of the same river basin.

Citation: Lin Li, Hui Xia, Zheng Li and Zhijun Zhang, Temporal-Spatial Evolution Analysis of Lake Size Distribution Using Landsat Imagery Data, doi:10.3390/rs70810364.

STUDY AREA

The four lakes—Dongting, Poyang, Chaohu, and Taihu Lakes, respectively located in northern Hunan, Northern Jiangxi, Central Anhui and Southern Jiangsu—are known as China's four largest lakes in the Yangtze River basin. All of them are natural lakes and are downstream of the Three Gorges Dam (Figure 1). The dam was built for flood control, irrigation, and electricity. In the event of a flood, there are 22 cubic kilometers of space behind the dam to hold back the flood waters. The flood waters are stored behind the dam for future use. In the dry season, the water is released down the river, and the reservoir level drops to prepare for the next flood.

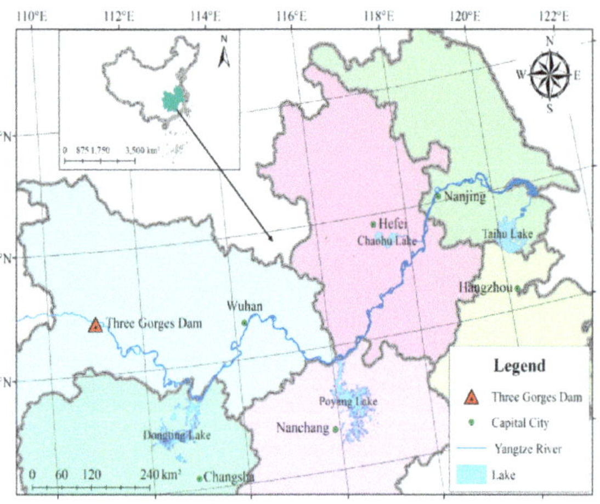

Figure 1: The study area and locations of the four lakes in China.

Dongting Lake, located in the area from 28°30′N to 30°20′N and from 112°25′E to 113°15′E, has a water storage capacity of approximately 17.8 billion·m^3 and plays an important role in flood storage. Its climate is mild and humid, and its MAAT (mean annual air temperature) is approximately 17° Celsius. Dongting Lake district has an agriculture-based economy, and the population of its surrounding is approximately 11 million (3.46 people/ha). It is a major source of drinking and irrigation water for surrounding areas.

Poyang Lake, located between 28°22′N and 29°45′N and between 115°47′E and 116°45′E, has a water storage volume of approximately 25 billion·m^3. The annual inflow into Poyang Lake is about 16% of the annual runoff of the Yangtze River. Its climate and MAAT are very similar to Dongting Lake. There are nearly 20 million people (3.81 people/ha) living around Poyang Lake and as a large bird sanctuary, Poyang Lake provides winter habitat to thousands of migratory birds from Siberia, Mongolia, Japan, North Korea, northeast China, and northwest China from late fall through early winter.

Chaohu Lake, 15 km southeast of the city of Hefei, is located between 31°25′N and 31°43′N and between 117°16′E and 117°51′E. The region is characterized by a northern subtropical monsoon moist climate with MAAT of approximately 15°–16° Celsius. The permanent population of its surrounding is about 11 million (4.56 people/ha), and the area of land under irrigation is nearly 3200 km^2.

Taihu Lake is geographically located between 30°55′N and 31°32′N and between 119°52′E and 120°36′E, with hilly mountains to the west. Its climate is mild and humid with MAAT is approximately 16°–18° Celsius. The permanent population of its surrounding is approximately 23 million (10.63 people/ha), and the area of land under irrigation is nearly 3240 km^2. The time of wet and dry season alternation is variable in years for the four lakes. Dongting and Poyang Lake are leaf-shaped lakes with a lot of shallow areas, and their outlets are close to the Yangtze River, allowing water from the lake and the river to complement each other during the dry and the wet season, in time. In contrast, Chaohu Lake and Taihu Lake are block-shaped lakes with less shallow areas, and their outlets are far from the Yangtze River.

DATA USE AND METHODOLOGY

Landsat Data

The Yangtze River was blocked off in 2002 after the completion of key parts of the Three Gorges Dam. The goal of this research is to estimate changes in lake size and centroid, downstream of the dam, and identify the probable cause of changes to lake morphology. In total, 287 Landsat images covering the four large lakes from 2002 to 2013 were obtained from the USGS Global Visualization Viewer [26] and were used to extract the water area of the target lakes (Table 1). To cover the data range of 2002 to 2013, both Landsat TM and ETM were used. All spatial resolutions are 30 m. Bands 1, 2, 4, 5, and 7 of TM and ETM were included.

Table 1: Data selection in the study area

Lakes	Number and Years of Landsat TM/ETM Data Used for Each of the Four Lakes												
	2002	2003	2004	2005	2006	2007	2008	2009	2010	2011	2012	2013	Total
Dongting Lake	4	8	6	8	5	4	4	2	3	5	0	9	58
Poyang Lake	5	6	10	2	6	5	5	6	7	4	5	1	61
Chaohu Lake	8	9	11	6	9	8	5	8	13	7	5	5	94
Taihu Lake	10	6	10	8	2	4	5	6	8	5	5	5	74
	TM					ETM							
Total	74					213							287

Methodology

Image Processing

Data obtained from the USGS Global Visualization Viewer are all from Level 1, meaning that the data have gone through the radiometric and geometric correction, but not terrain correction [26]. Since this study does not include a comparative analysis of water area with other topographic data, the terrain correction is not needed. Usually, satellite digital images provide a pixel's DN-values, with which only relative comparisons can be made within the same image. To quantitatively compare data obtained from different places and times with different types of sensors, the DN-values must be transformed into corresponding radiance pixel values via a conversion process called radiometric calibration [27,28].

Radiometric calibration methods for the TM and ETM sensor data are different. In this research, the ENVI™ (version 5.0 SP2) desktop software Landsat calibration tool was used to conduct radiometric calibration [29], and FLAASH (Fast Line-of-sight Atmospheric Analysis of Hypercubes) atmospheric correction tool was used to correct for atmospheric contamination [30].

Water Extraction Method

Compared with types of land cover such as vegetation or bare land, water shows special characteristics in the near-infrared band, which can be used to distinguish water from other object types. There are numerous methods for recognizing water with a single band or multi-band, including the Open Water Likelihood (OWL) water extraction index method [31,32], among others. One of the most commonly-used methods is the Normalized Difference Water Index (NDWI) proposed by McFeeters, using band 2 and band 4 of the Landsat images for water extraction [33]. However, NDWI cannot effectively distinguish between buildings and water, leading to misclassification of the two cover types. The NDWI method was improved by replacing band 4 with band 5, which effectively distinguishes water from buildings. The new water index is named the Modified Normalized Difference Water Index (MNDWI) [34]. However, neither of the above methods can distinguish water from shadows. In this study, a water index method, the Automated Water Extraction Index (AWEI) [23] was applied that distinguished water from shadows, buildings, asphalt, and other sources of spectral confusion. AWEI was defined as:

$$AWEI_{nsh} = 4 \times (\rho_{band2} - \rho_{band5}) - (0.25 \times \rho_{band2} + 2.75 \times \rho_{band5}) \quad (1)$$

$$AWEI_{sh} = \rho_{band1} + 2.5 \times \rho_{band2} - 1.5 \times (\rho_{band4} + \rho_{band5}) - 0.25 \times \rho_{band7} \quad (2)$$

where ρ represents spectral reflectance values of the band, and its subscript describes the specific band. The subscript "$_{nsh}$" represents the formula applied to situations where non-shadow surface features (such as building surfaces and asphalt roads) constitute the main problem for water extraction. In contrast, the subscript "$_{sh}$" indicates that this formula should be applied to remove shadows from water. Water usually absorbs almost all of the radiation of bands 4, 5, and 7, and it has the highest albedo in bands 1 and 2. Therefore, water can be effectively distinguished from other surface features by Equations (1) and (2). However, in some time series, lakes or parts of them are very turbid, and contain floating vegetation or intensive algal blooms. These are often inseparable from adjacent land. Visual interpretation was used to eliminate the error classified extraction results caused by algal blooms. Figure 2 is an example of the lake area extracted by AWEI from Landsat TM/ETM images, where (a) is the water area of Dongting Lake extracted on 19 March 2002; (b) is the surface area of Poyang Lake extracted on 8 January 2002; (c) is the lake body of Chaohu Lake extracted from Landsat data on 8 January 2002; and (d) is the water area of Taihu Lake extracted on 2 January 2002.

The omission error and commission error of the AWEI method of water detection are only half of those of the MNDWI, and a fourth of the errors associated with the Maximum Likelihood method [23].

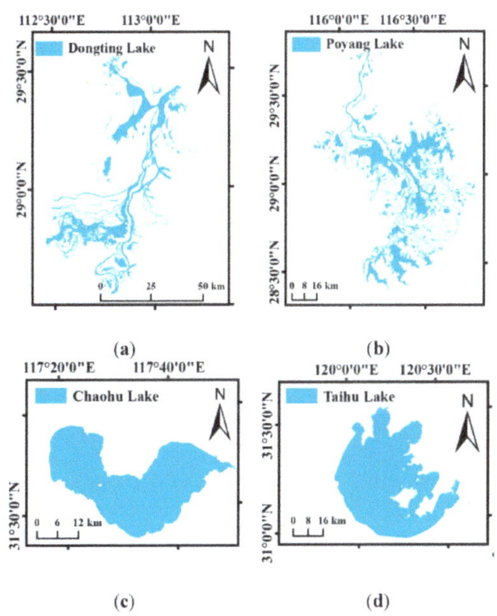

Figure 2: Extraction examples of (**a**) Dongting Lake; (**b**) Poyang Lake; (**c**) Chaohu Lake, and (**d**) Taihu Lake.

Lake Expansion Index (LEI)

To provide a quantitative analysis of the temporal and spatial changes in lake areas downstream of the Three Gorges Project, the annual rate of change in area relative to the previous period was estimated [11,35]. And an extended annual rate of change in area relative to the lake average size of the entire period was developed. The formulas for calculating the annual change and the extension in lake water area are defined as:

$$\eta = \frac{S_b - S_a}{S_a} \times \frac{1}{T} \times 100\% \quad (3)$$

$$\eta_{ext} = \frac{S_b}{S_{average}} \times 100\% \qquad (4)$$

where η is the annual percent change in lake area. The subscript "$_{ext}$" represents "extended". Sa represents the lake area during the current period. Sb represents the lake area of the next period. Saverage represents the lake average area of the entire period. T indicates the length of the analysis period, in years. If η is positive, the lake area increases. If η is negative, the area decreases. If ηext is larger than 1, the lake area is larger than average. If ηext less than 1, the area is smaller than average.

The possible number of Landsat acquisitions that could be used for lake extraction in the time-series is limited by factors such as temporal resolution of the Landsat sensor, thick cloud covering the surface features, large area of eutrophication, turbid water, *etc.* Seasonal variations in lake sizes could not be estimated reliably with the available Landsat data. However, the annual average for the four lakes can be calculated more reliably with selecting an equal number of lake sizes evenly distributed throughout the wet and dry seasons. The annual average is the mean value of the selected wet and dry season lake sizes.

Precipitation Analysis

To estimate the impact of rainfall on the lake areas, the monthly rainfall data of the four lakes from 2002 to 2013 were obtained from the China Meteorological Data Sharing System [36]. This data set consists of monthly surface precipitation gridded data with a 0.5° × 0.5° horizontal resolution since 1961. The latest precipitation data from approximately 2400 stations compiled by the National Meteorological Information Center were interpolated spatially by thin plate spline (TPS) using the ANUSPLIN software. The data were cross-validated, and error analysis was performed. We summed the monthly precipitation values to determine the annual surface precipitation. The average annual area and the annual surface precipitation were normalized [18,37] to improve visualization and comparison.

Centroid Determination Analysis

To examine the spatial-temporal characteristics, the lake centroid is extracted from each lake as a frame of reference, and the geographical space is divided into eight quadrants with intervals of 45 degrees. In this way, the offset can be estimated in every direction. The shifting distances and the shifting angles from the lake's previous centroid to the next are calculated by identifying the quadrant in which the deflection occurs [38,39]. All of the distance values in each quadrant in each time period are accumulated, and values in the opposite direction (such as north and south) cancel out.

RESULTS

Estimated Area and Surface Changes

To monitor the temporal and spatial changes in water area of the four downstream lakes, images of the lakes from 2002 to 2013 were used for data analysis. The statistical description of the water area extraction results by using AWEI was given in Table 2. Dongting Lake, with the shortest distance from the Three Gorges Project, had an average area of 1225 km^2 from 2002 to 2013. Its largest area was estimated on 3 September 2002, and the smallest area was estimated on 12 March 2011. In contrast, the average area of Poyang Lake was 2586 km^2 from 2002 to 2013. The largest area was estimated on 16 August 2010, and the smallest area was estimated on 11 December 2011. The extraction results of Dongting and Poyang Lakes were more discrete. The coefficients of variation, which were 0.46 and 0.28 separately, indicated that the values are much different from the mean value. Data sets with positive kurtosis tend to have a distinct peak near the mean, decline rather rapidly, and have heavy tails. However, the kurtoses of the two lakes were −1.23 and −1.45, respectively, which showed that the area values are more evenly distributed (Figure 3).

Table 2: Statistical description of four lake extraction areas in southeastern China

	Dongting Lake	Poyang Lake	Chaohu Lake	Taihu Lake
Number	52	61	73	67
Summation (km^2)	63,697	157,725	57,345	157,075
Minimum (km^2)	521	1390	769	2283
Maximum (km^2)	2232	3808	806	2423
Mean (km^2)	1225	2586	786	2344
Range (km^2)	1711	2418	38	140
SD	567.50	721.66	9.09	30.11
CV	0.46	0.28	0.01	0.01
Kurtosis	−1.45	−1.23	−0.50	−0.04

SD = standard deviation, CV = coefficient of variation.

Chaohu and Taihu Lakes are far away from the Three Gorges Project and have similar morphologies and temporal-spatial changes in water area (Figure 3). Overall, we can see that the CV of Chaohu and Taihu Lakes are both 0.01, which implies that the water areas of Chaohu and Taihu Lakes have not changed substantially. It could also be inferred from the kurtoses of Chaohu and Taihu Lakes (0.50 and 0.04, respectively) that the area results exhibited an approximate normal distribution. The average areas of Chaohu and Taihu Lakes were 786 km^2 and 2345 km^2, respectively, from 2002 to 2013. The seasonal water area changes of Chaohu and Taihu Lakes were within 3.7% and 5.1%, respectively, of the total lake areas.

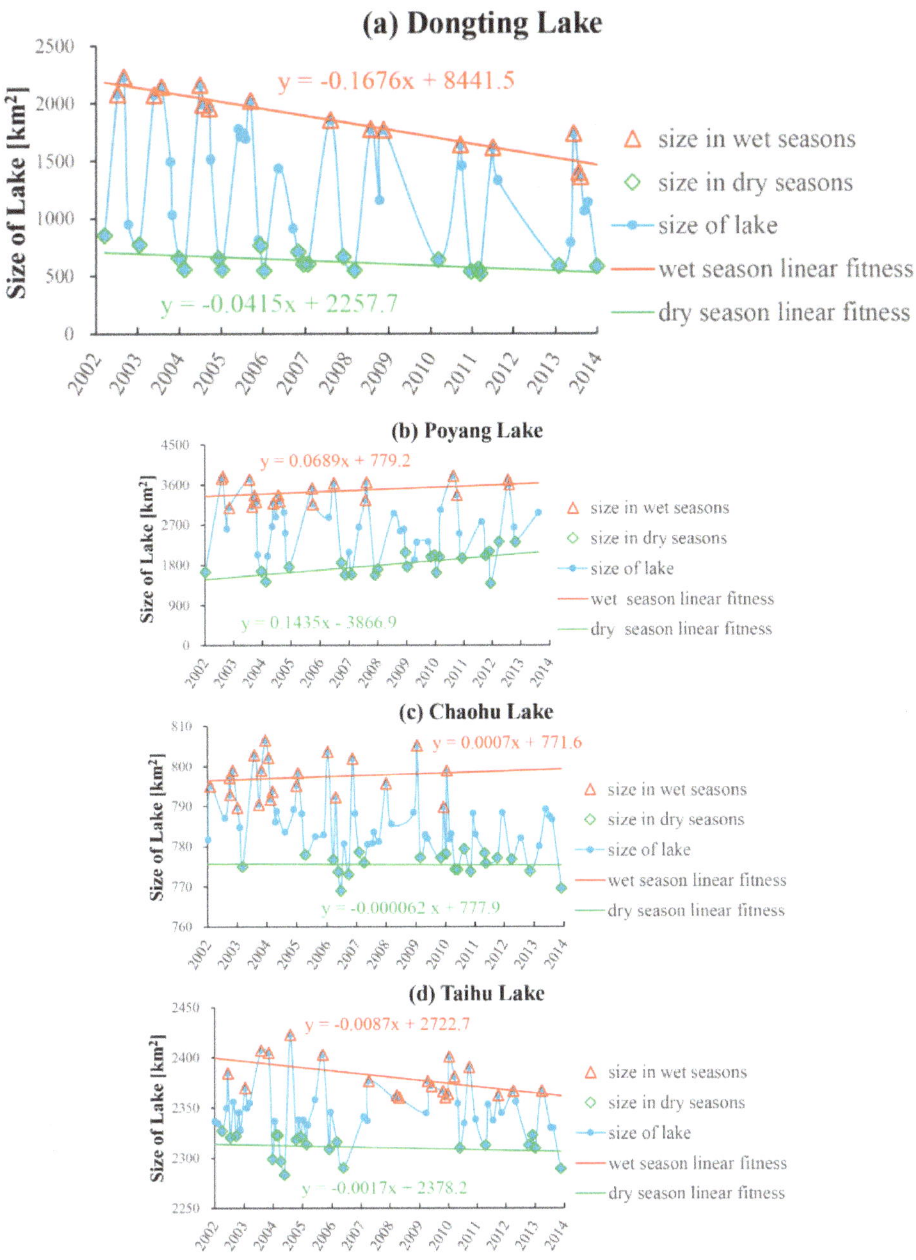

Figure 3: Seasonal sizes of four Yangtze River lakes from 2002 to 2013 for (**a**) Dongting Lake; (**b**) Poyang Lake; (**c**) Chaohu Lake, and (**d**) Taihu Lake.

Four images of the lakes were chosen for accuracy assessment. The method in assessment of the positional accuracy of AWEI method is a human visual inspection towards Landsat images. Water and non-water areas were distinguished by visual inspection in each pixel of the lake edges and the extraction result was used as the ground truth comparing to the result extracted at the same time period by AWEI method. The results of accuracy assessment were summarized in a confusion matrix and the specific information of classification accuracy for AWEI method of four lakes is shown in Table 3.

Table 3: Summary of classification accuracy of the four lakes

	Dongting Lake		Poyang Lake		Chaohu Lake		Taihu Lake	
Date	13 October 2002		20 August 2002		8 January 2002		9 October 2002	
Class	Non-Water	Water	Non-Water	Water	Non-Water	Water	Non-Water	Water
Commission (%)	1.09	1.26	0.8	0.79	0.06	0.6	2.08	1.76
Omission (%)	0.15	8.68	0.24	2.62	0.19	0.19	1.69	2.16
Prod. Acc. (%)	99.85	91.32	99.76	97.38	99.81	99.81	98.31	97.84
User Acc. (%)	98.91	98.74	99.2	99.21	99.94	99.4	97.92	98.24
Overall Accuracy (%)	98.89		99.2		99.81		98.08	
Kappa Coefficient	0.94		0.98		0.99		0.96	

The LEI and extended LEI for the four lakes from 2002 to 2013 were reported in Figure 4 and Figure 5. The largest percent change of water area increase for Dongting Lake was 0.4% from 2011 to 2013, when the water area increased by 3.8 km^2 per year. The largest annual decrease of the Dongting water area was 8.0% from 2005 to 2006, corresponding to a decrease of 101.2 km^2 per year. The largest percent change of water area increase for Poyang Lake was 45.1% from 2009 to 2010, an increase of 947.8 km^2 per year. The largest annual decrease in the water area was 24.7% during 2010 to 2011, which corresponded to a decrease of 752.8 km^2 per year. Area changes of Poyang Lake substantially fluctuated, which may be due to two factors. First, Poyang Lake is the largest of the four lakes. Under the condition of the same change rate, the change in area was much larger than the others. Second, severe drought occurred in 2009 and 2011, which also contributed to a large area change in the corresponding time period in Poyang Lake (Figure 4). The subsequent chapters will explain this in detail.

Figure 4: The LEI of the four lakes from 2002 to 2013.

The annual rates of area change for Chaohu Lake were most variable in 2002–2004. During 2002–2003, the water area of Chaohu Lake increased by 7.4 km^2 but then decreased by 8.2 km^2 in 2003–2004. The annual percentage changes of water area for the two years were 0.9% and −1.0%, respectively. Throughout the study period, the annual rate of water area change of Chaohu Lake was ±1.1%. This shows that the water area of Chaohu Lake was stable during the study period. Overall, the patterns of change for Taihu and Chaohu Lakes are similar. The annual percent change for Taihu Lake was ±0.4%. The minimum and maximum values were −0.4% and 0.1%, respectively.

During the 12 years, the difference between the maximum and the minimum areas of Dongting Lake was 460 km^2, and the difference for Poyang Lake was 948 km^2. However, the differences between the maximum and the minimum areas of Taihu Lake and Chaohu Lake were only 22 km^2 and 13 km^2, respectively. Comparing the annual average areas of the four lakes, the annual average for Chaohu and Taihu Lakes are very regular, whereas the seasonal differences for Poyang Lakes is great and Dongting Lake shows a steady decline (Figure 5).

Anomaly is the difference between estimated lake size values and statistical averages of the four lakes. And it is mainly used for volatility observation of sample data relative to the average. By calculating the anomalies of water areas from 2002 to 2013 [40], this study shows that the anomaly in Dongting Lake area declined from 2002 to 2013, even as the anomaly fluctuated considerably over that time. During the period from 2002 to 2006, the water area was generally larger than average, whereas from 2006 to 2013, the water area was statistically significantly smaller than average. Poyang Lake showed a similar trend in water area anomaly, but with a different timing (2007 *vs.* 2006) of the switch from larger to smaller than average anomalies (Figure 6).

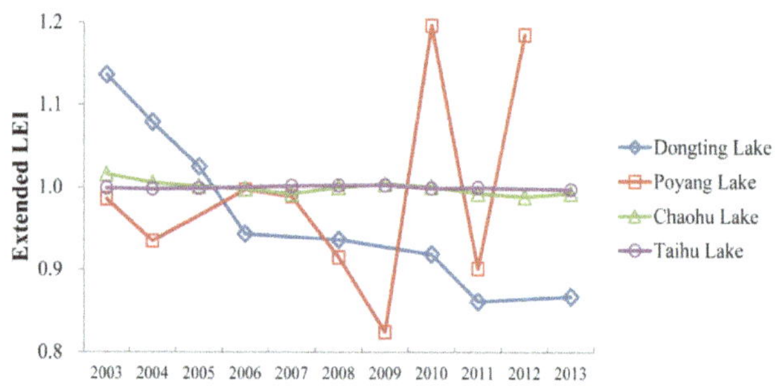

Figure 5: The extended LEI of the four lakes from 2002 to 2013.

The anomaly of Chaohu Lake changed little from 2002 to 2013. The water area fluctuated within 25 km^2 of the annual average area. Before April 2007, the water area tended to be larger than the 2002–2013 average area, and it was smaller than average after August 2007. The trend-line of the anomaly histogram of Taihu Lake shows that the water area changed so little that the trend-line almost overlaps the horizontal axis (Figure 6).

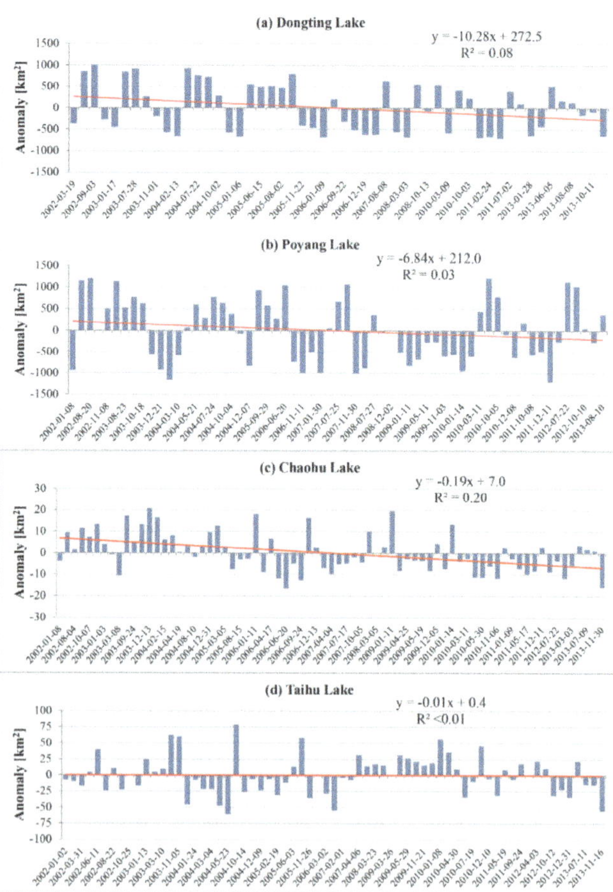

Figure 6: Anomaly of lower Yangtze River lakes from 2002 to 2013 for **(a)** Dongting Lake; **(b)** Poyang Lake; **(c)** Chaohu Lake, and **(d)** Taihu Lake.

Relationship between Rainfall and Lake Area

Annual precipitation and annual average area of the four lakes are significantly correlated. The p-values for Dongting and Poyang Lakes are all less than 0.05, indicating a significant correlation between rainfall and lake area. The p-values of Chaohu and Taihu Lakes are both less than 0.01, indicating highly-significant correlations. The Pearson correlation coefficient r-values of the four lakes are between 0.4 and 0.8, indicating that lake area is positively correlated with precipitation (Figure 7).

It is evident from Figure 7b that Poyang Lake suffered droughts in 2009 and 2011. Rainfall at Poyang Lake decreased, and the Yangtze River water level decreased at the same time, which led to the water of Poyang Lake pouring into the Yangtze River. Thus, in 2009 and 2011, the annual area reached a minimum, resulting in more fluctuation in the area change of Poyang Lake during the period from 2009 to 2013.

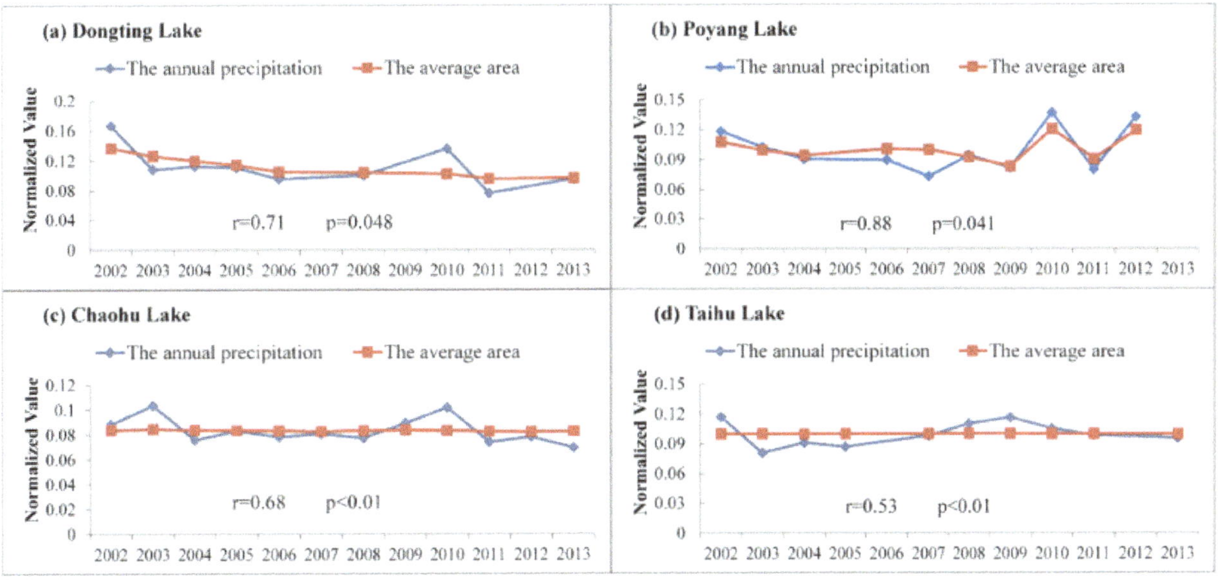

Figure 7: Normalized annual precipitation and average lake area correlation analysis for 2002–2013. (**a**) Dongting Lake; (**b**) Poyang Lake; (**c**) Chaohu Lake; (**d**) Taihu Lake.

Lake Centroid Displacement

The changes of a lake are evident not only in its area but also in its centroid. The spatial analysis shows some directionality in lake expansion and contraction during the study period, and the directionality depends on the centroid displacement of lakes. The results are shown in Figure 8. The initial positions of the centroid of each year are also extracted, connected as a time-sequence and attached with shifting directions. The four lakes are connected to the Yangtze River during all time periods considered in this study. The results show that the lake centroid of Dongting Lake mainly shifted towards the northeast and southwest from 2002 to 2013, whereas the lake centroid of Poyang Lake mainly shifted towards the northwest and southeast. However, the centroids of Chaohu and Taihu Lakes shifted less than 1 km over nearly 12 years, remaining relatively stable. The lake centroid of Chaohu Lake mainly shifted towards the west and the southeast, with distances of 305.7 m and 287.8 m, respectively. The centroid of Taihu Lake mainly shifted towards the west, the northeast, and the southeast, with distances of 989 m, 441.7 m, and 883.2 m, respectively (Figure 8). Although the movement of Poyang and Dongting Lakes seems to be much bigger than Chaohu and Taihu Lakes, comparing to the size of themselves, the changes are not significant (for example, centroid of Poyang Lake shifted about 8 km during 12 years, but this is approximately 1/80 of the length between one side to the other side of Poyang Lake). The centroids of the two lakes shift in the direction of the lakes' outlets, suggesting that the main driver of lake centroid displacement in Dongting and Poyang Lakes is the water complementation between the lakes and the Yangtze River. Unlike Dongting and Poyang Lakes, there is no regularity in the shifts of the centroids of Chaohu and Taihu Lakes.

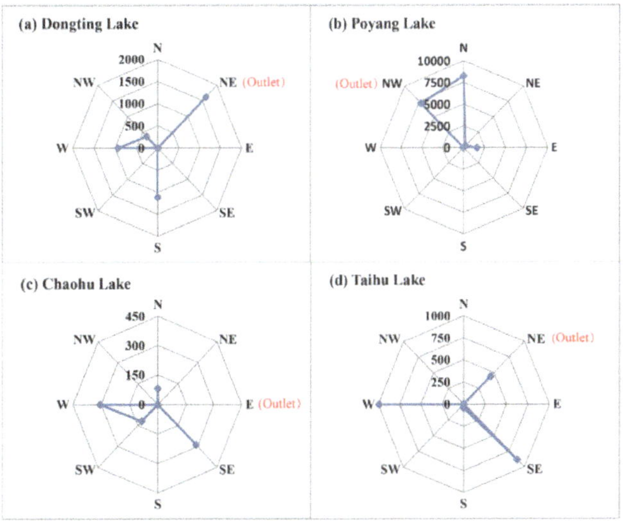

Figure 8: Rose diagrams of the movements of the centroids of the four lakes, (**a**) Dongting Lake; (**b**) Poyang Lake; (**c**) Chaohu Lake; (**d**) Taihu Lake. The units are in meters.

DISCUSSION

Different Size-Distribution Variations for Four Lakes

Both the flood season and the dry season water areas of Dongting Lake have been decreasing during the 12 years since the Three Gorges Project started to store water. However, Poyang Lake, which has the second shortest distance from the Three Gorges Project and has a similar morphology to Dongting Lake, is very different from Dongting Lake in the temporal and spatial patterns of water area. The water area of Poyang Lake in the flood and dry seasons has been increasing from 2002 to 2013. Moreover, there are only slight fluctuations of the wet and dry seasons for Chaohu Lake and Taihu Lake which can be seen clearly from the slopes of the wet and dry season fitting lines in Figure 3c,d, which are all less than 0.01. Additionally, Chaohu and Taihu Lakes have low shifts comparing to Dongting and Poyang Lakes.

Figure 6 shows that there are many anomaly values larger than 1000 km^2 or less than −1000 km^2 in Poyang Lake, while values in Dongting Lake are all within ±1000 km^2. This indicates that Poyang Lake might have different size change trends from Dongting Lake, although they are very similar in morphology. Related to the precipitation analysis, it could be argued that the different climate in these two lakes is a major cause for the different size change. Weather might be the main factor for the quite different variation of Poyang Lake from the other lakes. It also can be seen from the correlation analysis of precipitation and area that the area of Poyang Lake and rainfall are synchronous, especially during 2008 to 2013 (Figure 7). Furthermore, the annual inflows into Poyang Lake is about 16% of the annual runoff of the Yangtze River, which means the water inflows of Poyang Lake are the highest among the four lakes. In addition, Poyang Lake is an important protected area in China. From the end of the last century to the beginning of the 21st century, farmers returned to the fields in lakeside areas to Poyang Lake in large scales and a large number of embankments on the low-lying land were demolished. Overall, the unique character of size variation of Poyang Lake might be caused by precipitation, water inflows from the Yangtze River, and government policy.

The morphology and topography of the four lakes may also impact the centroid changes. Dongting Lake and Poyang Lake are leaf-shaped lakes with a lot of shallow areas, and their outlets are close to the Yangtze River, allowing water from the lake and the river to complement each other during the dry and the wet season in time. This results in the sizes of the two lakes being related to precipitation and the centroids of the two lakes moving back and forth towards or away from the outlets (Figure 9). In contrast, Chaohu Lake and Taihu Lake are block-shaped lakes, and their outlets are far from the Yangtze River. Therefore, they seem to show almost no response to precipitation, despite this relationship being significant and their lake centroids are not regularly displaced towards the outlet.

Dam Influence or Dissipate Influence on Lakes

Poyang Lake suffered large variations in rainfall during some periods of time, resulting in substantial lake area fluctuations, which can also cause substantial fluctuations in the surrounding ecological parameters, and cause losses of lake products. Therefore, lake area monitoring using multi-temporal remote sensing data can be used for assisting the management of upstream dams, such as increasing the drainage capacity of the dam in severe drought years but increasing the storage capacity of the dam in flood years. Implementation of such management of dams can preserve the surface area stability and sustainability of downstream lakes.

The impact of the Three Gorges Dam downstream has been a weakening in the Yangtze River forcing on the lakes downstream, allowing more lake flows to the river from July to March. This effect might also partially complete the dam's mission to mitigate flood risks in the lake basin, especially during the peak wet season of the Yangtze River basin from July to September. When the flood season faded away, the water would be kept from flowing downstream. When the dry season comes, the water storage, which was blocked before, could feed the lakes downstream for mitigating drought risks. These measures for dam management can make the downstream lake areas keep in a stable range.

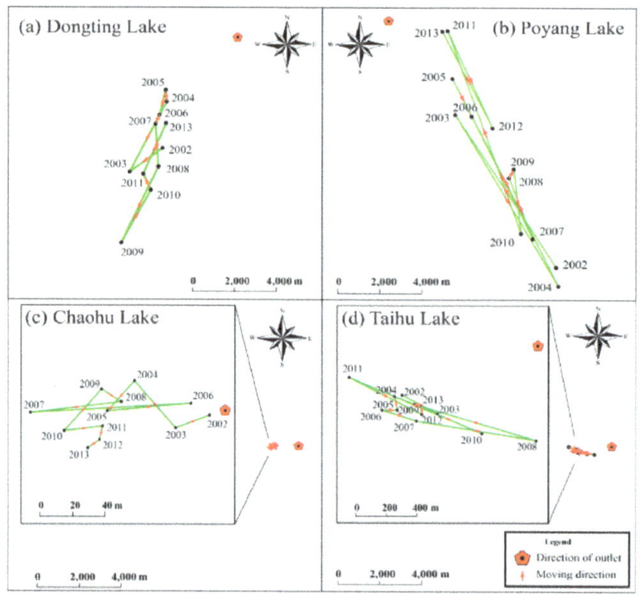

Figure 9: The movement of the centroids of the four lakes, (**a**) Dongting Lake; (**b**) Poyang Lake; (**c**) Chaohu Lake; (**d**) Taihu Lake.

Lake Monitoring and Other Applications

The Landsat TM and ETM sensors provided available measurements per 16 days, even though, as a result of high evaporation in the lake regions, there was sometimes only three or four cloud-free images available during a year. Although an insufficient number of images may lead to some uncertainties in the lake size statistics, the observation frequency of at least two images covering dry and wet seasons, after discarding cloudy scenes, could meet the requirements for this research. Higher frequency remote data can be applied to our research method for lake size monitoring, as well. For example, HJ-1A/1B CCD instruments provided one image with resolution of 30 m every two days [20], and MODIS instruments provided two daytime images per day [18,19]. These data could be better used in monitoring the long-term lake size and distribution changes in the future.

The results of the four lake size-distribution evolution presented here also have implications for water resource monitoring and management. This is very important because the lakes benefit the society and ecology in the surrounding regions [31,41], and lake water is the most important water supply in aquaculture, industry, and domestic water use around the four lakes regions [42,43]. The 12-year monitoring data record can serve as additional data to assess and quantify the seasonal water conditions around the lakes. For example, a comprehensive water pollution index (WPI) was used to describe water quality and identified the major pollutants in 28 Chinese large lakes and reservoirs [44]. Annual variations of water level in Dongting Lake during the period of 1961–2010 were analyzed to determine whether anthropogenic or climatic factor should be responsible for the variations [45].

CONCLUSIONS

China is rich in surface water, especially in the middle and lower Yangtze River Basin downstream of the Three Gorges Project. Dongting, Poyang, Chaohu, and Taihu Lakes are the largest natural lakes in this region. The temporal-spatial evolutions of these lakes largely represent the overall trends of other lakes in this basin [46,47]. Water extraction methods of remote sensing technology provide the scientific basis for long-term monitoring and analysis of variation for lakes in the middle and lower Yangtze River areas.

In this paper, the areas of the four lakes are extracted from 12 years of multi-temporal Landsat TM/ETM satellite image data with an overall accuracy of higher than 98%. The changes in the four natural lakes are analyzed from several approaches, including lake area statistics from 2002 to 2013, lake expansion index, correlation analysis between rainfall and lake area, and lake centroid displacement. Overall, this study provides the following conclusions:

The annual average areas of Dongting, Chaohu, and Taihu Lakes decreased respectively 452, 11, and 5 km^2 (29.6%, 1.4% and 0.2%) from 2002 to 2013, while Poyang Lake shows a increasing of 300 km^2 (11.0% of the total area in 2002), which is related to the different precipitation in great extent. If the decreasing trend continues, there is cause for concern for the environment of the three lakes in the future, especially the Dongting Lake. Therefore, timely and appropriate measures need to be taken by policy-makers to stop the drop in lake area and restore the original environment.

Furthermore, our study also revealed that regional climate and morphology of lake are two key factors for the spatio-temporal evolution of lake size-distribution in the middle and lower reaches of the Yangtze River. Size variation of leaf-shaped lakes with a lot of shallow areas, with outlets close to the river, is closely related to the precipitation (with correlation coefficient of 0.71 and 0.88, respectively, in Dongting and Poyang Lakes) and the inflow/outflow of the river. Meanwhile, size variation of block-shaped lakes with little shallow areas, with outlets far from the river, is much depended on the precipitation and the consumption of water.

However, in this study, size variation and centroid movement of the four lakes were monitored from the period after the dam's water storage. Variability in lake size and distribution showed certain trends over the time series, but climate change is a slow process that can be expected to occur over an extended time. Contextualizing lake behavior pre- and post-dam within long term trends in climate will require analysis over the total time span of the Landsat archive and is that we hope to investigate in follow-up studies.

ACKNOWLEDGMENTS

This work was supported by the National High Technology Research and Development Program (863 Program) of China (No. 2013AA12A202), the Special Fund for Surveying, Mapping and Geographical Information Scientific Research in the Public Interest (No. 201412014), and the National Natural Science Foundation of China (No. 41271458). Cartolab of Wuhan University is hereby acknowledged for providing the facilities used during this research and for supporting the research. The authors are also grateful to the U.S. Geological Survey (USGS) for providing the multi-temporal Landsat data and to the China Meteorological Data Sharing System for providing the precipitation data.

AUTHOR CONTRIBUTIONS

This research was mainly conducted and the article prepared by Lin Li and Hui Xia. Lin Li supervised this research, and his comments were considered throughout the paper. Hui Xia and Zheng Li collected, analyzed and processed data, including the satellite imagery, and contributed to the data analysis. Zhijun Zhang contributed to the correlation analysis.

REFERENCES

1. Milesi, C.; Samanta, A.; Hashimoto, H.; Krishna Kumar, K.; Ganguly, S.; Thenkabail, P.S.; Srivastava, A.N.; Nemani, R.R.; Myneni, R.B. Decadal variations in NDVI and food production in India. *Remote Sens.* 2010, *2*, 758–776.
2. Thenkabail, P.S.; Dheeravath, V.; Biradar, C.M.; Gangalakunta, O.R.P.; Noojipady, P.; Gurappa, C.; Velpuri, M.; Gumma, M.; Li, Y. Irrigated area maps and statistics of India using remote sensing and national statistics. *Remote Sens.*2009, *1*, 50–67.
3. Akdim, N.; Alfieri, S.M.; Habib, A.; Choukri, A.; Cheruiyot, E.; Labbassi, K.; Menenti, M. Monitoring of irrigation schemes by remote sensing: Phenology *versus* retrieval of biophysical variables. *Remote Sens.* 2014, *6*, 5815–5851.
4. Alderman, K.; Turner, L.R.; Tong, S. Floods and human health: A systematic review. *Environ. Int.* 2012, *47*, 37–47.
5. Bond, N.R.; Lake, P.; Arthington, A.H. The impacts of drought on freshwater ecosystems: An Australian perspective. *Hydrobiologia* 2008, *600*, 3–16.
6. Charron, D.F.; Thomas, M.K.; Waltner-Toews, D.; Aramini, J.J.; Edge, T.; Kent, R.A.; Maarouf, A.R.; Wilson, J. Vulnerability of waterborne diseases to climate change in Canada: A review. *J. Toxicol. Environ. Health Part A* 2004, *67*, 1667–1677.
7. Davranche, A.; Lefebvre, G.; Poulin, B. Wetland monitoring using classification trees and SPOT-5 seasonal time series. *Remote Sens. Environ.* 2010, *114*, 552–562.

8. Li, K.; Wu, S.; Dai, E.; Xu, Z. Flood loss analysis and quantitative risk assessment in China. *Nat. Hazards* 2012, *63*, 737–760.

9. Muala, E.; Mohamed, Y.A.; Duan, Z.; van der Zaag, P. Estimation of reservoir discharges from Lake Nasser and Roseires Reservoir in the Nile Basin using satellite altimetry and imagery data. *Remote Sens.* 2014, *6*, 7522–7545.

10. Muster, S.; Heim, B.; Abnizova, A.; Boike, J. Water body distributions across scales: A remote sensing based comparison of three arctic tundra wetlands. *Remote Sens.* 2013, *5*, 1498–1523.

11. Wei, W.; Pengfeng, X.; Xuezhi, F.; Hui, L.; Ronghua, M.; Hongtao, D. Remote sensing analysis for changes of lakes in the southeast of Qiangtang area, Qinghai-Tibet plateau in recent 30 years. *J. Lake Sci.* 2010, *22*, 874–881. (In Chinese).

12. Jiang, H.; Feng, M.; Zhu, Y.; Lu, N.; Huang, J.; Xiao, T. An automated method for extracting rivers and lakes from Landsat imagery. *Remote Sens.* 2014, *6*, 5067–5089.

13. Klein, I.; Dietz, A.; Gessner, U.; Dech, S.; Kuenzer, C. Results of the global waterpack: A novel product to assess inland water body dynamics on a daily basis. *Remote Sens. Lett.* 2015, *6*, 78–87.

14. Klein, I.; Dietz, A.J.; Gessner, U.; Galayeva, A.; Myrzakhmetov, A.; Kuenzer, C. Evaluation of seasonal water body extents in central Asia over the past 27 years derived from medium-resolution remote sensing data. *Int. J. Appl. Earth Obs. Geoinform.* 2014, *26*, 335–349.

15. Meng, L.; Guo, S.; Li, S. Summary on extraction of water body from remote sensing image and flood monitoring. *Water Resour. Informatiz.* 2012, *3*, 18–25.

16. Mizuochi, H.; Hiyama, T.; Ohta, T.; Nasahara, K.N. Evaluation of the surface water distribution in north-central Namibia based on MODIS and AMSR series. *Remote Sens.* 2014, *6*, 7660–7682.

17. Rokni, K.; Ahmad, A.; Selamat, A.; Hazini, S. Water feature extraction and change detection using multitemporal Landsat imagery. *Remote Sens.* 2014, *6*, 4173–4189.

18. Feng, L.; Hu, C.; Chen, X.; Cai, X.; Tian, L.; Gan, W. Assessment of inundation changes of Poyang Lake using MODIS observations between 2000 and 2010. *Remote Sens. Environ.* 2012, *121*, 80–92.

19. Pekel, J.F.; Vancutsem, C.; Bastin, L.; Clerici, M.; Vanbogaert, E.; Bartholomé, E.; Defourny, P. A near real-time water surface detection method based on HSV transformation of MODIS multi-spectral time series data. *Remote Sens. Environ.* 2014, *140*, 704–716.

20. Lu, S.; Ouyang, N.; Wu, B.; Wei, Y.; Tesemma, Z. Lake water volume calculation with time series remote-sensing images. *Int. J. Remote Sens.* 2013, *34*, 7962–7973.

21. Haas, E.M.; Bartholomé, E.; Combal, B. Time series analysis of optical remote sensing data for the mapping of temporary surface water bodies in Sub-Saharan western Africa. *J. Hydrol.* 2009, *370*, 52–63.

22. Jain, S.K.; Saraf, A.K.; Goswami, A.; Ahmad, T. Flood inundation mapping using NOAA AVHRR data. *Water Resour. Manag.* 2006, *20*, 949–959.

23. Feyisa, G.L.; Meilby, H.; Fensholt, R.; Proud, S.R. Automated water extraction index: A new technique for surface water mapping using Landsat imagery. *Remote Sens. Environ.* 2014, *140*, 23–35.

24. Gao, Y.; Su, F.; Sun, X.; Xue, Z.; He, Y. A study on spatial and temporal variations of coastal wetland in Pearl River estuary. In Proceedings of the 2010 IEEE International Geoscience and Remote Sensing Symposium (IGARSS'20), Honolulu, HI, USA, 25–30 July 2010; IEEE: New York, NY; USA, 2010; pp. 429–432.

25. Junhong, B.; Hua, O.; Baoshan, C.; Qinggai, W.; Hui, C. Changes in landscape pattern of alpine wetlands on the Zoige Plateau in the past four decades. *Acta Ecol. Sin.* 2008, *28*, 2245–2252.

26. Global Visualization Viewer. Available online: http://glovis.Usgs.Gov/ (accessed on 30 October 2014).

27. Chander, G.; Markham, B.L.; Barsi, J.A. Revised Landsat-5 thematic mapper radiometric calibration. *IEEE Geosci. Remote Sens. Lett.* 2007, *4*, 490–494.

28. Huang, C.; Wylie, B.; Yang, L.; Homer, C.; Zylstra, G. Derivation of a tasselled cap transformation based on Landsat 7 AT-satellite reflectance. *Int. J. Remote Sens.* 2002, *23*, 1741–1748.

29. Calibrating Images Tutorial. Available online: http://www.Exelisvis.Com/docs/radiometriccalibration.Html (accessed on 30 October 2014).

30. Atmospheric Correction. Available online: http://www.Exelisvis.Com/docs/atmosphericcorrection.Html#flaash(accessed on 30 October 2014).

31. Chen, Y.; Huang, C.; Ticehurst, C.; Merrin, L.; Thew, P. An evaluation of MODIS daily and 8-day composite products for floodplain and wetland inundation mapping. *Wetlands* 2013, *33*, 823–835.

32. Guerschman, J.P.; Warren, G.; Byrne, G.; Lymburner, L.; Mueller, N.; van Dijk, A.I. *MODIS-Based Standing Water Detection for Flood and Large Reservoir Mapping: Algorithm Development and Applications for the Australian Continent*; CSIRO: Canberra, ACT, Australian, 2011.

33. McFeeters, S. The use of the normalized difference water index (Ndwi) in the delineation of open water features. *Int. J. Remote Sens.* 1996, *17*, 1425–1432.

34. Xu, H. Modification of normalised difference water index (Ndwi) to enhance open water features in remotely sensed imagery. *Int. J. Remote Sens.* 2006, *27*, 3025–3033.

35. Karlsson, J.M.; Lyon, S.W.; Destouni, G. Temporal behavior of lake size-distribution in a thawing permafrost landscape in northwestern Siberia. *Remote Sens.* 2014, *6*, 621–636.

36. China Meteorological Data Sharing Service System. Available online: http://www.Cma.Gov.Cn/2011qxfw/2011qsjgx/(accessed on 30 October 2014).

37. Wen, X.; Hu, D.; Cao, B.; Shen, S.; Tang, X. Dynamics

change of Honghu Lake's water surface area and its driving force analysis based on remote sensing technique and topmodel model. In Proceedings of the IOP Conference Series: Earth and Environmental Science, Bendigo, VIC, Australia, 24–27 March 2014; IOP Publishing: Bristol, UK, 2014; p. 012130.

38. Zhang, F.; Tiyip, T.; Feng, Z.; Kung, H.T.; Johnson, V.; Ding, J.; Tashpolat, N.; Sawut, M.; Gui, D. Spatio-temporal patterns of land use/cover changes over the past 20 years in the middle reaches of the Tarim River, Xinjiang, China. *Land Degrad. Dev.* 2013, *26*.

39. Zhang, J.; Huang, F. Monitoring forest and its ecosystem services value changes in Tonghua City using Landsat TM images. In Proceedings of the 2011 4th International Congress on Image and Signal Processing (CISP), Shanghai, China, 15–17 October 2011; IEEE: New York, NY; USA, 2011; pp. 1641–1645.

40. Conway, D. Extreme rainfall events and lake level changes in East Africa: Recent events and historical precedents. In *The East African Great Lakes: Limnology, Palaeolimnology and Biodiversity*; Springer Netherlands: New York, NY, USA, 2004; pp. 63–92.

41. Ye, X.; Zhang, Q.; Liu, J.; Li, X.; Xu, C.-Y. Distinguishing the relative impacts of climate change and human activities on variation of streamflow in the Poyang Lake catchment, China. *J. Hydrol.* 2013, *494*, 83–95.

42. Elsawwaf, M.; Feyen, J.; Batelaan, O.; Bakr, M. Groundwater-surface water interaction in Lake Nasser, southern Egypt. *Hydrol. Process.* 2014, *28*, 414–430.

43. Li, X.-Y.; Xu, H.-Y.; Sun, Y.-L.; Zhang, D.-S.; Yang, Z.-P. Lake-level change and water balance analysis at Lake Qinghai, west china during recent decades. *Water Resour. Manag.* 2007, *21*, 1505–1516.

44. Meng, X.; Zhang, Y.; Yu, X.; Zhan, J.; Chai, Y.; Critto, A.; Li, Y.; Li, J. Analysis of the temporal and spatial distribution of lake and reservoir water quality in China and changes in its relationship with GDP from 2005 to 2010. *Sustainability*2015, *7*, 2000–2027.

45. Yuan, Y.; Zeng, G.; Liang, J.; Huang, L.; Hua, S.; Li, F.; Zhu, Y.; Wu, H.; Liu, J.; He, X. Variation of water level in Dongting Lake over a 50-year period: Implications for the impacts of anthropogenic and climatic factors. *J. Hydrol.*2015, *525*, 450–456.

46. Liu, W.; Zhang, Q.; Liu, G. Influences of watershed landscape composition and configuration on lake-water quality in the Yangtze River basin of China. *Hydrol. Process.* 2012, *26*, 570–578.

47. Long-sheng, C. Environment significance and pollution status of heavy metals in lake sediments of the middle and lower reaches of Yangtze River. *J. Anhui Agric. Sci.* 2013, *16*, 093.

CHAPTER 12 WALL CLIMBING ROBOTS: RECENT ADVANCES AND CHALLENGES

Shunsuke Nansai and Rajesh Elara Mohan

Keywords: Wall climbing robot, façade cleaning robot, nested reconfigurable design principle

INTRODUCTION

Robots have been advancing exponentially over the last three decades, moving beyond the traditional bounds of industrial applications into service missions sharing social spaces with humans. Frey and Osborne have estimated that 47% of total U.S. employment will be replaced by robots and/or artificial intelligence (AI) in the near future [1]. Early evidence points to significant improvement in productivity and safety over a number of service tasks that are dull, dirty and/or dangerous [2]. Façade cleaning of high-rise buildings and skyscrapers offers enormous opportunities for the use of robots. In recent decades, skyscrapers, as represented by the Burj Khalifa [3,4,5,6] in Dubai and Shanghai Tower [7,8,9] in Shanghai, have been built due to improvements in the construction technologies and processes. Even in such newfangled skyscrapers, the façades are generally cleaned by humans. In the case of some skyscrapers, including the Burj Khalifa, equipment, such as gondolas, is not installed, and humans are employed to clean the façade by hanging down from a rope. Use of manual workforce poses a high degree of the risk of falling, accidents and even resulting in human fatalities. Numerous incidents of accidents have been reported even with the use of gondolas in façade cleaning jobs. One such situation involved a gondola being uncontrollable due to a gust of wind at Shanghai World Financial Center [10]. In another instance, the gondola became suspended in mid-air at a 240-m height at One World Trade Center in New York [11]. Robotic solutions offer enormous potential in significantly minimizing risk to humans, as well improving productivity in façade cleaning jobs.

One of the fundamental components of any façade cleaning robot is its adhesive mechanism, which enables the robot to attach itself to the wall surface. Such a mechanism should also be adaptive enough to allow the robot to be mobile so as to cover the required area. This paper presents a detailed review of such adhesive mechanisms for wall climbing robots, categorizing them into six distinct classes. This paper concludes by expanding beyond adhesive mechanisms by discussing a set of the desirable design attributes of an ideal glass façade cleaning robot towards facilitating targeted future research with clear technical goals and well-defined design trade-off boundaries.

WALL CLIMBING METHOD

A number of literature works on wall climbing robots have been reported. This section categorizes the adhesion principles in wall climbing methods and summarizes the robot precedence for each of the adhesion principles. The six adhesion principles identified include suction cup adhesion, suction cup crawler, vacuum pump adhesion, magnetic adhesion and bio-inspired adhesion. Each of these principles presents a unique set of advantages and disadvantages viz-à-viz the nature of traversing terrain and application settings.

Suction Cup Adhesion

This method works effectively for fine-textured wall surfaces with less asperities, such as in the case of glass.

Citation: Shunsuke Nansai and Rajesh Elara Mohan, Wall Climbing Robots: Recent Advances and Challenges, doi:10.3390/robotics5030014.

A series of Skycleanershave been developed to clean and locomote on a glass wall by utilizing vacuum suction cups [12,13]. The Skycleaner is installed with the suction cups on both ends of actuators, like an X-Y stage, and these suction cups are connected to a vacuum pump [14]. Additionally, the robot adheres to the glass wall using the suction force. One of the important characteristics in this case is that the robot is totally driven by pneumatic actuators. The use of pneumatic actuators offer the advantage of payload reduction. The segment variable bang-bang control has been proposed in order to control the position of the pneumatic actuator precisely [13,15,16]. Another precedent involving a wall climbing robot with an ankle joint is presented [17]. Thus, even if the façade design of the building has curvature and a small angle exists between neighboring glasses, such robots would still be capable of adapting to such situations [18]. Furthermore, the trajectory generation method, which covers all areas of the wall, and its evaluation method have been proposed [19,20].

Wang et al. have proposed a new suction method utilizing the suction cup [21]. By applying a vacuum pump principle, the suction cup is capable of keeping its inward vacuum continuously, even if the air leaks from outside. In particular, sinusoidal or square wave vibration is added to the suction cup. The proposed method has been analyzed utilizing Bernoulli's principle; the suction pressure depending on vibrations with different parameters has been formulated, and the effectiveness of the proposed method has been verified through the experimental results. The experiments also have shown that the proposed method is effective on a glass wall, which allows the suction cup to keep the suction force on a variety of surfaces, including concrete, cabinets and wooden board.

Kawasaki et al. have developed a hexapod wall climbing robot with the suction cups on its toes [22]. The robot adopts a parallel linkage mechanism for the leg, which is optimized by the genetic algorithm (GA), and is able to realize wall climbing by using only one degree-of-freedom. In addition, the prototype has been implemented, and it has been verified that the wall climbing locomotion can be realized through a series of experiments.

Yano et al. have developed a biped-type wall climbing robot utilizing scanning type suction cups [23]. The adhesive mechanism consists of the suction cups, cam mechanism and a vacuum pump. A key characteristic of the suction system is to be able to keep the suction force, even if there is small steps or cracks, with the level change of the suction surface. As the mechanical feature, a valve that is activated by the cam mechanism is installed between the suction cups and the vacuum pump. With this cam mechanism, the valve is opened periodically, and the vacuum pump generates the suction force to provide the adhesive force needed to attach to the wall surface [24]. The effectiveness of the developed self-contained robot has been demonstrated through an exhaustive set of experiment [25].

Zhu et al. have developed a wall climbing robot that uses the suction cup principle. The developed platform is capable of self-localization and obstacle avoidance utilizing a CCD camera and a pair of laser diodes [26]. For achieving locomotion, the robot is equipped with an X-Y stage mechanism, as well as a rotational joint. Furthermore, small linear actuators are installed into the suction cup, which enables vertical motion. In addition, this work also presents a window frame crossing method. The effectiveness of the developed robot and the proposed strategies has been demonstrated through field trials at both of the City University of Hong Kong and the British Broadcasting Corporation (BBC) [27]. The efficacy and validity of the developed system have been demonstrated through a range of experiments [28].

From the above set of literature works, it is evident that leg-based mechanisms are often preferred while using the suction cup adhesion principle. Both the Skycleaner robot series, as well as the robot developed by Zhu et al. adopt a multi-legged tripod gait with respect to the robot's locomotion [29,30,31]. Another robot platform developed by Kawasaki et al. possesses a morphology of hexapod robots [32,33,34]. Furthermore, the robot developed by Yano et al. uses a biped structure.

The literature presented also points to this method being effective for a fine textured wall surface with less asperities. Suction cups possess an internal structure that easily lets air out, but restricts air incursion, since its edge faces outside. Thereby, it generates high levels of negative pressure inside the chamber; i.e., a highly normal force is seamlessly generated against a wall surface. It prevents the suction cups from slipping, as it is equivalent to generating a high friction force between the robot and the wall surface. With this mechanisms, the control of legged robots as summarized above on the vertical wall surface becomes very close to that in the case of the horizontal floor. Given that the suction cups are constantly moving together with the robot platform, the performance of the suction mechanism greatly depends on the nature of the traversed terrain. One critical deficiency of this mechanisms is the depression of the negative pressure due to air leaking. To overcome this particular issue, both the Skycleaner series and the robot developed by Zhu et al. connect vacuum pumps to the suction cups to generate negative pressures. Wang et al. have proposed a method that generates negative pressure by pressing the suction cup periodically based on the principle of the vacuum pump. Yano et al. have proposed the scanning type suction cup and demonstrated the ability of the developed robot to maintain the suction force on a bubblysurface. Another common disadvantage while using suction cup adhesion principles is that of slow moving speed. In the case of ground-based legged robots, they can switch the idling leg and the supporting leg smoothly as the platform guarantees the friction forces of the toes by utilizing its own weight. However, in the case of wall climbing robots that use the suction cup adhesion principle, there is an additional time delay arising from the need to vary the negative pressure inside the suction cups for its control.

Suction Cup Crawler Adhesion

This method involves installing the suction cups on the tracks of a crawler. The mobility of the robot is achieved through the use of the crawler rotation while maintaining the suction forces. Hence, this method overcomes the low speed issue faced in the case of the suction cup adhesion principle.

Kim et al. have developed a wall climbing robot using the suction cup crawler principle wherein mechanical valves are installed onto the suction cups [35]. The robot platform uses two tracks with twelve suction cups on each track, and these suction cups are activated by a novel mechanical valve design [36]. The mechanical valve is composed of a spring and a cam mechanism. The valve is opened mechanically while the suction cup moves to the bottom of the track. This work also features the optimization of the vacuum pressure using the Taguchi method [37,38]. In addition, efforts have been taken to modularize the robot design, and a platform with three core modules has been developed [39]. Each module is equipped with the above-mentioned suction cup crawlers, by controlling each of these modules based on kinematic analysis, a transition control between two orthogonal surfaces has been realized and validated through field experiments with the developed robot [40].

Serbot AG, an industrial company in Switzerland, has developed a wall and solar panel cleaning robot, GEKKO [41]. GEKKO is equipped with a vacuum suction cup crawler, which rotates horizontally. The crawler has a semi-circular shape, which has a straight line on one side and an arc line on the other side, with both linear and rotational motions being realized. By changing the number of brushes and the crawlers, the developed robot can be manually reconfigured to adapt to different wall areas.

Zhu et al. have developed a wall climbing robot with suction cup installed crawlers [42]. The robot has two crawlers, and each crawler has installed two columns of suction cups. A steering mechanism is installed using a front pulley, and the robot is capable of limited turning motion.

Shafts supporting the suction cup are equipped with springs that enable the robot to overcome small obstacles, such as the window frame. In addition, a vision system utilizing a CCD camera is installed to perceive the external environment. This work also presents the kinematics and the stability analysis of the developed robot mechanism. Experiments have been conducted to analyze the payload capacity and safety factor for the developed robot [43].

The suction cup crawler principle greatly improves the slow speed issue present in the case of Section 2.1 by varying negative pressure while the suction cups are passing under the crawlers. This principle allows for increased payload and stability, as the suction force is maintained even if air leaking occurs in some suction cups due to asperity. One common deficiency with this approach is the difficulty in turning. This is due to the high frictional force, which prevents slipping, as mentioned in Section 2.1, while the latter is needed in order to turn. To overcome this issue, Kim et al. have adopted a modular-type design and installed a waist joint. In the case of Zhu, a steering mechanism on the front pulley has been designed to overcome the turning issue. GEKKO has a horizontal crawler track, which is semi-circular in shape, where both linear and turning motions have been realized by switching the suction mechanism accordingly.

Vacuum Pump Adhesion

This method uses a negative pressure between the robot and the wall generally through the use of a pump. Numerous precedents exist for this principle demonstrating very high levels of mobility. Gao et al. have developed two kinds of wall climbing robots utilizing the wheel system with negative pressure generated by a vacuum pump which acts both as an impeller or a propeller [44]. The robots basically consist of one sealed central body, two wheels for mobility and two vacuum pumps [45]. For propeller systems, the generated negative pressure from the propeller and interspace between the seal and the wall is calculated and analyzed based on the dynamics and the relationships between the robot parameters and the air leaking into vacuum chamber [46]. A time constant and a responsiveness curve are also simulated utilizing an analogy between fluid dynamics and the electric circuit [47]. As for the impeller system, the suction principle of the impeller is elucidated, and the stability of the robot is discussed based on the static and dynamic analysis on the wall [48]. The technical specifications of the robot is decided based on these analyses, and the minimum suction force needed to adhere the robot to the wall is calculated [49].

Miyake et al. have developed a small wall climbing robot that is capable of being used in domestic indoor environments by improving the sealing ability and decreasing the friction forces of the suction cup through the use of a liquid substance [50,51]. Experiments have been performed to validate the relationships between the friction forces and the suction force of the vacuum system on a wet surface [52,53]. A novel suction system based on this method has been developed in order to be integrated with the robot [54,55]. Robot demonstrations involving successful integration of the vacuum pump adhesion onto a wheeled robot and locomotion trials have been performed. A trajectory to clean a rectangle-shaped window has been generated utilizing a trajectory generation method based on patterns, and the effectiveness of the robot has been demonstrated through real experiments [56,57].

Zhao et al. have developed a wall climbing robot with a vacuum pump and a circular body [58]. The robot consists of a vacuum pump, circular body, four wheels and a sealing module. A mechanism that uses both air and regulating springs has been designed to achieve adjustable negative pressure in a vacuum chamber. In addition, an airflow model has been developed by utilizing fluid network theory, and its characteristics have been discussed in detail [59]. It has been demonstrated that the developed platform reliably adheres itself to wall surfaces, as well as achieves mobility [60].

This method offers high levels of mobility by utilizing commonly-used mobility mechanisms, such as wheels. Such an approach greatly reduces the control complexity as the locomotion control theories from general purpose wheeled robots can be directly applied. Such robots have to be able to maintain optimal negative pressure, as lower values of negative pressure result in the robot slipping off, whereas higher values of negative pressure inhibit robot mobility. To overcome such issues, Gao et al. and Zhao et al. have modeled the negative pressure generated by the vacuum pumps by utilizing the fluid network theory. Simulation studies has been run to study the performance and stability of the robot under varying negative pressure conditions. Miyake et al. have decreased the friction force of the suction cup by utilizing a liquid substance. The use of a liquid-based substance also increases the sealing ability of the suction cup.

Magnetic Adhesion

This method is often adopted for walls that have high levels of magnetic permeability. Since most of the prior work in this area uses permanent magnets, this eliminates the need for any additional devices, such as power supply. This results in improved payload capacity.

Gao et al. have developed a wall climbing robot for the maintenance of boiler water-cooling tubes [44]. Theoretical studies have been undertaken to analyze the situations that lead to robots falling off based on which a physical robotic platform using permanent magnets has been designed [61]. Experiments have been performed at a thermal power plant to validate the merits of the proposed approach [62].

Lee et al. have developed a modular wall climbing robot with a crawler utilizing permanent magnets [35,63,64]. This platform is characterized by its high payload capacity and wall transition ability. A robot platform comprising six links has been developed based on rigorous static and dynamic analysis [65]. A second iterated robot platform has been developed taking a modular approach comprising three torsos, a head-mounted arm and a tail-mounted arm [66]. Links connecting each module consist of a passive joint with torsion springs and an active joint [67]. The use of passive joints and tail-mounted arms to achieve the transition between two wall surfaces has been demonstrated with rigorous static and dynamic analysis [68].

Xu et al. have developed a wall climbing robot with magnetic tracks [69]. The permanent magnets and metal blocks are placed alternatively in the crawlers mounted on the robot. A novel elastic torso has been designed that allows for traversing concave-convex wall surfaces. In addition, a parallelogram mechanism has been adopted to further improve the adaptability of the robot. A rigorous study on system safety has been done as part of this research work.

This method does not require any additional devices, such as a power supply, to navigating wall surfaces, as only permanent magnets are needed to generate the required suction force. This load saving helps with improving the payload capacity, as witnessed in the literature summarized above. The major drawback with this approach is that the robot can only operate on wall surfaces with high levels of magnetic permeability. Even with metal surfaces, there is a high risk of damage due to robot slippage. In order to overcome this problem, Gao et al., Lee et al. and Xu et al. have extensively studied the slippage and stability in robots that use the magnetic adhesion principle.

Rope and/or Rail Gripping

This method utilizes a rope ascender from the upper section or specialized equipment installed on the wall to support the navigating robot platform. Humans generally adopt this technique when they clean façade windows. This approach ensures a high degree of safety, as the robot is always secured to the wall or roof through a rope or cable, which at times even helps with robot mobility.

Zhang et al. have developed a wall climbing robot in order to clean a spherically-shaped façade [70]. The developed robot is aimed at use at the International Grand Theatre in China. The developed robot consists of two frames, and by use of which, it can climb up the spherically-shaped façade by gripping the sliding rods, which are equipped on the wall surface [71,72,73]. Stability and safety have been studied through extensive kinematic and dynamic analysis both during climbing and cleaning phases [74]. The effectiveness of the developed robot and the proposed climbing approach has been verified through a number of practical tests.

Wang et al. have developed a cleaning robot for a reversed inclining glass façade [75]. The targeted infrastructure is the airport control tower at the Guangzhou Airport, where wall mounted rail lines and a conveyor belt mechanism have been put in place to support the mobility of the robot, as well as deliver power, detergent liquid and water needed for cleaning. The cleaning unit adheres and cleans the glass wall based on the vacuum pump principle, while the robot has two suction areas to overcome the asperity of the surface and gaps between the windows. Thus, the robot is capable of adhering safely to the glass surface even if the air leaks from either one of the suction areas.

Seo et al. has developed ROPE RIDE [76]. The ROPE RIDE enables climbing up a vertical surface by utilizing a rope dropped down from the top of the building, and two additional propeller thrusters are installed to adhere on the wall with certainty. Triangular-shaped crawlers are used for achieving mobility, navigating over or around obstacles. A rigorous quantitative stability analysis has been performed, and potential cases of instability have been identified. ROPE RIDE has been controlled based on kinematics analysis and task principles, and the effectiveness of the developed robot has been shown through real-world experiments. In addition, a novel control system has been developed to handle the cleaning task [77]. A new impedance control system has been proposed, and the effectiveness of the force control system to press brushes against the wall with a constant force has been shown through repeated experiments [78].

SIRIUS is a wall cleaning robot developed by Fraunhofer IFF, which is an industrial company in Germany [79,80,81,82]. SIRIUS realizes up-and-down motion by utilizing a crane that is installed on buildings for façade maintenance, and it adheres on the wall by a suction system that navigates like a linear actuator [83,84,85,86].

Fraunhofer IFF has also developed another wall cleaning robot, Filius [87]. Filius cleans arch-like roofs and navigates using wheels with a rope that secures the robot platform from the top of the building. The robot cleans the roof utilizing a brush installed on the front.

Qian et al. have developed a lifted type wall climbing robot via a rope from the roof without using any actuators on board [60,88]. The robot does not have independent actuation force; instead, it can move up and down by utilizing its own weight and a lifting force from a crane at the rooftop. The developed robot uses two vacuum pumps to adhere to the glass curtain surface. Two vacuum chambers are installed, one each at the upside and the downside, respectively, and a washing brush is wedged between the vacuum chambers. A rigorous stability analysis has been performed. Further analysis on negative pressure and fluids in the vacuum pumps has been done using the fluid network theory [59]. A closed loop fuzzy control system has been put in place to handle the negative pressure, and its effectiveness has been validated through well-designed experiments.

This approach ensures a high degree of safety from falling off the wall, as the robot is always tied to a rope attached to the rooftop, which also aids in improving the mobility performance. Due to the reduced need for additional components that provide active suction power, the developed robot platform is often compact in size and lighter weight. On the other hand, this method requires particular equipment, such as cranes and ropes. However, in most cases, rail-based wall cleaning robots are chosen due to the existence of this specific equipment as part of the concerned infrastructure. Even though, the use of rope facilitates movement of the robot in the vertical direction. It still poses a number of issues with respect to robot movement in the horizontal direction. Both SIRIUS and the robot developed by Qian et al. initially possessed only the ability to move in the vertical direction, while ROPE RIDE has installed crawlers for traversing in the horizontal direction. If the robots work on vertical surfaces or a reversed inclining wall, grip mechanisms, as well as suction mechanisms are necessary for adhering to the wall surfaces. Both ROPE RIDE and the robot developed by Qian et al. have additional installed mechanisms to adhere to the wall surfaces.

Bio-Inspired Adhesion

This approach mimics the characteristics of living species in nature that can navigate over vertical wall surfaces. This method uses bio-inspiration as its core principle for synthesizing new artificial mechanisms for wall climbing. Since species in nature have evolved over millions of years, artificial systems mimicking their locomotion over wall surfaces have the potential to offer a high degree of efficiency.

Funatsu et al. have developed a climbing robot that is able to climb up by hooking its claws [89]. This robot has been designed to be light in weight and compact in size, so that the robot climbs up walls only using its claws without any suction forces. Stability analysis has been performed in relation to the asperity of the wall surface, and the stable operating limits have been identified through simulations. In addition, experiments have been performed to validate the success of the developed prototype.

Mini-WhegsTM is a climbing robot with insect-inspired wheel legs [90,91]. The wall climbing ability has been experimented with several design variants of the Mini-Whegs platform, and the results have been discussed in detail. These platforms make use of adhesion means, such as claws, Scotch tape and micro-structured tapes, as well as sophisticated wheel-leg designs [92,93]. Experiments have been performed to verify the ability of these platforms to navigate over wall surfaces. In addition, experiments involving wall transitions, as well as having a complex angular relationship have been documented [94]. Furthermore, a robot with a body joint has been developed, and it has been shown that the robot is capable of transitioning between two orthogonal surfaces [95]. A legged robot driven by both spine and linkage mechanism has also been developed, and it has been shown that the robot can move over meshed surfaces [96].

Sitti et al. have proposed a gecko foot-inspired suction method [97]. Fiber arrays in varying morphologies have been developed using templates. Two popular approaches to this end include the use of a nanoprobe indented flat wax surface and the use of nanopore membranes [97]. A fiber array fabricated via micromolding followed by spatula tip formation via dipping has been made [98]. In addition, a directional and controllable fiber array has also been realized [99,100]. Using fiber arrays, adhesive pads that generate relatively higher adhesion force in one direction and lower in the opposite direction have been developed. By utilizing these adhesive pads, wall climbing robots called Geckobot and Waalbot have been implemented, and the effectiveness of the robots has been demonstrated through experiments [101,102,103,104,105,106,107,108].

This method is expected to develop robots of high ability by mimicking the characteristics of animals who can climb up walls, like geckos. Since no large-sized mechanisms are necessary for adhesion with bio-inspired approaches, the robots can be designed to be small and light weight. Standing examples of this case are the robot developed by Funatsu et al., Mini-WhegsTM and the robots developed by Sitti et al., which are relatively lighter and smaller than the robots that use other adhesion principles for wall climbing. Since most of the precedents use legged mechanisms deriving inspiration from living species, the locomotion of these platforms is often realized based on ground-based legged robotic platforms. Given that most living species that traverse over wall surfaces only carry their weight, robots developed deriving inspiration from them often follow suit to have a lighter payload capacity. However, with the advent of nanotechnology-based materials for adhesion, there is a renewed interest in the field of bio-inspired wall climbing robots.

Discussion

This paper presents a detailed review of wall climbing robots categorizing them into six distinct classes based on the adhesive mechanism that they use. These categories are suction cup adhesion, suction cup crawler adhesion, vacuum pump adhesion, magnetic adhesion, rope and/or rail gripping and bio-inspired adhesion. This paper also discusses the specific design approaches in synthesizing each of the six adhesion methods. The advantages and disadvantages of the identified six classes of adhesion methods for enabling robots to navigate wall surfaces are listed. Even though significant progress has been made towards the development of wall climbing robots and their component technologies, there still remains numerous challenges towards deploying them in a real-world setting for applications such as glass façade cleaning. This is mainly attributed to the diverse nature of the building types and the uncertain environment in which these robots are expected to operate [71,72,73,74]. One of the grand challenges associated with the design of wall climbing robots is that of developing a general purpose robot that can navigate over any type of wall.

Desirable Design Attributes

With most of the developed world aging rapidly and the simultaneous push for productivity improvement in job processes, these have significantly contributed to the advancement of robotics in a number of application domains. Glass façade cleaning is one such application domain that is witnessing increased interest due to the nature of the task being dull, dangerous and dirty. The glass façade cleaning tasks could be subdivided into the following phases:

- Navigating over the defined area.
- Adapting to window shapes.
- Washing/brushing windows.

- Wiping washing liquid.

However, not all of the above phases are used for every cleaning exercise. The choice of the phases greatly depends on the user requirements and prevailing environmental conditions. For instance, in hot and dry conditions, all of the phases are generally carried out. On the other hand, the washing phase may not be performed in rainy conditions. Furthermore, fugitive dust should be swept from buildings that are close to a desert by brushing.

Typically, it is easier for robotic platforms to complete each of these tasks individually. However, it is highly difficult for robots to possess all of these capabilities integrated into a single platform while being able to adapt in response to the façade cleaning tasks in real-world environments. The absence of a clearly-defined set of performance indicators for such robots further adds to the uncertainty in designing an efficient platform while making optimal trade-offs between competing design criteria. By rigorous analysis of 28 reported research papers on glass façade cleaning robots, we derived the following set of desirable design attributes for an ideal glass façade cleaning robot:

- Minimize total time to cover a given wall area
- Maximize the total wall area covered
- Minimize the wall area cleaned multiple times
- Minimize wall slippage
- Minimize energy consumption
- Maximize fault tolerance
- Maximize perceptible information
- Minimize noise
- Maximize dust removal
- Maximize safety
- Maximize user interaction
- Minimize infrastructure changes needed for robot deployment

With the rising interest in façade cleaning robots and related research topics, the identification of desirable design attributes, such as the ones presented above, becomes highly critical in facilitating targeted research with clear technical goals and well-defined design trade-off boundaries. This set of design attributes can also be used as metrics for comparing robot performance in glass façade cleaning tasks.

CONCLUSIONS

In the current paper, we presented a detailed survey of adhesion principles adopted in wall climbing robots. We categorized the adhesion principles found in the literature into six broad categories, namely: suction cup adhesion, suction cup crawler, vacuum pump adhesion, magnetic adhesion and bio-inspired adhesion. We further summarized the impact of each of these adhesion mechanisms on various robot design considerations, including locomotion, stability, infrastructure needs and perception by deriving specific cases from the literature. Going beyond the adhesion issues in wall climbing robots, we also put forward a set of desirable design attributes for glass façade cleaning robots for the first time derived from rigorous analyses that has the potential to facilitate targeted future research with clear technical goals and well-defined design trade-off boundaries. Such a set of design attributes also offers significant potential for use as performance metrics in evaluating façade cleaning robots, as well as comparing competing designs. A novel nested reconfigurable approach to the design of glass façade cleaning robots is currently being studied where the proposed set of desirable design attributes drive the core of the concept formulation. Our long-term aim is to develop highly efficient glass façade cleaning robots that are able to seamlessly adapt to the dynamic needs of the operating environment, task and occurrence of faults.

ACKNOWLEDGMENTS

This research is supported by the SUTD-JTC Industrial Infrastructure Innovation Center under grants IPJTCT 31501.

AUTHOR CONTRIBUTIONS

S.N. surveyed the literatures. S.N. and R.E.M contributes discussions/design attributes; S.N. and R.E.M. wrote the paper.

REFERENCES

1. Frey, C.B.; Osborne, M.A. The future of employment: How susceptible are jobs to computerisation. *Retriev. Sept.* 2013, *7*, 2013.
2. International Federation of Robotics. Service Robot Statistics. Available online: http://www.ifr.org/service-robots/statistics/ (accessed on 28 June 2016).
3. Aldred, J. Burj Khalifa—A new high for high-performance concrete. *Proc. Inst. Civ. Eng.* 2010, *163*, 66–73.
4. Baker, W.F. The World's tallest building. Available online: http://www.structuremag.org/wp-content/uploads/2014/08/D-Spotlight-Burj-June111.pdf (accessed on 28 June 2016).
5. Baker, W.F.; Korista, D.S.; Novak, L.C. Burj Dubai: Engineering the world's tallest building. *Struct. Des. Tall Spec. Build.* 2007, *16*, 361–375.
6. Weismantle, P.A.; Smith, G.L.; Sheriff, M. Burj Dubai: An architectural technical design case study. *Struct. Des. Tall Spec. Build.* 2007, *16*, 335–360.
7. ZELJIC, A.S. Shanghai Tower Façade Design Process. International Conference of Building Envelope Systems. Available online: http://www.gensler.com/uploads/documents/Shanghai_Tower_Facade_Design_Process_11_10_2011.pdf (accessed on 28 June 2016).
8. Xia, J.; Poon, D.; Mass, D. Case study: Shanghai Tower. *CTBUH J.* 2010, *2*, 12–18.
9. Zhaoa, X.; Ding, J.; Suna, H. Structural design of shanghai tower for wind loads. *Procedia Eng.* 2011, *14*, 1759–1767.
10. BBC. Shanghai Window Cleaning Cradle Swings Out of Control. Available online: http://www.bbc.com/news/world-asia-china-32176401 (accessed on 28 June 2016).
11. BBC. Window Washers Rescued from High up World

Trade Center. Available online: http://www.bbc.com/news/world-us-canada-30028969 (accessed on 28 June 2016).

12. Zhang, H.; Zhang, J.; Zong, G.; Wang, W.; Liu, R. Sky cleaner 3: A real pneumatic climbing robot for glass-wall cleaning. *Robot. Autom. Mag. IEEE* 2006, *13*, 32–41.
13. Zhang, H.; Zhang, J.; Zong, G. Effective nonlinear control algorithms for a series of pneumatic climbing robots. In Proceedings of the 2006 IEEE International Conference on Robotics and Biomimetics, ROBIO'06, Pisa, Italy, 20–22 February 2006; pp. 994–999.
14. Zhang, H.; Zhang, J.; Zong, G. Requirements of glass cleaning and development of climbing robot systems. In Proceedings of the 2004 International Conference on Intelligent Mechatronics and Automation, Chengdu, China, 26–31 August 2004; pp. 101–106.
15. Zhang, H.; Zhang, J.; Liu, R.; Zong, G. A novel approach to pneumatic position servo control of a glass wall cleaning robot. In Proceedings of the 2004 IEEE/RSJ International Conference on Intelligent Robots and Systems (IROS 2004), Sendai, Japan, 28 September–2 October 2004; Volume 1, pp. 467–472.
16. Tokhi, O.; Zhang, H.; Zhang, J.; Wang, W.; Liu, R.; Zong, G. A series of pneumatic glass-wall cleaning robots for high-rise buildings. *Ind. Robot Int. J.* 2007, *34*, 150–160.
17. Zhang, H.; Zhang, J.; Zong, G. Realization of a service climbing robot for glass-wall cleaning. In Proceedings of the 2004 IEEE International Conference on Robotics and Biomimetics, ROBIO 2004, Shenyang, China, 22–26 August 2004; pp. 395–400.
18. Zhang, H.; Zhang, J.; Liu, R.; Wang, W.; Zong, G. Pneumatic Climbing Robots for Glass Wall Cleaning. In *Climbing and Walking Robots*; Springer: Berlin, Germany, 2005; pp. 1061–1069.
19. Zhang, H.; Wang, W.; Zhang, J.; Zong, G. Control Hierarchy Realization and Cleaning Trajectory Evaluation of a Wall Cleaning Robot. In Proceedings of the 2006 IEEE Conference on Robotics, Automation and Mechatronics, Bangkok, Thailand, 7–9 June 2006; pp. 1–6.
20. Zhang, H.; Zhang, J.; Zong, G. Cleaning Trajectory Evaluation of a Wall Cleaning Robot Based on Synthesis Standards. In Proceedings of the IMACS Multiconference on Computational Engineering in Systems Applications, Beijing, China, 4–6 October 2006; Volume 2, pp. 1695–1700.
21. Wang, W.; Wang, K.; Zong, G.H.; Li, D.Z. Principle and experiment of vibrating suction method for wall-climbing robot. *Vacuum* 2010, *85*, 107–112.
22. Kawasaki, S.; Kikuchi, K. Development of a small legged wall climbing robot with passive suction cups. In Proceedings of the 3rd International Conference on Design Engineering and Science—ICDES, Pilsen, Czech Republic, 31 August–3 September 2014; pp. 112–116.
23. Yano, T.; Suwa, T.; Murakami, M.; Yamamoto, T. Development of a semi self-contained wall climbing robot with scanning type suction cups. In Proceedings of the 1997 IEEE/RSJ International Conference on Intelligent Robots and Systems, IROS'97, Grenoble, France, 7–11 September 1997; Volume 2, pp. 900–905.
24. Ikeda, K.; Yano, T. Fundamental study of a scanning-type suction cup. *Adv. Robot.* 1990, *5*, 309–320.
25. Yano, T.; Numao, S.; Kitamura, Y. Development of a self-contained wall climbing robot with scanning type suction cups. In Proceedings of the 1998 IEEE/RSJ International Conference on Intelligent Robots and Systems, Victoria, BC, Canada, 13–17 October 1998; Volume 1, pp. 249–254.
26. Zhu, J.; Sun, D.; Tso, S.K. Application of a service climbing robot with motion planning and visual sensing. *J. Robot. Syst.* 2003, *20*, 189–199.
27. Sun, D.; Zhu, J.; Lai, C.; Tso, S. A visual sensing application to a climbing cleaning robot on the glass surface. *Mechatronics* 2004, *14*, 1089–1104.
28. Sun, D.; Zhu, J.; Tso, S.K. *A Climbing Robot for Cleaning Glass Surface with Motion Planning and Visual Sensing*; INTECH Open Access Publisher: Rijeka, Croatia, 2007.
29. Ota, Y.; Inagaki, Y.; Yoneda, K.; Hirose, S. Research on a reduced-DOF walking robot with parallel mechanism that has high homogeneity of output force. *J. Robot. Soc. Jpn.* 2000, *18*, 66–74.
30. Ota, Y.; Inagaki, Y.; Yoneda, K.; Hirose, S. Research on a six-legged walking robot with parallel mechanism. In Proceedings of the 1998 IEEE/RSJ International Conference on Intelligent Robots and Systems, Victoria, BC, Canada, 13–17 October 1998; Volume 1, pp. 241–248.
31. Lee, T.T.; Liao, C.M.; Chen, T.K. On the stability properties of hexapod tripod gait. *IEEE J. Robot. Autom.* 1988, *4*, 427–434.
32. Haynes, G.C.; Rizzi, A.A. Gaits and gait transitions for legged robots. In Proceedings of the 2006 IEEE International Conference on Robotics and Automation, ICRA 2006, Orlando, FL, USA, 15–19 May 2006; pp. 1117–1122.
33. Chen, W.; Yao, S.; Low, K. Modular formulation for dynamics of multi-legged robots. In Proceedings of the 8th International Conference on Advanced Robotics, ICAR'97, Monterey, CA, USA, 7–9 July 1997; pp. 279–284.
34. Inagaki, S.; Yuasa, H.; Suzuki, T.; Arai, T. Wave CPG model for autonomous decentralized multi-legged robot: Gait generation and walking speed control. *Robot. Auton. Syst.* 2006, *54*, 118–126.
35. Lee, G.; Kim, H.; Seo, K.; Kim, J.; Sitti, M.; Seo, T. Series of Multilinked Caterpillar Track-Type Climbing Robots. *J. Field Robot.* 2014.
36. Kim, H.; Kim, D.; Yang, H.; Lee, K.; Seo, K.; Chang, D.; Kim, J. A wall climbing robot with vacuum caterpillar wheel system operated by mechanical valve. In Proceedings of the 9th International Conference on Climbing and Walking Robots, Brussels, Belgium, 12–14 September 2006; pp. 28–33.
37. Roy, R.K. *A Primer on the Taguchi Method*; Society of Manufacturing Engineers: Dearborn, MI, USA, 2010.
38. Kim, H.; Kim, D.; Yang, H.; Lee, K.; Seo, K.; Chang, D.; Kim, J. Development of a wall-climbing robot

using a tracked wheel mechanism. *J. Mech. Sci. Technol.* 2008, *22*, 1490–1498.

39. Kim, H.; Seo, K.; Lee, K.; Kim, J.; Kim, H. Development of a multi-body wall climbing robot with tracked wheel mechanism. In Proceedings of the 13th International Conference on Climbing and Walking Robots (CLAWAR), Nagoya, Japan, 31 August–3 September 2010; pp. 439–446.
40. Lee, G.; Kim, H.; Seo, K.; Kim, J.; Kim, H.S. MultiTrack: A multi-linked track robot with suction adhesion for climbing and transition. *Robot. Auton. Syst.* 2015, *72*, 207–216.
41. Serbot-AG. Available online: http://www.serbot.ch/en/ (accessed on 28 June 2016).
42. Tso, S.; Fung, Y.; Chow, W.; Zong, G.; Liu, R. Design and implementation of a glass-wall climbing robot for high rise building. In Proceedings of the World Automation Congress, Kobe, Japan, 19–23 September 2000.
43. Zhu, J.; Sun, D.; Tso, S.K. Development of a tracked climbing robot. *J. Intell. Robot. Syst.* 2002, *35*, 427–443.
44. Wang, Y.; Liu, S.; Xu, D.; Zhao, Y.; Hao, S.; Gao, X. Development and application of wall-climbing robots. In Proceedings of the 1999 IEEE International Conference on Robotics and Automation, Detroit, AL, USA, 10–15 May 1999; Volume 2, pp. 1207–1212.
45. Gao, X.; Kikuchi, K. Study on a kind of wall cleaning robot. In Proceedings of the IEEE International Conference on Robotics and Biomimetics, ROBIO 2004, Shenyang, China, 22–26 August 2004; pp. 391–394.
46. Liu, S.; Gao, X.; Li, K.; Li, J.; Duan, X. A small-sized wall-climbing robot for anti-terror scout. In Proceedings of the IEEE International Conference on Robotics and Biomimetics, ROBIO 2007, Sanya, China, 16–18 September 2007; pp. 1866–1870.
47. Xu, D.; Gao, X.; Wu, X.; Fan, N.; Li, K.; Kikuchi, K. Suction ability analyzes of a novel wall climbing robot. In Proceedings of the 2006 IEEE International Conference on Robotics and Biomimetics, ROBIO'06, Pisa, Italy, 20–22 February 2006; pp. 1506–1511.
48. Li, J.; Gao, X.; Fan, N.; Li, K.; Jiang, Z. BIT Climber: A centrifugal impeller-based wall climbing robot. In Proceedings of the 2009 International Conference on Mechatronics and Automation, ICMA 2009, Changchun, China, 9–12 August 2009; pp. 4605–4609.
49. Li, J.; Gao, X.; Fan, N.; Li, K.; Jiang, Z.; Jiang, Z. Adsorption performance of sliding wall climbing robot. *Chin. J. Mech. Eng.* 2010, *23*, 1.
50. Miyake, T.; Ishihara, H. Mechanisms and basic properties of window cleaning robot. In Proceedings of the 2003 IEEE/ASME International Conference on Advanced Intelligent Mechatronics, AIM 2003, Kobe, Japan, 20–24 July 2003; Volume 2, pp. 1372–1377.
51. Miyake, T.; Ishikawa, H. Window Wiping System. U.S. Patent 8,099,818, 24 January 2012.
52. Miyake, T.; Ishihara, H.; Yoshimura, M. Basic studies on wet adhesion system for wall climbing robots. In Proceedings of the 2007 IEEE/RSJ International Conference on Intelligent Robots and Systems, IROS 2007, San Diego, CA, USA, 29 October–2 November 2007; pp. 1920–1925.
53. Miyake, T.; Ishihara, H.; Yoshimura, M. Application of Wet Vacuum-based Adhesion System for Wall Climging Mechanism. In Proceedings of the 2007 International Symposium on Micro-Nanomechatronics and Human Science, MHS'07, Nagoya, Japan, 11–14 November 2007; pp. 532–537.
54. Miyake, T.; Ishihara, H.; Tomino, T. Basic research on vacuum-based wet adhesion system for wall climbing robots-measurement of lubricating action and seal action by the liquid. In Proceedings of the 2008 IEEE International Conference on Mechatronics and Automation, ICMA 2008, Takamatsu, Japan, 5–8 August 2008; pp. 616–621.
55. Miyake, T.; Ishihara, H.; Tomino, T. Vacuum-based wet adhesion system for wall climbing robots-lubricating action and seal action by the liquid. In Proceedings of the 2008 IEEE International Conference on Robotics and Biomimetics, ROBIO 2008, Bangkok, Thailand, 22–25 February 2009; pp. 1824–1829.
56. Miyake, T.; Ishihara, H.; Shoji, R.; Yoshida, S. Development of small-size window cleaning robot a traveling direction control on vertical surface using accelerometer. In Proceedings of the 2006 IEEE International Conference on Mechatronics and Automation, Luoyang, China, 25–28 August 2006; pp. 1302–1307.
57. Miyake, T.; Ishihara, H.; Shoji, R.; Yoshida, S. Development of small-size window cleaning robot by wall climbing mechanism. In Proceedings of the 23rd International Symposium on Automation and Robotics in Construction, Tokyo, Japan, 3–5 October 2006; pp. 215–220.
58. Zhao, Y.; Fu, Z.; Cao, Q.; Wang, Y. Development and applications of wall-climbing robots with a single suction cup. *Robotica* 2004, *22*, 643–648.
59. Qian, Z.; Zhao, Y.; Fu, Z.; Wang, Y. Fluid model of sliding suction cup of wall–climbing robots. *Int. J. Adv. Robot. Syst.*2006, *3*, 275–284.
60. Qian, Z.Y.; Zhao, Y.Z.; Fu, Z. Development of wall-climbing robots with sliding suction cups. In Proceedings of the 2006 IEEE/RSJ International Conference on Intelligent Robots and Systems, Beijing, China, 9–15 October 2006; pp. 3417–3422.
61. Gao, X.; Jiang, Z.; Gao, J.; Xu, D.; Wang, Y.; Pan, H. Boiler maintenance robot with multi-operational schema. In Proceedings of the 2008 IEEE International Conference on Mechatronics and Automation, ICMA 2008, Takamatsu, Japan, 5–8 August 2008; pp. 610–615.
62. Gao, X.; Xu, D.; Wang, Y.; Pan, H.; Shen, W. Multifunctional robot to maintain boiler water-cooling tubes. *Robotica*2009, *27*, 941–948.
63. Lee, G.; Park, J.; Kim, H.; Seo, K.; Kim, J.; Seo, T. Wall Climbing Robots with Track-Wheel Mechanism. In Proceedings of the 3rd International Conference on Machine Learning and Computing (ICMLC 2011), Guillin, China, 10–14 July 2011; pp. V3-334–V3-337.
64. Seo, T.; Sitti, M. Tank-like module-based climbing robot

using passive compliant joints. *IEEE/ASME Trans. Mechatron.*2013, *18*, 397–408.

65. Lee, G.; Seo, K.; Lee, S.; Park, J.; Kim, H.; Kim, J.; Seo, T. Compliant track-wheeled climbing robot with transitioning ability and high-payload capacity. In Proceedings of the 2011 IEEE International Conference on Robotics and Biomimetics (ROBIO), Phuket Island, Thailand, 7–11 December 2011; pp. 2020–2024.

66. Lee, G.; Wu, G.; Kim, S.H.; Kim, S.H.; Seo, T. Combot: Compliant climbing robotic platform with transitioning capability and payload capacity. In Proceedings of the 2012 IEEE International Conference on Robotics and Automation (ICRA), Saint Paul, MI, USA, 14–18 May 2012; pp. 2737–2742.

67. Lee, G.; Wu, G.; Kim, J.; Seo, T. High-payload climbing and transitioning by compliant locomotion with magnetic adhesion. *Robot. Auton. Syst.* 2012, *60*, 1308–1316.

68. Nam, S.; Oh, J.; Lee, G.; Kim, J.; Seo, T. Dynamic analysis during internal transition of a compliant multi-body climbing robot with magnetic adhesion. *J. Mech. Sci. Technol.* 2014, *28*, 5175–5187.

69. Xu, Z.; Ma, P. A wall-climbing robot for labelling scale of oil tank's volume. *Robotica* 2002, *20*, 209–212.

70. Zhang, H.; Zhang, J.; Liu, R.; Zong, G. Mechanical design and dynamcis of an autonomous climbing robot for elliptic half-shell cleaning. *Int. J. Adv. Robot. Syst.* 2007, *4*, 437–446.

71. Zhang, H.; Zhang, J.; Liu, R.; Wang, W.; Zong, G. Design of a climbing robot for cleaning spherical surfaces. In Proceedings of the 2005 IEEE International Conference on Robotics and Biomimetics (ROBIO), Hong Kong/ Macau, 29 June–3 July 2005; pp. 375–380.

72. Zhang, H.; Zhang, J.; Liu, R.; Zong, G. Realization of a service robot for cleaning spherical surfaces. *Int. J. Adv. Robot. Syst.* 2005, *2*, 53–58.

73. Zhang, H.; Zhang, J.; Liu, R.; Zong, G. Climbing technique of the cleaning robot for a spherical surface. In Proceedings of the 2005 IEEE International Conference on Mechatronics and Automation, Niagara Falls, NY, Canada, 29 July–1 August 2005; Volume 4, pp. 2061–2066.

74. Zhang, H.; Wang, W.; Liu, R.; Zhang, J.; Zong, G. Locomotion realization of an autonomous climbing robot for elliptic half-shell cleaning. In Proceedings of the 2nd IEEE Conference on Industrial Electronics and Applications, ICIEA 2007, Harbin, China, 23–25 May 2007; pp. 1220–1225.

75. Wang, W.; Tang, B.; Zhang, H.; Zong, G. Robotic cleaning system for glass facade of high-rise airport control tower. *Ind. Robot Int. J.* 2010, *37*, 469–478.

76. Seo, K.; Cho, S.; Kim, T.; Kim, H.S.; Kim, J. Design and stability analysis of a novel wall-climbing robotic platform (ROPE RIDE). *Mech. Mach. Theory* 2013, *70*, 189–208.

77. Kim, T.Y.; Kim, J.H.; Seo, K.C.; Kim, H.M.; Lee, G.U.; Kim, J.W.; Kim, H.S. Design and Control of a Cleaning Unit for a Novel Wall-Climbing Robot. In *Applied Mechanics and Materials*; Trans Tech Publications Inc.: Pfaffikon, Switzerland, 2014; Volume 541, pp. 1092–1096.

78. Kim, T.; Seo, K.; Kim, J.H.; Kim, H.S. Adaptive impedance control of a cleaning unit for a novel wall-climbing mobile robotic platform (ROPE RIDE). In Proceedings of the 2014 IEEE/ASME International Conference on Advanced Intelligent Mechatronics (AIM), Besancon, France, 8–11 July 2014; pp. 994–999.

79. Fraunhofer-IFF. Available online: http://www.iff.fraunhofer.de/en.html (accessed on 28 June 2016).

80. Böhme, T.; Schmucker, U.; Elkmann, N.; Sack, M. Service robots for facade cleaning. In Proceedings of the 24th Annual Conference of the IEEE Industrial Electronics Society, IECON'98, Aachen, Germany, 31 August–4 September 1998; Volume 2, pp. 1204–1207.

81. Elkmann, N.; Schmucker, U.; Scharfe, H.; Schoop, C.; Kubbe, I. Drive Device for Moving a Robot or Vehicle on Flat, Inclined or Curved Surfaces, Particularly of a Glass Construction and Robot with Drive Device. U.S. Patent 5,959,424, 28 September 1999.

82. Elkmann, N.; Hortig, J.; Fritzsche, M. Cleaning automation. In *Springer Handbook of Automation*; Springer: Berlin, Germany, 2009; pp. 1253–1264.

83. Elkmann, N.; Felsch, T.; Sack, M.; Saenz, J.; Hortig, J. Innovative service robot systems for facade cleaning of difficult-to-access areas. In Proceedings of the 2002 IEEE/RSJ International Conference on Intelligent Robots and Systems, Lausanne, Switzerland, 30 September–4 October 2002; Volume 1, pp. 756–762.

84. Elkmann, N.; Kunst, D.; Krueger, T.; Lucke, M.; Böhme, T.; Felsch, T.; Stürze, T. SIRIUSc–Façade cleaning robot for a high-rise building in munich, germany. In *Climbing and Walking Robots*; Springer: Berlin, Germany, 2005; pp. 1033–1040.

85. Elkmann, N.; Lucke, M.; Krüger, T.; Kunst, D.; Stürze, T.; Hortig, J. Kinematics, sensors and control of the fully automated facade-cleaning robot SIRIUSc for the Fraunhofer headquarters building, Munich. *Ind. Robot Int. J.* 2008, *35*, 224–227.

86. Bridge, B.; Elkmann, N.; Lucke, M.; Krüger, T.; Kunst, D.; Stürze, T.; Hortig, J. Kinematics, sensors and control of the fully automated façade-cleaning robot SIRIUSc for the Fraunhofer headquarters building, Munich. *Ind. Robot Int. J.*2008, *35*, 224–227.

87. Nof, S.Y.; Ceroni, J.; Jeong, W.; Moghaddam, M. E-Service Industry. In *Revolutionizing Collaboration through E-Work, E-Business, and E-Service*; Springer: Berlin, Germany, 2015; pp. 315–356.

88. Qian, Z.Y.; Zhao, Y.Z.; Fu, Z.; Cao, Q.X. Design and realization of a non-actuated glass-curtain wall-cleaning robot prototype with dual suction cups. *Int. J. Adv. Manuf. Technol.* 2006, *30*, 147–155.

89. Funatsu, M.; Kawasaki, Y.; Kawasaki, S.; KikuchiI, K. Development of cm-Scale Wall Climbing Hexapod Robot with Claws. In Proceedings of the 3rd International Conference on Design Engineering and Science—ICDES, Pilsen, Czech Republic, 31 August–3 September 2014; pp. 101–106.

90. Daltorio, K.A.; Horchler, A.D.; Gorb, S.; Ritzmann, R.E.; Quinn, R.D. A small wall-walking robot with compliant, adhesive feet. In Proceedings of the 2005 IEEE/RSJ International Conference on Intelligent Robots and Systems (IROS 2005), Edmonton, KY, Canada, 2–6 August 2005; pp. 3648–3653.
91. Gorb, S.N.; Sinha, M.; Peressadko, A.; Daltorio, K.A.; Quinn, R.D. Insects did it first: A micropatterned adhesive tape for robotic applications. *Bionspir. Biomim.* 2007, *2*, S117. [PubMed]
92. Daltorio, K.A.; Gorb, S.; Peressadko, A.; Horchler, A.D.; Ritzmann, R.E.; Quinn, R.D. A robot that climbs walls using micro-structured polymer feet. In *Climbing and Walking Robots*; Springer: Berlin, Germany, 2006; pp. 131–138.
93. Daltorio, K.A.; Wei, T.E.; Gorb, S.N.; Ritzmann, R.E.; Quinn, R.D. Passive foot design and contact area analysis for climbing mini-whegs. In Proceedings of the 2007 IEEE International Conference on Robotics and Automation, San Diego, CA, USA, 29 October–2 November 2007; pp. 1274–1279.
94. Daltorio, K.A.; Wei, T.E.; Horchler, A.D.; Southard, L.; Wile, G.D.; Quinn, R.D.; Gorb, S.N.; Ritzmann, R.E. Mini-whegs TM climbs steep surfaces using insect-inspired attachment mechanisms. *Int. J. Robot. Res.* 2009, *28*, 285–302.
95. Daltorio, K.A.; Witushynsky, T.C.; Wile, G.D.; Palmer, L.R.; Malek, A.A.; Ahmad, M.R.; Southard, L.; Gorb, S.N.; Ritzmann, R.E.; Quinn, R.D. A body joint improves vertical to horizontal transitions of a wall-climbing robot. In Proceedings of the 2008 IEEE International Conference on Robotics and Automation, ICRA 2008, Pasadena, CA, USA, 19–23 May 2008; pp. 3046–3051.
96. Wile, G.D.; Daltorio, K.A.; Diller, E.D.; Palmer, L.R.; Gorb, S.N.; Ritzmann, R.; Quinn, R.D. Screenbot: Walking inverted using distributed inward gripping. In Proceedings of the 2008 IEEE/RSJ International Conference on Intelligent Robots and Systems, IROS 2008, Nice, France, 22–26 September 2008; pp. 1513–1518.
97. Sitti, M.; Fearing, R.S. Synthetic gecko foot-hair micro/nano-structures for future wall-climbing robots. In Proceedings of the 2003 IEEE International Conference on Robotics and Automation, ICRA'03, Taipei, Taiwan, 14–19 September 2003; Volume 1, pp. 1164–1170.
98. Aksak, B.; Murphy, M.P.; Sitti, M. Gecko inspired micro-fibrillar adhesives for wall climbing robots on micro/nanoscale rough surfaces. In Proceedings of the 2008 IEEE International Conference on Robotics and Automation, ICRA 2008, Pasadena, CA, USA, 19–23 May 2008; pp. 3058–3063.
99. Murphy, M.P.; Aksak, B.; Sitti, M. Gecko-Inspired Directional and Controllable Adhesion. *Small* 2009, *5*, 170–175. [PubMed]
100. Mengüç, Y.; Yang, S.Y.; Kim, S.; Rogers, J.A.; Sitti, M. Gecko-Inspired Controllable Adhesive Structures Applied to Micromanipulation. *Adv. Funct. Mater.* 2012, *22*, 1246–1254.
101. Menon, C.; Sitti, M. Biologically inspired adhesion based surface climbing robots. In Proceedings of the 2005 IEEE International Conference on Robotics and Automation, ICRA 2005, Barcelona, Spain, 18–22 April 2005; pp. 2715–2720.
102. Murphy, M.P.; Sitti, M. Waalbot: An agile small-scale wall-climbing robot utilizing dry elastomer adhesives. *IEEE/ASME Trans. Mechatron.* 2007, *12*, 330–338.
103. Murphy, M.P.; Tso, W.; Tanzini, M.; Sitti, M. Waalbot: An agile small-scale wall climbing robot utilizing pressure sensitive adhesives. In Proceedings of the 2006 IEEE/RSJ International Conference on Intelligent Robots and Systems, Beijing, China, 9–15 October 2006; pp. 3411–3416.
104. Murphy, M.P.; Kute, C.; Mengüç, Y.; Sitti, M. Waalbot II: Adhesion recovery and improved performance of a climbing robot using fibrillar adhesives. *Int. J. Robot. Res.* 2011, *30*, 118–133.
105. Unver, O.; Uneri, A.; Aydemir, A.; Sitti, M. Geckobot: A gecko inspired climbing robot using elastomer adhesives. In Proceedings of the 2006 IEEE International Conference on Robotics and Automation, ICRA 2006, Orland, CA, USA, 15–19 May 2006; pp. 2329–2335.
106. Unver, O.; Murphy, M.; Sitti, M. Geckobot and waalbot: Small-scale wall climbing robots. In Proceedings of the AIAA 5th Aviation, Technology, Integration, and Operations Conference, Arlington, MA, USA, 26–28 September 2005.
107. Menon, C.; Murphy, M.; Sitti, M. Gecko inspired surface climbing robots. In Proceedings of the 2004 IEEE International Conference on Robotics and Biomimetics, ROBIO 2004, Shenyang, China, 22–26 August 2004; pp. 431–436.
108. Carlo, M.; Metin, S. A biomimetic climbing robot based on the gecko. *J. Bion. Eng.* 2006, *3*, 115–125.

CHAPTER 13 BRIDGE DECK LOAD TESTING USING SENSORS AND OPTICAL SURVEY EQUIPMENT

Hubo Cai, Osama Abudayyeh, Ikhlas Abdel-Qader, Upul Attanayake, Joseph Barbera, and Eyad Almaita,

INTRODUCTION

Bridges are a critical component of a nation's ground transportation infrastructure system that allows the movement of people and goods from one place to another. Bridges are exposed to load effects and weather impacts that lead to their deterioration. Federal Highway Administration (FHWA) statistics show that about 20% of the National Highway System (NHS) bridges and 27% of non-NHS bridges in US are either structurally deficient or functionally obsolete [1], resulting in a total of about 152,000 bridges requiring immediate attention in the form of repair or replacement. Therefore, routine inspection and load testing are needed to obtain good estimates on bridge response and to support making decisions of appropriate maintenance and rehabilitation activities to ensure their structural integrity and safety [2, 3].

A recent trend in bridge management is the utilization of sensors embedded in the bridge structure to monitor the long-term performance of bridges, under various loads and environmental impacts. Given the monitored data, meaningful information regarding the bridge reliability can be extracted [4]. This is particularly true with the growing adoption of the prefabrication technology in bridge construction. This technology allows the prefabrication of structural components to be conducted offsite under a well-controlled environment with the benefit of higher-quality precast structural components that are expected to perform better with lower maintenance needs, compared to conventional cast-in-place concrete bridges [5]. By reducing the onsite construction time, this prefabrication technology can also save construction time and minimize traffic disruption [6]. Due to its relatively short history in bridge construction, long-term performance data of prefabricated bridges are rarely existent, and sensor technology is many times utilized to monitor their long-term performance. This scenario of embedding sensors in bridge structure is also the case with the Parkview Bridge in Kalamazoo, Michigan, which was constructed in 2006 as a totally prefabricated system. Vibrating-wire strain gages (sensors) were embedded in the precast deck panels to form a sensor network for continuous health monitoring and condition assessment.

While the original intention of embedding sensors into bridges is to continuously monitor their long-term performance, these sensors may serve as an alternative method for measuring the structural response of a bridge under nondestructive load testing. This paper presents the results of a load testing study to evaluate the response of the Parkview Bridge under varying load configurations via nondestructive load testing and the utilization of an embedded sensor network. In this study, a load testing method that incorporates optical surveying systems, the embedded sensor network, and deflection analysis was designed and implemented. The major contribution of this study is its demonstration of how an embedded sensor network might be utilized to acquire real-time equivalent static stresses under various loading scenarios.

BRIDGE LOAD TESTING

The actual response of a bridge to loads is usually better than what the theory dictates [7]. Factors that contribute to the

Citation: Hubo Cai, Osama Abudayyeh, Ikhlas Abdel-Qader, Upul Attanayake, Joseph Barbera, and Eyad Almaita, "Bridge Deck Load Testing Using Sensors and Optical Survey Equipment," Advances in Civil Engineering, vol. 2012, Article ID 493983, 11 pages, 2012. doi:10.1155/2012/493983.

load capacity difference include unintended composite action, load distribution effects, participation of parapets, railings, curbs, and utilities, material property differences, unintended continuity, participation of secondary members, effects of skew, portion of load carried by deck, and unintended arching action due to frozen bearings [7].

Load testing is recommended by AASHTO [8] as an "effective means of evaluating the structural response of a bridge." The purpose of conducting load testing on existing bridges is to evaluate their structural response without causing damages. Therefore, load testing is usually conducted in a nondestructive manner and is sometimes referred to as nondestructive load testing. The goal of this type of testing is to compare field response of the bridge under test loads with its theoretical response [7]. Nondestructive load testing can be further categorized into diagnostic testing and proof testing. Diagnostic testing methods provide the measurements necessary to analyze differential loading effects (i.e., moment, shear, axial force, deflection, etc.) present in various structural members due to applied loads [9]. Proof-load testing aims at determining the magnitude and configuration of loads that cause critical structural components to approach their elastic limit.

Tasks involved in a program of load testing typically include the determination of testing objectives and load configuration, the selection and placement of instrumentation, the adoption of appropriate analysis techniques, and the evaluation and comparison of test results and analytical results [10]. Load testing is being carried out widely to evaluate and rate bridge response on a case-by-case manner to evaluate new construction materials and technologies [10, 11]. While a number of bridge load testing studies can be found in the literature, such studies on bridges using embedded sensors, which is the focus of this paper, are almost nonexistent.

BRIDGE UNDER TEST: PARKVIEW BRIDGE

The Parkview Bridge is located in Kalamazoo, MI. The 249-foot long bridge over US-131 is particularly relevant concerning the advancement of both bridge construction technology and structural monitoring. Being the first of its kind in Michigan, the concrete superstructure of the Parkview Bridge is almost entirely precast, including girders and deck panels. The substructure also has precast concrete piers. This configuration allowed for the implementation of rapid bridge construction (RBC) technology. A second feature of the structure is a network of sensors that was embedded in the deck panels during casting. This sensor network will continuously monitor the response of the bridge to various loads and environmental effects (temperature changes).

The Parkview Bridge consists of four spans, which from east to west are approximately 38, 83, 83, and 45 feet, respectively. The three lanes of traffic are supported by seven precast prestressed AASHTO type III girders spaced at 8′7″. Figure 1(a) provides a cross-sectional view of the bridge to illustrate the girder spacing and the deck-panel dimensions. The bridge deck comprises 48 precast nonprestressed 9-inch thick deck panels, skewed at 23 degrees and posttensioned longitudinally; see Figure 1(b). The sensors within the deck panels are placed at varying locations and orientations with the objective of measuring all relevant stresses that are predicted to develop in the bridge deck. Figure 2 illustrates the locations and orientations of the sensors embedded in the bridge deck panels.

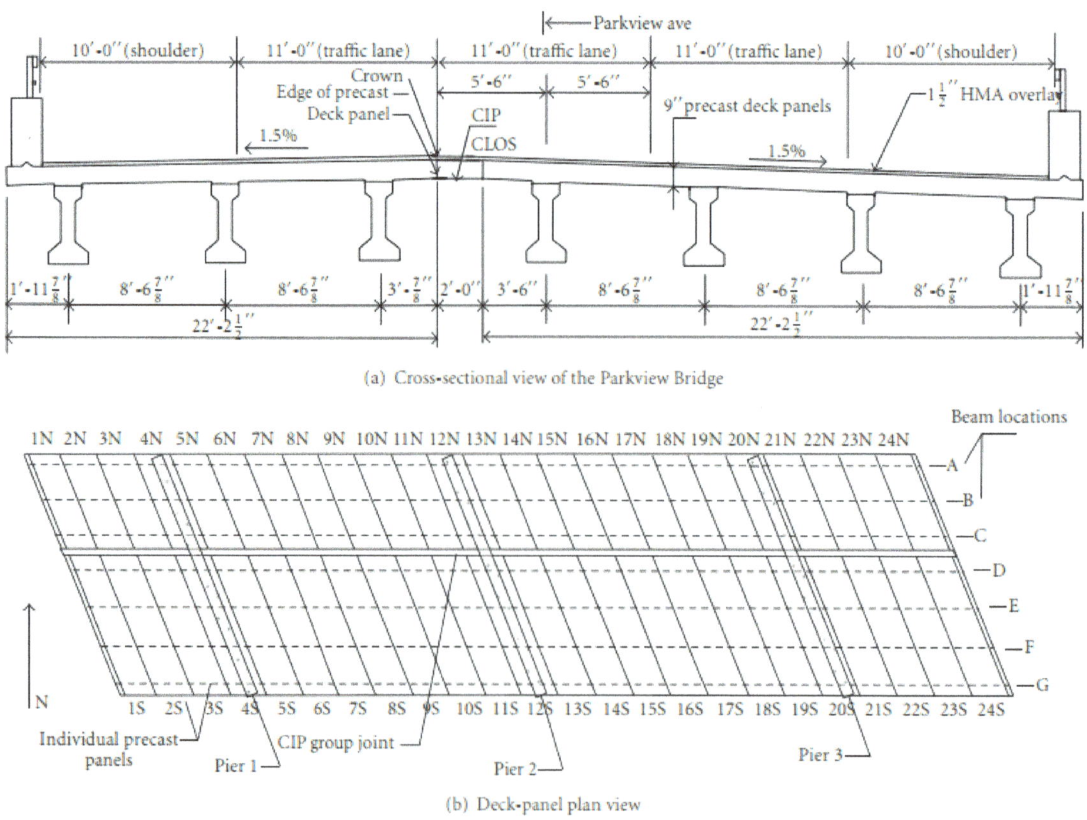

(a) Cross-sectional view of the Parkview Bridge

(b) Deck-panel plan view

Figure 1: Parkview bridge schematic.

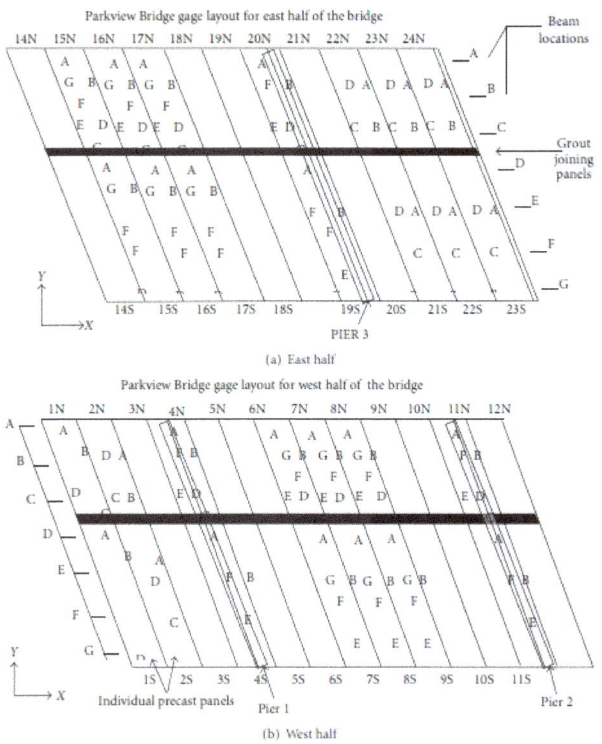

Figure 2: Gage layout and orientation for the Parkview Bridge.

Figure 3 presents a flow chart that describes the approaches and tasks used to carry out the testing. The first task is to determine load configurations that should produce the maximum moments in the bridge girders and the maximum deflections of the bridge deck surface. The next step is to compute stresses using calculated girder deck composite-section properties and measured deflections. Roadway-surface deflections are measured by optical surveying using a Trimble Dini level with a precision of 0.0012 inch (0.003048 cm). Stresses derived from measured deflections are compared to stresses derived from sensor strain readings for verification and further analysis.

LOAD TESTING DESIGN AND IMPLEMENTATION METHODOLOGY

This section describes the objectives, approaches, testing scenarios, load configurations, and testing procedures related to the design and implementation of the load testing program.

Load Testing Objectives and Approaches

The overall goal of conducting load tests on the Parkview Bridge is to quantify its response to various loading conditions. Specific objectives include measuring roadway-surface deflections due to various live loads, deriving stresses from measured deflections via analytical models, and comparing analytical results with results derived from sensor readings.

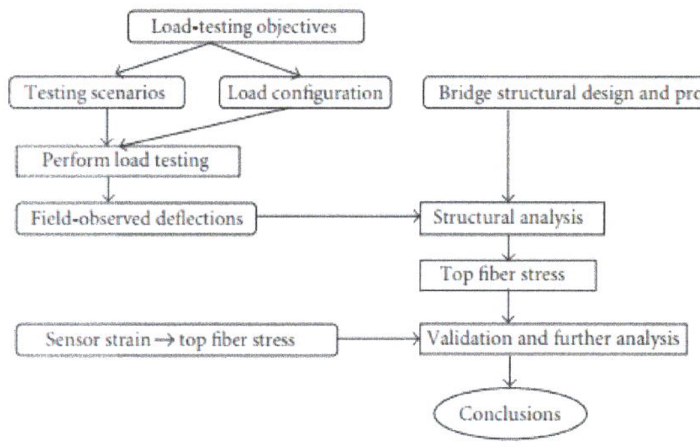

Figure 3: Flow chart of the bridge load testing.

Testing Scenarios

A total of ten load scenarios, including four single directional (one truck) and six bidirectional (two trucks with opposite heading directions), were designed and implemented in this study. Figure 4 illustrates the midspan locations, where the test loads were placed. Table 1 lists the testing scenarios with their load locations. Before the load testing, roadway surface elevations at these midspan locations, and also at locations above the bridge piers (represented by triangles), were measured before the loads were in place to provide a surface baseline. During the load tests, when the loads were in place, surface elevations at these locations were measured again to determine surface deflections due to specific test loads.

Table 1: Testing scenarios

Scenarios	Truck location (single direction—1 truck)	Truck location (bidirectional—2 trucks)
1	47	—
2	42	—
3	49	—
4	40	—
5	—	45, 44
6	—	47, 42
7	—	49, 40
8	—	51, 38
9	—	47, 40
10	—	45, 38

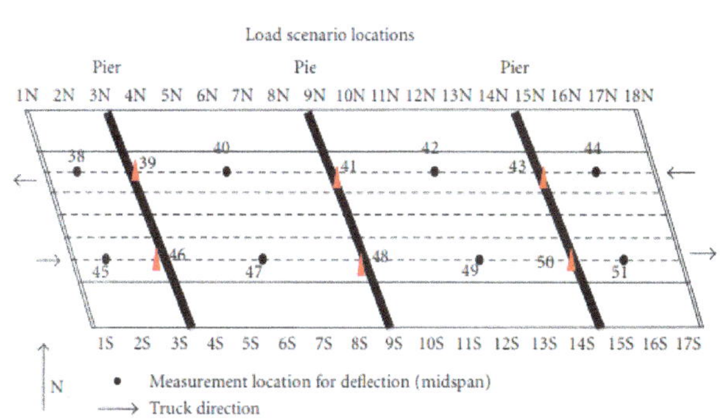

Figure 4: Load locations.

Load Configuration

Two types of trucks were used to provide test loads. Figure 5 illustrates the configuration of the trucks used for the single-directional and the bidirectional testing. In this study, single-directional testing means using only one truck, while bidirectional testing means using two trucks simultaneous and when these two trucks were driven to their intended locations facing each other. These configurations were chosen to closely match two trucks that are included in Michigan's set of legal truck-load configurations. The type I truck illustrated in Figure 5 was chosen to approximately match the HS20 design truck used in the design of the bridge. Table 2provides the measured axle weights for all three trucks (one type I truck and two type II trucks) used in this study.

Table 2: Actual loaded truck weights

	Loaded truck weight configurations		
Axle no.	Single directional truck type 1 weights (pounds*)	Bidirectional truck type 2 weights (pounds*)	
Front axle	9,640	17,850	18,350
No. 2 axle	35,540	18,050	18,600
No. 3 axle		17,800	18,250
No. 4 axle	34,580	—	—
No. 5 axle		—	—
Gross weight	79,760	53,700	55,200

*1 pound of mass = 0.4536 kg; 1 pound of force = 4.4482 N.

Type I truck for single-directional testing

(a)

Type II truck for bidirectional testing

(b)

Figure 5: General configurations of trucks.

Load Testing Procedure

The steps below were followed.

Step 1. Load trucks to approximate AASHTO HS20 and legal Michigan truck configuration and utilize truck weigh stations to determine actual axle weights.

Step 2. Based upon actual axle weight, mark a point on each truck in such a way that when this point is aligned with midspan locations of deck panels, the truck load causes the maximum moment in deck panels.

Step 3. While the trucks are being loaded prior to testing, the survey crew sets up and determines girder midspan locations and pier centerline locations as shown in Figure 4. These locations are marked for future alignment of truck marks and midspan locations in Step 5.

Step 4. Measure surface elevations of points illustrated in Figure 4 to establish a baseline for testing scenario 1.

Step 5. Move type I truck to point 47 in Figure 4 (scenario 1). After the truck is in place, measure surface elevations to determine bridge deflection due to the truck load.

Step 6. Record the time the truck is in position and ensure that the truck stays in position for a minimum of 10 minutes to allow the sensors to register their readings (sensors collect data at 5-second increments during the 10-minute time period, and these readings were averaged).

Step 7. Move the truck off the bridge.

Step 8. Repeat Steps 4 to 7 for the remaining testing scenarios.

Control of Temperature Variations

It is well recognized that environmental factors, particularly temperature variations, can lead to large strains, and when the structure is restrained as in bridges, they can cause large stresses, even though the wire strain gages are self-temperature compensating. Considering the extreme weather conditions that are common in Michigan during winter, load testing studies shall at least plan for temperature variations. In this load testing study, the original plan of controlling the effects of temperature variation was to start the testing in the morning and monitor the ambient temperature, and when it falls out of the 5° range, the testing pauses and resumes either later that day when the temperature cools down or next morning. In reality, it rained on the day of testing, and the temperature variation was a way below the 5° range to allow all the 10 scenarios to be tested on a single day.

LOAD TESTING RESULTS

The load-test program was performed twice: first in September 2008 before the bridge was opened to traffic and again in June 2009. During the first load test, the sensor network was not yet operational. During the second load test, the sensor network was functional and was used in the load test of the bridge. This section presents the results from the second load test only.

Table 3 summarizes surface deflections measured in the field during the second load testing, under the ten loading scenarios.

Table 3: Measured surface deflections

Location	Surface deflections in 10 scenarios (inches*)									
	1	2	3	4	5	6	7	8	9	10
38	0.01	0.05	−0.05	0.04	0.01	0.00	0.01	−0.06	0.00	−0.05
39	0.01	−0.04	−0.02	0.02	0.02	−0.01	−0.02	0.00	0.00	−0.01
40	0.00	−0.05	−0.02	−0.04	0.00	−0.01	−0.06	0.01	−0.07	0.02
41	0.02	−0.02	−0.04	−0.03	0.00	0.00	−0.02	0.00	0.00	−0.01
42	0.00	−0.09	−0.02	0.04	0.01	−0.07	−0.01	0.02	0.02	0.00
43	0.00	−0.02	−0.03	0.01	0.04	−0.01	0.00	0.01	0.01	−0.01
44	−0.02	−0.01	−0.03	0.00	−0.03	0.00	0.01	0.01	0.00	0.01
45	0.01	−0.05	−0.04	0.02	0.02	0.00	0.00	−0.02	0.01	−0.02
46	0.10	−0.02	−0.04	0.04	0.01	0.02	−0.01	−0.01	0.02	0.21
47	−0.04	−0.04	−0.02	0.01	−0.01	−0.07	0.00	0.02	−0.06	0.01
48	0.02	−0.01	−0.05	0.00	−0.02	0.00	0.02	0.01	−0.02	0.00
49	0.01	−0.01	−0.07	0.01	−0.01	0.01	−0.04	0.03	−0.01	0.00
50	0.00	−0.01	−0.03	−0.01	−0.02	−0.02	0.01	0.01	0.00	0.00
51	−0.02	−0.01	0.00	−0.02	0.00	0.01	0.00	−0.01	−0.01	0.01

*1 inch = 2.54 cm.

Field Measurements Validation

Three-dimensional plots of surface-deflection measurements were developed and compared with the finite element (FE) simulation results. Figure 6 illustrates the comparison for testing scenario 1. The deflected shape of the surface obtained from field observations matchs the deflected shape from analytical results, providing confidence in the measurements from testing scenario 1 observations. Similar comparisons were conducted to validate results in all the 10 testing scenarios.

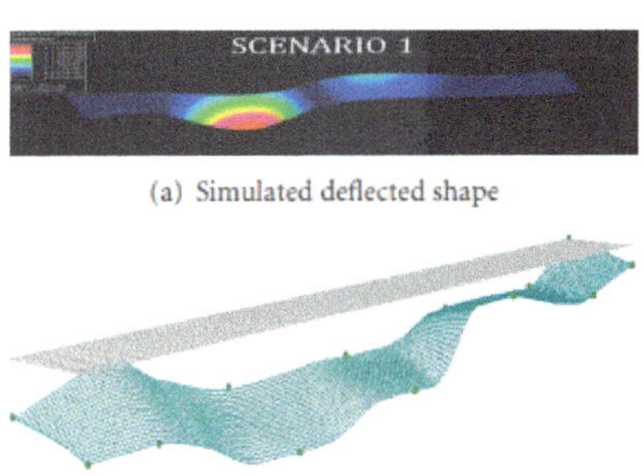

(a) Simulated deflected shape

(b) Actual deflected shape

Figure 6: Validating scenario 1 results (not drawn to scale).

Top Fiber Live Load Stresses from Deflection Measurements

After validating field observations, moments were derived from surface deflections, and top fiber stresses were calculated utilizing the two PCI equations below and using the simply supported moments at midspans (i.e., assuming zero moments at the piers for conservative results at midspans) [12]. The moments at the piers were then computed from the mid span moments using distribution factors obtained from simulated, unit-force loadings that mimic the truck loads from the 10 scenarios,

$$M_{\mathrm{LL}} = D\left(\frac{48EI_c}{5L^2}\right),$$

$$\sigma_{\mathrm{LL}} = \frac{M_{\mathrm{LL}}\,y}{I_c}, \qquad (1)$$

where M_{LL}: live load moment; D: deflection (inches); E: section modulus of elasticity (4,600,000 psi); I_c: moment of inertia of composite section (438,913 in^4); M_{LL}: moment (lb-in); L: span length (inches); y: distance from the top fiber to the neutral axis (18 inches); σ_{LL}: stress (psi).

The live-load stress results derived from deflection measurements are presented in Tables 4 and 5 (negative values indicate compression).

Table 4: Live load stresses from deflections (analytical)—south side panels (psi*)

Scenarios	Midspan (45)	Pier 1 (46)	Midspan (47)	Pier 2 (48)	Midspan (49)	Pier 3 (50)	Midspan (51)
Scenario 1	47	40	−34	12	10	−9	−50
Scenario 2	−186	−159	−29	−10	−11	−10	−17
Scenario 3	−140	−119	−19	−7	−57	−55	−7
Scenario 4	93	80	6	2	6	−6	−63
Scenario 5	70	60	−7	−2	−6	−6	−10
Scenario 6	0	0	−58	21	5	−5	33
Scenario 7	0	0	0	0	−35	34	−3
Scenario 8	−93	−80	14	5	20	19	−30
Scenario 9	47	40	−51	−19	−8	−7	−23
Scenario 10	−96	82	9	−3	−2	2	36

*1 psi = 6,895 kPa.

Table 5: Live load stresses from deflections (analytical)—north side panels (psi*)

Scenarios	Midspan (38)	Pier 1 (39)	Midspan (40)	Pier 2 (41)	Midspan (42)	Pier 3 (43)	Midspan (44)
Scenario 1	47	0	0	0	0	−57	−66
Scenario 2	186	−37	−38	−25	−68	−23	−27
Scenario 3	−186	−18	−19	−5	−12	−60	−70
Scenario 4	140	27	−28	−13	35	9	10
Scenario 5	47	4	−4	−2	5	81	−95
Scenario 6	0	−9	−10	21	−58	−9	10
Scenario 7	47	−5	−5	−4	−12	31	36
Scenario 8	−233	6	7	−7	19	23	27
Scenario 9	0	57	−59	4	10	9	−10
Scenario 10	−143	−15	15	−1	−3	−20	23

* 1 psi = 6,895 kPa.

Top Fiber Live Load Stresses from Sensor Readings

During the load testing, top fiber strains were recorded by the embedded sensors in the bridge deck panels and downloaded to the laboratory computer for analysis. Even though the sensors are installed throughout the entire bridge deck, only those sensors located along the longitudinal load path were used to derive live load stresses and to compare them to the stresses derived from deflections. Tables 6 and 7 present the stresses derived from the sensors for the south-side and the north-side panels, respectively (negative values indicate compression).

Table 6: Live load stresses from sensors—south side panels (psi*)

Scenarios	Midspan (45)	Pier 1 (46)	Midspan (47)	Pier 2 (48)	Midspan (49)	Pier 3 (50)	Midspan (51)
Scenario 1	1.48	10.60	−11.68	6.52	5.38	−6.82	−0.33
Scenario 2	0.80	−1.50	2.52	1.00	9.57	−14.05	0.45
Scenario 3	1.55	0.65	2.48	11.77	−24.14	8.97	2.80
Scenario 4	3.00	11.70	−2.43	11.95	2.75	−0.65	−0.20
Scenario 5	−2.80	−2.05	0.28	−0.20	−0.27	−0.55	−0.15
Scenario 6	−0.40	6.30	−13.50	22.10	−3.65	11.30	−0.15
Scenario 7	0.93	−0.28	−0.69	11.30	−30.68	4.72	1.18
Scenario 8	−5.95	0.00	0.85	−0.55	0.93	3.35	2.65
Scenario 9	3.50	−2.00	15.63	−21.35	−5.12	3.70	−1.60
Scenario 10	−8.55	−2.40	0.20	0.25	1.38	−0.55	0.05

* 1 psi = 6,895 kPa.

Table 7: Live load stresses from sensors—north side panels (psi*)

Scenarios	Midspan (38)	Pier 1 (39)	Midspan (40)	Pier 2 (41)	Midspan (42)	Pier 3 (43)	Midspan (44)
Scenario 1	1.60	8.85	−8.88	4.55	5.47	−0.87	−0.10
Scenario 2	1.95	−1.95	2.32	1.45	14.63	−19.70	0.40
Scenario 3	1.95	0.70	3.82	10.63	−11.06	4.93	4.02
Scenario 4	3.95	18.75	−26.33	17.75	3.53	0.25	0.50
Scenario 5	−2.75	−1.30	−0.10	−0.65	0.65	0.10	0.15
Scenario 6	−0.15	4.45	−6.53	21.90	−25.93	16.60	0.20
Scenario 7	1.55	5.45	−10.87	11.82	−3.51	0.02	1.70
Scenario 8	−6.05	0.35	−0.60	−0.95	−0.92	−2.65	2.55
Scenario 9	4.45	8.65	10.67	−21.30	−10.75	3.45	−0.95
Scenario 10	−6.90	3.10	0.17	−1.25	−1.22	−1.95	0.20

* 1 psi = 6,895 kPa.

Total Stresses

Live load stresses are added to dead load stresses calculated based upon structural design and material properties of this bridge to obtain the total top fiber stresses, which are then compared to the allowable compression and tension stresses to ensure that the structure is within design limits under varying loading scenarios. In calculating stresses from load test deflections, we conservatively assumed simply supported span moments as follows:

$$\sigma_{LL} = \frac{M_{LL} y}{I_c},$$

$$M_{LL} = D\left(\frac{48EI_c}{5L^2}\right), \tag{2}$$

where M_{LL}: live load moment (lb-in); D: deflection (inches); E: section modulus of elasticity (4,600,000 psi); I: moment of inertia of composite section (438,913 in^4); L: span length (inches); y: distance from the top fiber to the neutral axis (18 in); σ_{LL}: stress (psi).

The average 28-day compression strength (f' c) was recorded as approximately 8,000 psi from concrete samples taken during the casting. Therefore, maximum allowable stresses in the concrete are as follows:

$$\text{compression } (fc): \quad fc \le 0.45\, f'c \implies 3{,}600 \text{ psi},$$

$$\text{tension } (ft): \quad ft \le 6\sqrt{f'c} \implies 537 \text{ psi}. \tag{3}$$

Table 8 provides the deck dead-load stresses that are combined with the live-load stresses given in Tables 4through 7 to compute total stresses presented in Tables 9, 10, 11, and 12 for all scenarios using the deflection measurements and sensor-readings methods. It is clear from these tables that the total stresses are within the allowable limits for all testing scenarios for both the south and north panels.

Table 8: Deck dead load stresses (psi*)

Top fiber dead load stresses—south								
West abut	Midspan (45)	Pier 1 (46)	Midspan (47)	Pier 2 (48)	Midspan (49)	Pier 3 (50)	Midspan (51)	East abut
0	−436	−466	−773	−430	−760	−379	−428	0
Top fiber dead load stresses—north								
East abut	Midspan (44)	Pier 3 (43)	Midspan (42)	Pier 2 (41)	Midspan (40)	Pier 1 (39)	Midspan (38)	West abut
0	−428	−379	−760	−430	−773	−466	−436	0

*1 psi = 6,895 kPa.

Table 9: Total stresses based on deflection measurements—south (psi*)

Scenarios	Midspan (45)	Pier 1 (46)	Midspan (47)	Pier 2 (48)	Midspan (49)	Pier 3 (50)	Midspan (51)
Scenario 1	−390	−426	−806	−418	−751	−388	−478
Scenario 2	−623	−626	−801	−440	−771	−389	−444
Scenario 3	−576	−586	−792	−437	−817	−433	−434
Scenario 4	−343	−387	−767	−428	−755	−384	−491
Scenario 5	−366	−407	−779	−432	−766	−384	−438
Scenario 6	−436	−466	−830	−409	−755	−383	−395
Scenario 7	−436	−466	−773	−430	−796	−344	−431
Scenario 8	−529	−546	−758	−425	−740	−359	−458
Scenario 9	−390	−426	−824	−448	−768	−386	−451
Scenario 10	−532	−385	−764	−433	−762	−377	−391

*1 psi = 6,895 kPa.

Table 10: Total stresses based on deflection measurements—north (psi*)

Scenarios	Midspan (38)	Pier 1 (39)	Midspan (40)	Pier 2 (41)	Midspan (42)	Pier 3 (43)	Midspan (44)
Scenario 1	−390	−466	−773	−430	−760	−435	−494
Scenario 2	−250	−503	−811	−455	−828	−401	−454
Scenario 3	−623	−485	−792	−434	−773	−438	−498
Scenario 4	−296	−439	−801	−442	−726	−370	−418
Scenario 5	−390	−463	−776	−432	−755	−298	−522
Scenario 6	−436	−475	−782	−409	−819	−387	−418
Scenario 7	−390	−471	−777	−434	−772	−347	−391
Scenario 8	−669	−460	−766	−437	−741	−356	−401
Scenario 9	−436	−409	−832	−426	−750	−370	−438
Scenario 10	−579	−481	−757	−431	−763	−398	−405

*1 psi = 6,895 kPa.

Table 11: Total stresses based on sensor readings—south (psi*)

Scenarios	Midspan (45)	Pier 1 (46)	Midspan (47)	Pier 2 (48)	Midspan (49)	Pier 3 (50)	Midspan (51)
Scenario 1	−435	−455	−785	−423	−755	−386	−428
Scenario 2	−435	−468	−770	−429	−750	−393	−428
Scenario 3	−434	−465	−771	−418	−784	−370	−425
Scenario 4	−433	−454	−775	−418	−757	−380	−428
Scenario 5	−439	−468	−773	−430	−760	−380	−428
Scenario 6	−436	−460	−787	−408	−764	−368	−428
Scenario 7	−435	−466	−774	−419	−791	−374	−427
Scenario 8	−442	−466	−772	−431	−759	−376	−425
Scenario 9	−433	−468	−757	−451	−765	−375	−430
Scenario 10	−445	−468	−773	−430	−759	−380	−428

* 1 psi = 6,895 kPa.

Table 12: Total stresses based on sensor readings—north (psi*)

Scenarios	Midspan (38)	Pier 1 (39)	Midspan (40)	Pier 2 (41)	Midspan (42)	Pier 3 (43)	Midspan (44)
Scenario 1	−426	−370	−769	−425	−768	−467	−436
Scenario 2	−426	−381	−758	−429	−758	−486	−436
Scenario 3	−426	−378	−756	−419	−784	−461	−432
Scenario 4	−424	−360	−786	−412	−769	−466	−436
Scenario 5	−431	−380	−760	−431	−772	−466	−436
Scenario 6	−428	−375	−767	−408	−799	−449	−436
Scenario 7	−426	−374	−771	−418	−777	−466	−434
Scenario 8	−434	−379	−761	−431	−774	−469	−433
Scenario 9	−424	−370	−749	−451	−784	−463	−437
Scenario 10	−435	−376	−760	−431	−774	−468	−436

* 1 psi = 6,895 kPa.

DISCUSSIONS AND CONCLUDING REMARKS

Overall, the load tests were effective in providing information about the bridge's structural response. Stresses in all of the scenarios were well under the maximum allowable limits. The stress values derived from the strains acquired from the sensor network were compared to those derived from the measured deflections at every location and every scenario. An example comparison of stresses due to live loads for the south deck panels under scenario 1 is illustrated in Figure 7. Such a comparison leads to the following observations:(i) live load stresses at the deck top fiber are consistently small (both compression and tension stresses are less than 50 psi for scenario 1). In general, stresses due to live load are relatively smaller than stresses due to dead load;(ii)for a given location in testing scenario 1, the deflection-derived results and sensor-derived results are consistent from the perspective of whether the top-fiber stress is in compression or tension;(iii) at a given location under a specific testing scenario, the deck-panel top-fiber stresses derived from the sensor readings are consistently smaller than the stresses derived from deflection measurements;(iv)the stress distribution reveals compression on top fiber (positive moment) at the midspan location where the truck load was located in testing scenario 1 and tension on top fiber (negative moment) at the neighboring pier location. The stresses of the midpoints of nearby spans and pier locations are consistent with the loading configuration.

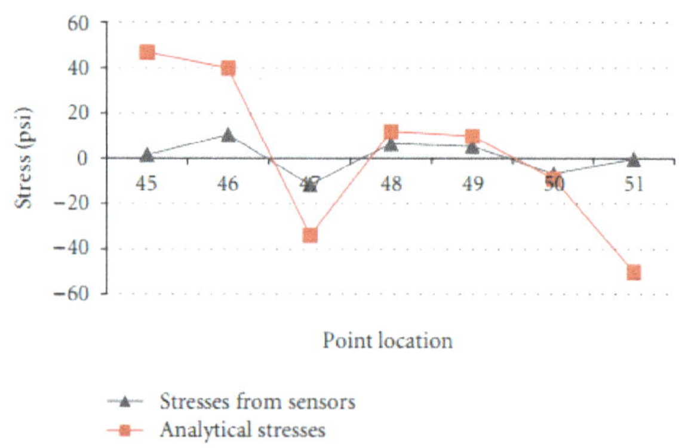

Figure 7: Comparison of analytical stresses and sensor stresses—south panels in testing scenario 1.

These comparisons were conducted for all 10 scenarios. It was observed from both sets of results that top fiber stresses due to live load are relatively small when compared to the stresses caused by dead loads and temperature variation. However, the difference between the two sets of calculated stresses is significant, considering the small stress values. Stress values derived from deflection measurements are consistently larger than those derived from sensor readings. Also, the stress types (tension/compression) derived from the deflection measurements do not always match those derived from sensor readings. In other words, some locations may have compression strains based on deflection measurements when the sensors are reading tension strains. These differences may be explained as follows:(i)the current practice of the analytical methods in calculating stresses is conservative, which confirms the reality that bridges perform better than what theoretical methods predict;(ii)the conservative assumptions used for computing stresses from deflections, namely using simply supported span moments when the bridge is continuously supported over four spans, result in large computed stresses;(iii)the surface deflection measures include effects of compression in bearings, slack in component fit, shear deformation, and pier/pile/cap deformations, leading to a larger surface deflection than that of the concrete deck itself;(iv)the bridge has a 3-inch flexible asphalt overlay on top of the concrete deck panels, possibly resulting in larger surface deflections than what the concrete deck is actually experiencing;(v)even though the surveying equipment can reach an accuracy of 0.0012 inch (0.003048 cm), the human error cannot be eliminated with methods used. That is, any measured deflections less than 0.01 inch (0.0254 cm) may be considered inaccurate due to the difficulty of eliminating human errors in reading the targets or holding the rod vertically. Therefore, it is believed that sensors provide much more accurate results than the optical survey instrument and whenever possible, sensors, instead of optical survey instrument, shall be used in bridge load testing;(vi) sensors are not located at the top fiber (18 inches (45.72 cm) from the neutral access of the composite section). Rather, they are 2.5 inches (6.35 cm) below the surface (due to protective-cover requirements) or 15.5 inches (39.37) from the neutral access, resulting in a slightly smaller stress than would exist at the top fiber. Sensor readings and resulting stresses could be adjusted to accommodate this situation and perhaps reduce the differences, but they will remain different.

Advanced deflection-measuring instruments and techniques may eliminate some of the human error, but at a premium cost. In our case, however, since the sensors already existed for the purpose of monitoring the health of the bridge deck, they provided a low-cost, accurate, and quick alternative to the deflection measurement load testing method. Additionally, with sensors installed throughout the bridge, a stress surface can be constructed to provide a complete view of the bridge structure's response to varying loading scenarios, a feature that is difficult to achieve using conventional surveying methods. Therefore, this study concludes that embedded sensors provide a more accurate and precise measuring alternative to measuring the structural response of bridges under various loads.

ACKNOWLEDGMENTS

The authors are grateful to the Michigan Department of Transportation (MDOT) for funding this study through MDOT Contract no. 04-0090/Z3. Any opinions, findings, conclusions, or recommendations expressed in this paper are those of the authors and do not necessarily reflect the views of MDOT or Western Michigan University.

REFERENCES

1. Federal Highway Administration (FHWA), "FHWA Bridge Programs NBI Data," 2008, http://www.fhwa.dot.gov/bridge/defbr07.cfm.
2. H. A. Capers and M. M. Valeo, "FHWA's international scan on ensuring bridge safety and serviceability," Transportation Research Record, no. 2202, pp. 117–123, 2010. ·
3. Federal Highway Administration (FHWA), Bridge Preservation Guide, Maintaining a State of Good Repair Using Cost Effective Investment Strategies, FHWA, 2011.
4. D. M. Frangopol, A. Strauss, and S. Kim, "Bridge reliability assessment based on monitoring," Journal of Bridge Engineering, vol. 13, no. 3, pp. 258–270, 2008. · · View at Scopus
5. W.-W. Wang, J.-G. Dai, G. Li, and C.-K. Huang, "Long-term behavior of prestressed old-new concrete composite beams," Journal of Bridge Engineering, vol. 16, no. 2, pp. 275–285, 2011. ·
6. O. Abudayyeh, H. Cai, B. Mellema, and S. Yehia, "Quantifying time and user cost savings for rapid bridge construction technique," Transportation Research Record, no. 2151, pp. 11–20, 2010. ·
7. NCHRP-234, "Manual for Bridge Rating Through Load-testing," National Cooperative Highway Research Program, Research Results Digest, Number 234, Transportation Research Board, Washington, DC, USA, 1998.
8. AASHTO, Manual for Condition Evaluation of Bridges, Washington, DC, USA, AASHTO, 2nd edition, 2000.
9. B. Phares, T. Wipf, L. Greimann, and Y. Lee, "Health Monitoring of Bridge Structures and Components Using Smart-Structure Technology," Tech. Rep. 0092-04-14, Wisconsin Highway Research Program, 2005.
10. D. D. Kleinhans, J. J. Myers, and A. Nanni, "Assessment of load transfer and load distribution in bridges utilizing FRP panels," Journal of Composites for Construction, vol. 11, no. 5, pp. 545–552, 2007. ·
11. T. Hou and P. J. Lynch, "Rapid-to-deploy wireless monitoring systems for static and dynamic load-testing of bridges: validation on the grove street bridge," in 13th Annual International Symposium on Smart Structures and Materials, Proceedings of SPIE, San Diego, Calif, USA, February 2006.
12. PCI, Bridge Design Manual, Precast/Prestressed Concrete Institute, Chicago, Ill, USA, 2003.

CHAPTER 14

LASER-BASED PEDESTRIAN TRACKING IN OUTDOOR ENVIRONMENTS BY MULTIPLE MOBILE ROBOTS

Masataka Ozaki, Kei Kakimuma, Masafumi Hashimoto and Kazuhiko Takahashi

Keywords: Pedestrian tracking, multi-mobile robots, laser range scanner, Bayesian filter, decentralized multi-sensor fusion

INTRODUCTION

Tracking (i.e., estimating the motion) of pedestrians is important to ensure safe navigation of mobile robots and vehicles. There has been much interest in the use of stereo vision or a laser range scanner (LRS) in mobile robotics and vehicle automation [1–5]. We previously presented a pedestrian tracking method using LRS mounted on mobile robots and automobiles [6–8].

Recently, many studies related to multi-robot coordination and cooperation have been conducted [9,10]. When these robots and vehicles are located near each other, they can share their sensing data. This implies that the robots and vehicles are considered to be a multi-sensor system. Therefore, even if pedestrians are located outside the sensing area of any individual robot or vehicle, it can detect pedestrians using the tracking data received from other robots and vehicles in the vicinity, and thus, multiple robots can improve the accuracy and reliability of pedestrian tracking.

In an intelligent transport system (ITS), if the tracking data is shared with neighboring vehicles through vehicle-to-vehicle communication, each vehicle can detect pedestrians efficiently. This facilitates the construction of an advanced driver-assistance-system. Even if pedestrians suddenly run into roads, the vehicles can detect them, and hence drivers can stop their vehicles to prevent an accident.

This paper presents a pedestrian tracking method employing multiple mobile robots and vehicles. Most studies of cooperative tracking by multiple mobile robots focus on motion planning and controlling issues [11–13]. These studies attempt to keep many moving objects visible to the mobile robots at all times while consuming as little motion energy as possible. In this paper, we address sensor-data fusion, through which pedestrian tracking is achieved by combining the tracking data from multiple mobile robots located in their vicinity.

There has been considerable research in cooperative pedestrian tracking using multiple static sensors located in the environment [14–18] and multi-sensors on robots [19,20]. Our previous work [8] presented a pedestrian tracking method using in-vehicle multi-laser range scanners; pedestrians were tracked by each LRS based on a Kalman filter. In order to enhance the tracking performance, the tracking data were blended based on covariance intersection (CI) method [21].

In this paper, we extend our previous method to pedestrian tracking with multiple mobile robots in the proximity to each other. As illustrated in Figure 1, our method contributes toward building a cooperative pedestrian tracking system using vehicles such as mobile robots, cars, and electric personal assistive mobility devices (EPAMD) in future urban city environments.

Recent studies [22,23] in cooperative pedestrian tracking by multiple mobile robots require centralized data fusion with a central server; sensing data captured by each robot are sent to a central server for subsequent data fusion. The centralized data fusion reduces system robustness and scalability. Our cooperative tracking system proposed in this paper functions in a decentralized manner without any central server. This paper is organized as follows: in Section 2, we present an overview

Citation: Masataka Ozaki, Kei Kakimuma, Masafumi Hashimoto and Kazuhiko Takahashi, Laser-Based Pedestrian Tracking in Outdoor Environments by Multiple Mobile Robots, doi:10.3390/s121114489.

of our experimental system. In Sections 3 and 4, we present methods of pedestrian tracking and robot localization. In Section 5, we describe simulation and experiment of pedestrian tracking to validate our method, followed by our conclusions.

Figure 1: Pedestrian tracking system using multiple vehicles such as mobile robots, cars, and EPAMD.

EXPERIMENTAL MOBILE ROBOTS

Figure 2 shows our mobile robot system used in the experiments. We use a Okatech Mecrobot wheeled mobile robot platform. Three robots each have two independent drive-wheels. A wheel encoder is attached to each of the drive wheels to measure the wheel velocity. A fiber-optic yaw rate gyro (Tamagawa Seiki, TA7319N3) is attached to the robot›s chassis to measure the turn velocity. This information is used to estimate the robot›s posture based on dead reckoning. Moreover, each robot is equipped with an RTK-GPS (Novatel ProPak-V3 GPS receiver) to identify its own posture in outdoor environments. The RTK-GPS provides three types of solution: fixed, float, and single solutions. Fixed solution offers range accuracy of less than 0.2 m, and float solution achieves range accuracy of about 0.2 to 1 m. In outdoor environments causing GPS multipath problems and bad weather conditions, we get single solutions with range accuracies of several meters.

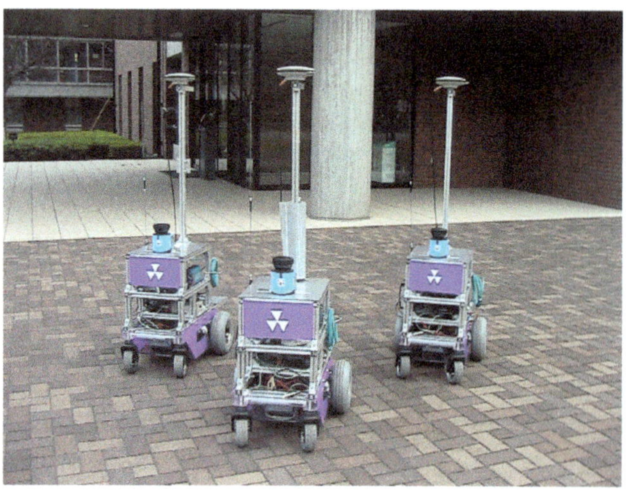

Figure 2: Overview of the mobile robot system.

The robot is equipped with a single-layered LRS (Sick LMS100). The LRS captures laser scan images that are represented by a sequence of distance samples in a horizontal plane of 270 deg. The angular resolution of the LRS is 0.5 deg, and the number of distance samples is 541 in one scan image. The onboard computer is a Lenovo ThinkPad R500 with a 2.4 GHz Intel core 2 duo processor, and the operating system used is Microsoft Windows Vista. The sampling frequency of the sensors is 10 Hz.

Broadcast communication via a wireless LAN is used to exchange information among the robots. It takes approximately 40 ms to exchange information between robots. We employ a ring-type network structure in which the robots transmit information in the sequence: robot #1, #2, and #3.

PEDESTRIAN TRACKING

Overview

We define two coordinate frames: the world coordinate frame, $\Sigma_w(O_w : X_wY_w)$ and the i-th robot coordinate frame, $\Sigma_i(O_i : X_iY_i)$ attached at the robot body, where i = 1, 2, 3. Each robot independently detects pedestrians using its own laser image based on an occupancy-grid-based method. Table 1 briefly shows our occupancy grid algorithm in the pseudo-code format. Our detection method is detailed in [6,7].

Table 1: Occupancy grid algorithm

1. Let $C[X_{max}, Y_{max}]$ be a two dimensional array of cells counting the number of observations, where X_{max} and Y_{max} are the maximum X and Y coordinates.
2. Intialize all cells in C to zero.
3. Make an observation with laser range scanner.
4. Determine which cells in C are occupied in the current laser scan image, and increment the occupied cells $C[X, Y]$.
5. If $C[X, Y] == 0$, then we have no information on the cell—free space.
6. If $C[X, Y] \geqq 7$, then the cell is "static cell"—static object.
7. If $0 < C[X, Y] < 7$, then the cell is "moving cell"—moving object; pedestrian.
8. Repeat from step 3.

The detected pedestrians are tracked using the following two tracking modes (Figure 3):

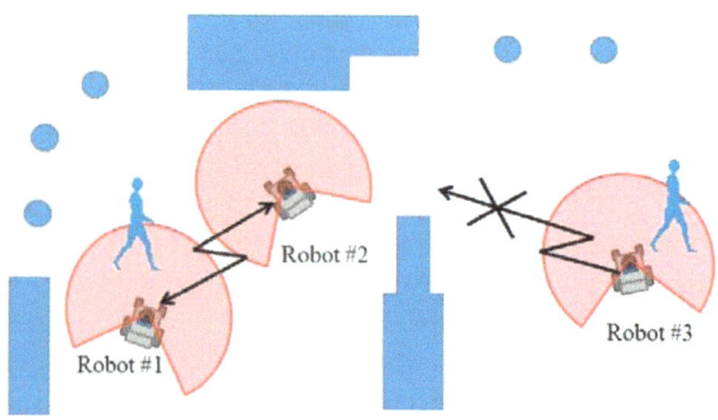

Figure 3: Tracking mode; robots #1 and #2 track a pedestrian in cooperative tracking mode, and robot #3 tracks a pedestrian in individual tracking mode. The red arc indicates the LRS sensing area.

- Individual tracking by a single robot: Each robot individually tracks pedestrians without any tracking data from other robots. The robot can only track pedestrians inside its LRS sensing area.
- Cooperative tracking by multiple robots: The robots track pedestrians by sharing their own tracking data so that each robot can track pedestrians both inside and outside its LRS sensing area.

Individual Tracking

A pedestrian position in Σ_w is denoted by (x,y). If the pedestrian is assumed to move at almost constant velocity, the rate kinematics is given by:

$$\boldsymbol{x}_{(t)} = \boldsymbol{F}\boldsymbol{x}_{(t-1)} + \boldsymbol{G}\Delta\boldsymbol{x}_{(t-1)}$$
$$= \begin{pmatrix} 1 & \tau & 0 & 0 \\ 0 & 1 & 0 & 0 \\ 0 & 0 & 1 & \tau \\ 0 & 0 & 0 & 1 \end{pmatrix} \boldsymbol{x}_{(t-1)} + \begin{pmatrix} \tau^2/2 & 0 \\ \tau & 0 \\ 0 & \tau^2/2 \\ 0 & \tau \end{pmatrix} \Delta\boldsymbol{x}_{(t-1)} \qquad (1)$$

where $\boldsymbol{x} = (x, \dot{x}, y, \dot{y})^T$. $\Delta\boldsymbol{x} = (\Delta\ddot{x}, \Delta\ddot{y})^T$ is an unknown acceleration (plant noise). τ is a sampling period of sensors; in our experimental system, τ is 0.1 s.

The measurement model related to the pedestrian is then:

$$\boldsymbol{z}_{(t)} = \boldsymbol{H}_{i(t)}\boldsymbol{x}_{(t)} + \boldsymbol{H}'_{i(t)}\boldsymbol{u}_{i(t)} + \Delta\boldsymbol{z}_{(t)}$$
$$= \begin{pmatrix} \cos\psi_{i(t)} & 0 & -\sin\psi_{i(t)} & 0 \\ \sin\psi_{i(t)} & 0 & \cos\psi_{i(t)} & 0 \end{pmatrix} \boldsymbol{x}_{(t)} + \begin{pmatrix} \cos\psi_{i(t)} & -\sin\psi_{i(t)} \\ \sin\psi_{i(t)} & \cos\psi_{i(t)} \end{pmatrix} \boldsymbol{u}_{i(t)} + \Delta\boldsymbol{z}_{(t)} \qquad (2)$$

where $\boldsymbol{z} = (z_x, z_y)^T$ is the measurement represented in Σ_i. $\Delta\boldsymbol{z}$ is the measurement noise. $\boldsymbol{u}_i = (x_i, y_i)^T$ is the position of the i-th robot in Σ_w. ψ_i is the orientation of the i-th robot in Σ_w. The posture (position and orientation) $\boldsymbol{x}_i = (x_i, y_i, \psi_i)^T$ is determined using the localization system described in Section 4.

From Equation (1), the pedestrian›s posture $\hat{\boldsymbol{x}}$ and its associated error covariance $\boldsymbol{P}$ are predicted using a Kalman filter [24]:

$$\begin{cases} \hat{\boldsymbol{x}}_{(t/t-1)} = \boldsymbol{F}\hat{\boldsymbol{x}}_{(t-1)} \\ \boldsymbol{P}_{(t/t-1)} = \boldsymbol{F}\boldsymbol{P}_{(t-1)}\boldsymbol{F}^T + \boldsymbol{G}\boldsymbol{Q}_{(t-1)}\boldsymbol{G}^T \end{cases} \qquad (3)$$

where $\boldsymbol{Q}$ is the covariance of the plant noise $\Delta\boldsymbol{x}$.

To track multiple pedestrians, as shown in Figure 4(a), a validation region with a constant radius is set around the predicted position $(\hat{x}, \hat{y})$ of each tracked pedestrian. The measurements inside the validation region are considered to be obtained from the tracked pedestrian, and it is applied to the track updated with the Kalman filter. On the other hand, the measurements outside the validation region are considered to be false alarms, and are therefore, discarded. From Equations (2) and (3), the posture of the tracked pedestrian and its associated error covariance are updated by:

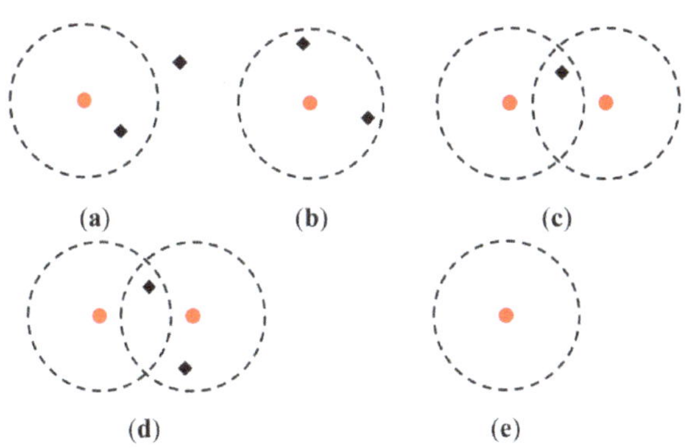

Figure 4: Tracking condition; the red circle and black diamond indicate the tracked pedestrian and the measurement, respectively. The dashed circle indicates the validation region. (**a**) Case 1. (**b**) Case 2. (**c**) Case 3. (**d**) Case 4. (**e**) Case 5.

$$\begin{cases} \hat{\boldsymbol{x}}_{(t)} = \hat{\boldsymbol{x}}_{(t/t-1)} + \boldsymbol{K}_{(t)}(\boldsymbol{z}_{(t)} - \boldsymbol{H}_{i(t)}\hat{\boldsymbol{x}}_{(t/t-1)} + \boldsymbol{H}'_{i(t)}\boldsymbol{u}_{i(t)}) \\ \boldsymbol{P}_{(t)} = \boldsymbol{P}_{(t/t-1)} + \boldsymbol{K}_{(t)}\boldsymbol{H}_{i(t)}\boldsymbol{P}_{(t/t-1)} \end{cases} \qquad (4)$$

where $\boldsymbol{K}_{(t)} = \boldsymbol{P}_{(t/t-1)}\boldsymbol{H}_{i(t)}^{T}\boldsymbol{S}_{(t/t-1)}^{-1}$, and $\boldsymbol{S}_{(t/t-1)} = \boldsymbol{H}_i^{(t)}\boldsymbol{P}_{(t/t-1)}\boldsymbol{H}_i^{(t)T} + \boldsymbol{R}_{(t)}$. $\boldsymbol{R}$ is the covariance of the measurement noise $\Delta\boldsymbol{z}$.

In our simulation and experiment described in Section 5, the radius of the validation region is set at 1.0 m. The covariances of the plant and measurement noises in Equations (3) and (4) are set at $\boldsymbol{Q}$ = diag (1.0 m^2/s^4, 1.0 m^2/s^4) and $\boldsymbol{R}$ = diag (0.01 m^2, 0.01 m^2), respectively.

In crowded environments, as shown in Figures 4(b–d), multiple measurements exist inside a validation region; multiple tracked pedestrians also compete for measurements. To achieve a reliable data association (matching of tracked pedestrians and measurements), we apply a global-nearest-neighbor (GNN) algorithm [25].

We consider that, in a validation region, J pedestrians exist and K measurements are received, where J does not necessarily equal K. We then define the distance measure λ_{jk} from the j-th tracked pedestrian to the k-th measurement, where j = 1,2, …, J and k = 1,2, …, K as:

$$\lambda_{jk} = \sqrt{(\boldsymbol{z}_{k(t)} - \hat{\boldsymbol{u}}_{j(t/t-1)})^T (\boldsymbol{S}_{j(t/t-1)})^{-1} (\boldsymbol{z}_{k(t)} - \hat{\boldsymbol{u}}_{j(t/t-1)})} \qquad (5)$$

where $\boldsymbol{S}_j^{(t/t-1)} = \boldsymbol{H}_j^{(t)}\boldsymbol{P}_j^{(t/t-1)}\boldsymbol{H}_j^{(t)T} + \boldsymbol{R}(t)$, and $\boldsymbol{u}_j^{(t/t-1)}$ is the predicted position of the j-th tracked pedestrian.

We then define the following cost matrix $\boldsymbol{\Lambda}$:

$$\boldsymbol{\Lambda} = \begin{pmatrix} \lambda_{11} & \lambda_{12} & \cdots & \lambda_{1K} \\ \lambda_{21} & \lambda_{22} & \cdots & \lambda_{2K} \\ \vdots & \vdots & \ddots & \vdots \\ \lambda_{J1} & \lambda_{J2} & \cdots & \lambda_{JK} \end{pmatrix} \qquad (6)$$

We assume that the a(j)-th measurement is assigned to the j-th pedestrian. The data association is achieved by finding the a(j) based on the Munkres algorithm [26] so that $\sum_{j=1}^{J} \lambda_{ja(j)}$ can be minimized. It is noted that if the k-th measurement does not exist inside the validation of the j-th tracked person, we set the distance measure at $\lambda_{jk} = \infty$.

Pedestrians always appear in and disappear from the LRS sensing area. They also face interaction and occlusion issues. In order to handle such conditions, we implement a tracking management system based on the following rules:

a. Track initiation: As shown in Figures 4(a–c), the measurements that are not matched with any tracked pedestrians are considered to come from new pedestrians or false alarms, which disappear soon. Therefore, we tentatively initiate tracking of the measurements with Kalman filter. If the measurements are always visible in more than N_1 s, they are considered to come from new pedestrians, and the tracking is continued. If the measurements disappear within N_1 s, they are considered to be the false alarms, and the tentative tracking is terminated.

b. Track termination: When the tracking pedestrians exit the sensing area of the LRS or they meet occlusion, no measurements exist within their validation regions. If no measurements arise from the temporal occlusion, the measurements appear again. We thus predict the positions of the tracking pedestrians with the Kalman filter. If the measurements appear again within N_2 s, we proceed with the tracking. Otherwise (see Figure 4(e)), we terminate the tracking. In our simulation and experiment described in Section 5, we set $N_1 = 1.5$ and $N_2 = 3.0$ by trial and error.

For simplicity, in this paper, pedestrians are assumed to move at an almost constant velocity, and they are tracked using the usual Kalman filter. If the pedestrians move randomly, such as walking, running, going or stopping suddenly, and turning suddenly, using multi-model-based tracking can improve the tracking performance [14,15].

Cooperative Tracking

When the robots are located near to each other, the tracking mode is switched to cooperative tracking. They communicate with each other and exchange their own tracking data, which consist of estimated positions and velocities of tracked pedestrians and their associated error covariances. Because tracking data are shared, each individual robot constantly tracks pedestrians both inside and outside its own LRS sensing area.

To elucidate cooperative tracking in detail, we consider two robots #1 and #2, as shown in Figure 5. The tracking data for the m-th pedestrian tracked by robot #1 is denoted by $\boldsymbol{I}_m^{(1)} = \{\hat{\boldsymbol{x}}_m^{(1)}, \boldsymbol{P}_m^{(1)}\}$, where m = 1,2, ... $\hat{\boldsymbol{x}}_m^{(1)}$ denotes the estimate (position estimate $\hat{\boldsymbol{q}}_m^{(1)}$ and velocity estimate $\hat{\dot{\boldsymbol{q}}}_m^{(1)}$); $\boldsymbol{P}_m^{(1)}$ is its associated error covariance. Similarly, the tracking data for the n-th pedestrian tracked by robot #2 is denoted by $\boldsymbol{I}_n^{(2)} = \{\hat{\boldsymbol{x}}_n^{(2)}, \boldsymbol{P}_n^{(2)}\}$, where n = 1,2, ... We consider that robot #1 combines the tracking data sent from robot #2 with its own tracking data. Combining the tracking data of robot #1 with that of robot #2 can be achieved similarly.

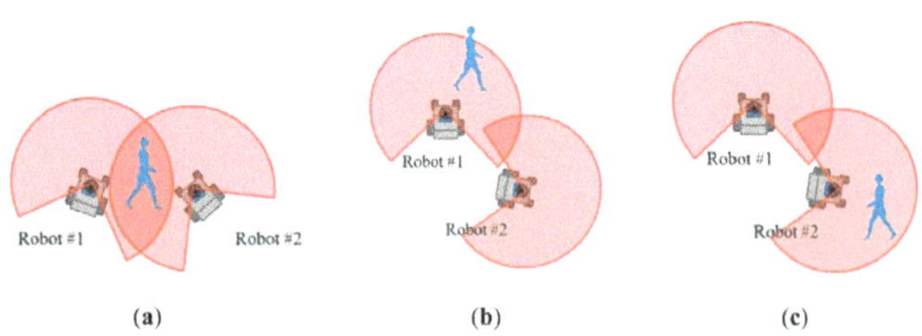

Figure 5: Conditions in cooperative tracking. (**a**) Case 1. (**b**) Case 2. (**c**) Case 3.

First, we set a validation region with a constant radius around the position estimate $\hat{\boldsymbol{q}}_m^{(1)}$ of the m-th pedestrian tracked by robot #1. We consider the position estimate $\hat{\boldsymbol{q}}_n^{(2)}$ of the n-th pedestrian tracked by robot #2 as the measurement, and then, we can determine data association (one-to-one matching of pedestrians tracked by robots #1 and #2) using the GNN algorithm. The GNN-based data association in cooperative tracking is similar as one in individual tracking mentioned in Section 3.1. In our simulation and experiment described in Section 5, the radius of the validation region is set at 1.2 m.

As shown in Figure 5(a), when a pedestrian is detected inside the sensing areas of both robots #1 and #2, the two estimates, $\hat{\boldsymbol{q}}^{(1)}$ and $\hat{\boldsymbol{q}}^{(2)}$, of the pedestrian can be matched. For the matched pedestrian, robot #1 updates its own tracking data by the CI method [21]:

$$\begin{cases} \hat{\boldsymbol{x}}_{(t)}^{(1)+} = \boldsymbol{P}_{(t)}^{(1)+}\left\{\omega \boldsymbol{P}_{(t)}^{(1)-1}\hat{\boldsymbol{x}}_{(t)}^{(1)} + (1-\omega)\boldsymbol{P}_{(t)}^{(2)-1}\hat{\boldsymbol{x}}_{(t)}^{(2)}\right\} \\ \boldsymbol{P}_{(t)}^{(1)+} = \left\{\omega \boldsymbol{P}_{(t)}^{(1)-1} + (1-\omega)\boldsymbol{P}_{(t)}^{(2)-1}\right\}^{-1} \end{cases} \tag{7}$$

where $\boldsymbol{I}^{(1)} = \{\hat{\boldsymbol{x}}^{(1)}, \boldsymbol{P}^{(1)}\}$ and $\boldsymbol{I}^{(2)} = \{\hat{\boldsymbol{x}}^{(2)}, \boldsymbol{P}^{(2)}\}$ denote the tracking data of the matched pedestrian. $\hat{\boldsymbol{x}}^{(1)+}$ and $\boldsymbol{P}^{(1)+}$ denote the updated tracking data and its associated error covariance, respectively. The weight ω is selected using the Golden selection search (GSS) method so that the determinant of $\boldsymbol{P}^{(1)+}$ can be minimized under the constraint $0 \leq \omega \leq 1$. In simulation and experiment described in Section 5, the convergence threshold of the weight ω is set at $1.0*10^{-4}$. In this case, we can determine the appropriate weight ω by iterative calculation of less than twenty times.

As shown in Figure 5(b,c), for non-matched pedestrian, robot #1 updates its own tracking data as follows:

- When a pedestrian appears inside the sensing area of robot #1 but outside that of robot #2, as shown in Figure 5(b), robot #1 has the tracking data $\boldsymbol{I}^{(1)}$, but robot #2 does not have $\boldsymbol{I}^{(2)}$. Then, robot #1 sets $\boldsymbol{I}^{(1)+} = \boldsymbol{I}^{(1)}$.
- When a pedestrian appears inside the sensing area of robot #2 but outside that of robot #1 as shown in Figure 6(c), robot #2 has the tracking data $\boldsymbol{I}^{(2)}$, but robot #1 does not have $\boldsymbol{I}^{(1)}$. Then, robot #1 sets $\boldsymbol{I}^{(1)+} = \boldsymbol{I}^{(2)}$.

Cooperative tracking with three or more robots can be achieved in a similar manner. Decentralized data fusion provides better system scalability and reliability than centralized data fusion [21]. Therefore, we combine the tracking data in a decentralized manner. Statistically, the tracking data are highly correlated. The conventional Kalman filter-based fusion hampers the development of a decentralized system because it needs to calculate the degree of their correlation. The CI method allows accurate fusion of the tracking data in a decentralized manner without the knowledge of the degree of their correlation. Therefore, we apply the CI algorithm.

Data association is important in pedestrian tracking. In this paper, we apply GNN-based data association to match the current measurement scan to the existing tracks. An alternative effective data association algorithm is multiple hypothesis tracking (MHT) [27,28]. In MHT, the feasible measurement-to-track association hypotheses are enumerated and evaluated up to a certain time depth. The MHT-based data association may outperform the GNN data association in crowded environments; however, in our experience, MHT data association makes real-time tracking difficult in crowded environments because it requires the evaluation of an exponentially increasing number of feasible data association hypotheses. The MHT data association also requires centralized data fusion with a central server [22]. Therefore, we apply GNN data association.

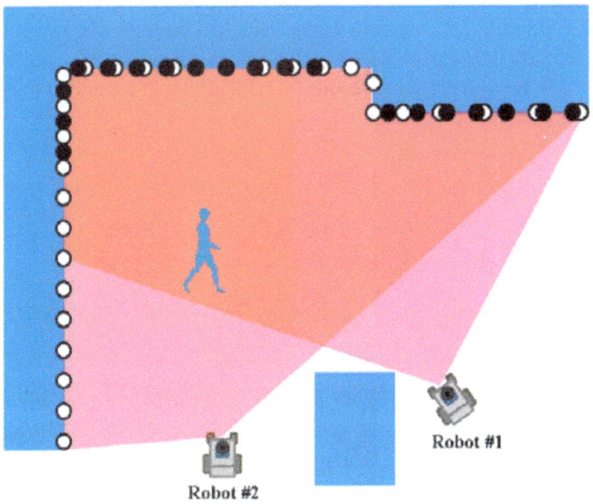

Figure 6: Robot localization by relative-scan matching; the solid and open circles indicate images scanned by robots #1 and #2, respectively.

ESTIMATION OF ROBOT POSTURE

To achieve cooperative tracking, each robot must always identify its own posture (position and orientation) with a high degree of accuracy in a world coordinate frame Σ_w and map the tracking data onto Σ_w, for which we apply RTK-GPS. The robot also determines its own posture by a scan matching based localization to improve the accuracy of its posture. If the robot cannot retrieve RTK-GPS information, only the scan matching based localization is applied to determine its own posture.

RTK-GPS Based Localization

The robot estimates its own velocity (linear/turning velocity) based on dead reckoning using the wheel encoders and gyro. The robot is assumed to move at nearly constant velocity. Motion and measurement models of the i-th robot are then given by Equations (8) and (9), respectively:

$$V_{i(t)} = V_{i(t-1)} + \Delta V_{i(t-1)} \tag{8}$$

$$z_{i(t)} = \begin{pmatrix} 1 & -b/2 \\ 1 & b/2 \\ 0 & 1 \end{pmatrix} V_{i(t)} + \Delta z_{i(t)} \tag{9}$$

where $V_i = (v_i, \dot{\psi}_i)^T$; v_i is the linear velocity and $\dot{\psi}_i$ is the turning velocity. $z_i = (z_l, z_r, z_\psi)^T$; z_l and z_r are the velocities of the left and right wheels, respectively, measured by the wheel encoders , and z_ψ is the gyro output. ΔV_i and Δz_i are unknown acceleration (disturbance) and the sensor noise, respectively. b is the tread length of the robot.

From Equations (8) and (9), the robot velocity V_i is estimated using Kalman filter. Based on the velocity estimate $\hat{V}_i$, we can determine the posture of the i-th robot $x_i = (x_i, y_i, \psi_i)^T$ and its associated covariance by Equations (10) and (11), respectively:

$$\hat{x}_{i(t/t-1)} = \begin{pmatrix} \hat{x}_{i(t-1)} + \hat{V}_{i(t-1)} \tau \cos(\hat{\psi}_{i(t-1)} + \frac{\hat{\dot{\psi}}_{i(t-1)}}{2} \tau) \\ \hat{y}_{i(t-1)} + \hat{V}_{i(t-1)} \tau \sin(\hat{\psi}_{i(t-1)} + \frac{\hat{\dot{\psi}}_{i(t-1)}}{2} \tau) \\ \hat{\psi}_{i(t-1)} + \hat{\dot{\psi}}_{i(t-1)} \tau \end{pmatrix} \tag{10}$$

$$P_{i(t/t-1)} = \nabla f_{(t-1)} P_{i(t-1)} \nabla f_{(t-1)}^T + \nabla f'_{(t-1)} Q_{i(t-1)} \nabla f'^T_{(t-1)} \tag{11}$$

where $\hat{x}_{i(t/t-1)}$ and $P_{i(t/t-1)}$ are the posture estimate and its error covariance, respectively. Q_i is the error covariance of $\hat{V}_i$. ∇f and $\nabla f'$ are the Jacobian matrices of Equation (10) at $\hat{x}_{i(t/t-1)}$ and $\hat{V}_i$, respectively. τ is a sampling period of sensors.

The measurement model related to the RTK-GPS is given by:

$$z_{GPS(t)} = Hx_{i(t)} + \Delta z_{GPS(t)} = \begin{pmatrix} 1 & 0 & 0 \\ 0 & 1 & 0 \end{pmatrix} x_{i(t)} + \Delta z_{GPS(t)} \tag{12}$$

where z_{GPS} is the measurement; position of the i-th robot in Σ_w, and Δz_{GPS} is the measurement noise.

If the robot obtains posture information from the RTK-GPS, the robot can update its own posture and its associated covariance using Kalman filter as follows:

$$\begin{cases} \hat{x}_{i(t)} = \hat{x}_{i(t/t-1)} + K_{(t)}[z_{GPS(t)} - H\hat{x}_{i(t/t-1)}] \\ P_{i(t)} = P_{i(t/t-1)} - K_{(t)} HP_{i(t/t-1)} \end{cases} \tag{13}$$

where $K_{(t)} = P_i^{(t/t-1)} H_{(t)}^T (HP_i^{(t/t-1)} H^T + R_{GPS}^{(t)})^{-1}$, and R_{GPS} is the covariance of the measurement noise Δz_{GPS}.

Scan Matching Based Localization

When multiple robots are located near each other, they have an overlapping sensing area. They improve their own posture accuracy by exchanging their laser-scan images and matching them in their overlapping sensing area.

To elucidate scan matching based localization in detail, we consider two robots #1 and #2, as shown in Figure 6, where the two robots are located near each other and their sensing areas partially overlap. We define the posture of robot #2 relative to robot #1 by ${}^2z_1 = ({}^2x_1, {}^2y_1, {}^2\psi_1)^T$ in Σ_w. Robot #1 broadcasts its own posture and a laser scan image obtained by its own LRS to robot #2. Robot #2 determines the relative posture, ${}^2z_1$, by matching its own laser scan image with that sent from robot #1 (Appendix). Hereafter, we call the laser scan matching for estimating relative posture as relative-scan matching.

Measurement model related to the relative-scan matching is given by:

$${}^2z_{1(t)} = g(\hat{x}_{1(t)}, x_{2(t)}) + \Delta^2 z_{1(t)} = \begin{pmatrix} \cos\psi_{2(t)} & -\sin\psi_{2(t)} & 0 \\ \sin\psi_{2(t)} & \cos\psi_{2(t)} & 0 \\ 0 & 0 & 1 \end{pmatrix} \begin{pmatrix} \hat{x}_{1(t)} - x_{2(t)} \\ \hat{y}_{1(t)} - y_{2(t)} \\ \hat{\psi}_{1(t)} - \psi_{2(t)} \end{pmatrix} + \Delta^2 z_{1(t)} \tag{14}$$

where $\hat{x}_1$ is the posture of robot #1 estimated by the RTK-GPS-based localization. $\Delta^2 z_1$ is the error of the relative posture. From Equations (13) and (14), robot #2 can determine its own posture $\hat{x}_2$ and its associated error covariance P_2 using Kalman filter:

$$\begin{cases} \hat{\boldsymbol{x}}_{2(t)} = \hat{\boldsymbol{x}}_{2(t/t-1)} + \boldsymbol{K}_{(t)}[{}^{2}z_{1(t)} - \boldsymbol{g}(\hat{\boldsymbol{x}}_{1(t)}, \hat{\boldsymbol{x}}_{2(t/t-1)})] \\ \boldsymbol{P}_{2(t)} = \boldsymbol{P}_{2(t/t-1)} - \boldsymbol{K}_{(t)} \nabla \boldsymbol{g}_{(t)} \boldsymbol{P}_{2(t/t-1)} \end{cases} \quad (15)$$

where $\boldsymbol{K}_{(t)} = \boldsymbol{P}_2^{(t/t-1)} \nabla \boldsymbol{g}_{(t)}^{T} (\nabla \boldsymbol{g}_{(t)} \boldsymbol{P}_2^{(t/t-1)} \nabla \boldsymbol{g}_{(t)}^{T} + {}^{2}\boldsymbol{R}_1^{(t)})^{-1}$. $\nabla \boldsymbol{g}$ is the Jacobian matrix of $\boldsymbol{g}$ in Equation (14) at $\hat{x}_2^{(t/t-1)}$, and ${}^{2}\boldsymbol{R}_1$ is covariance of $\Delta^2 z_1$.

SIMULATION AND EXPERIMENTAL RESULTS

Simulation Results

In the experiment described in Section 5.2, it is very difficult to recognize the true positions of the tracked pedestrians. Therefore, we evaluate the performance of the proposed method by simulating four-pedestrian tracking by two robots. As shown in Figure 7, two robots stops at the coordinates (x,y)=(−8.0,3.0)m and (7.0,−1.0)m, and pedestrians move at the velocity of 0.1–1.7 m/s; pedestrians #1 and #2 move side-by-side at the distance of 0.8 m from start point A, and pedestrians #3 and #4 move side-by-side at the distance of 0.8 m from start point B. Four pedestrians meet each other at point C. In the simulation, the pedestrians are assumed to be always detected correctly, and measurement noise of LRS is assumed to be uniform distribution between −0.05 m and 0.05 m. Simulation tool/software is self-produced using C++ language.

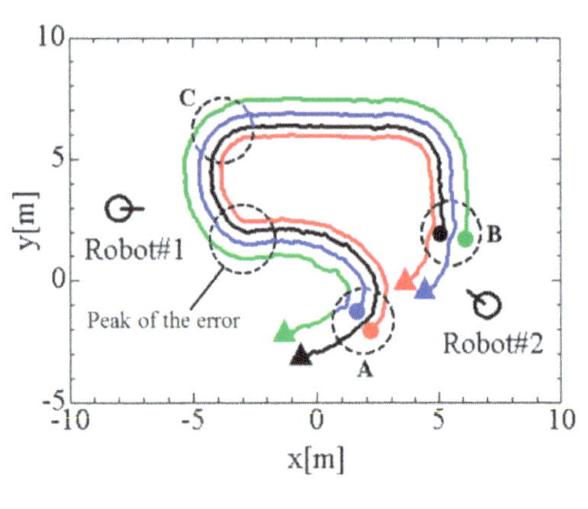

Figure 7: Simulation condition. Red, blue, green and black lines indicate moving path of pedestrians #1, #2, #3 and #4, respectively.

Figure 8(a) shows the effect of tracking mode and data association methods to the tracking error; GNN and conventional nearest neighbor (NN) [24] methods are applied for the data association. Tracking error is evaluated by the following root mean squared error (RMS):

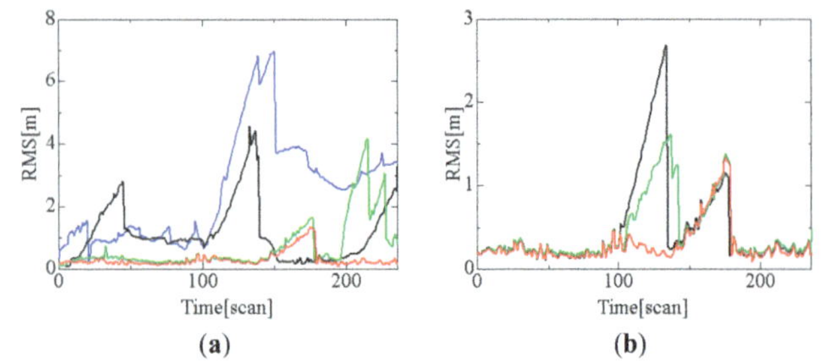

Figure 8: Tracking error. (**a**) Effect of data association method and tracking mode. (**b**) Effect of data fusion method. In (a), red, green, blue and black lines indicate the results by cooperative tracking using GNN, individual tracking using GNN, cooperative tracking using NN, and individual tracking using NN, respectively. In (b), red, green and black lines indicate the results by CI, Kalman filter and averaging method, respectively.

$$RMS_{(t)} = \sqrt{\frac{1}{4}\sum_{i=1}^{4}(\hat{\boldsymbol{u}}_{i(t)} - \boldsymbol{u}_{i(t)})^{T}(\hat{\boldsymbol{u}}_{i(t)} - \boldsymbol{u}_{i(t)})} \quad (16)$$

where $\hat{\boldsymbol{u}}_i^{(t)}=(\hat{x}_i^{(t)}, \hat{y}_i^{(t)})^T$ and $\boldsymbol{u}_i^{(t)}=(x_i^{(t)}, y_i^{(t)})^T$ denote the position estimate and the true position, respectively, of the i-th pedestrian, where i = 1, 2, 3, 4, at the t-th laser scan.

First of all, we compare the result of cooperative tracking by GNN data association with that of individual tracking by GNN data association. Both tracking modes have the similar tracking error before 200 scan (20 s); however, individual tracking mode causes large tracking error after 200 scan. Because pedestrian #2 is shadowed by pedestrian #1 around 183 scan, robot #1 loses pedestrian #2 and large tracking error occurs in individual tracking mode after 200scan. However, pedestrian #2 is visible by robot #2 even around 183 scan, and thus cooperative tracking can maintain the accurate tracking after 200scan.

Both tracking modes temporarily cause large tracking error around 160 scan (around (x,y) = (−3.0,0.7)m in Figure 7). This reason is why pedestrian #3 is temporarily shadowed by pedestrian #4 and track lost then occurs.

NN-based data association very often causes incorrect matching of tracked pedestrians and LRS measurements, and this results in the track being lost. On the other side, GNN data association reduces the track lost. Therefore, the tracking error by GNN data association becomes smaller than that by NN data association. As the result, it is clear from Figure 8(a) that cooperative tracking by GNN data association provides better tracking performance than other methods.

Next, we simulate the effect of the data fusion methods for the cooperative tracking to the tracking error. For comparison purpose, we consider three data fusion methods: CI method, Kalman filter, and averaging method. In the Kalman filter, data fusion is achieved by considering the tracking data sent from other robots to be measurements. Based on the averaging method, each robot tracks pedestrians by simply averaging its own tracking data with the tracking data sent from other robots; the averaging method equals CI method by setting the weight ω = 0.5. In this simulation, GNN data association is always applied for the data association.

Figure 8(b) shows the results. The Kalman filter and averaging method cause large tracking errors around 120 scans (around point C in Figure 7). The data fusion method is closely related to the data association method; the performance in data fusion affects that in data association, and vice versa. Compared to Kalman filter and average methods, CI method maintains the accurate tracking performance. From these simulations, we confirmed that cooperative tracking based on CI and GNN methods provides tracking performance better.

Experimental Results

To evaluate the tracking method, we conducted an experiment in an outdoor environment shown in Figure 9. Three robots and three pedestrians move around in the environment as shown in Figure 10. The moving speed of the robots is less than 0.3 m/s. The walking speeds of pedestrians #1, #2, and #3 are less than 1.5 m/s, 1.5 m/s, and 3.7 m/s, respectively: At first, pedestrian #3 walks at the same speed of pedestrians #1 and #2, and he runs at a speed of 3.7 m/s on the way. The experimental time is 188 scans (18.8 s).

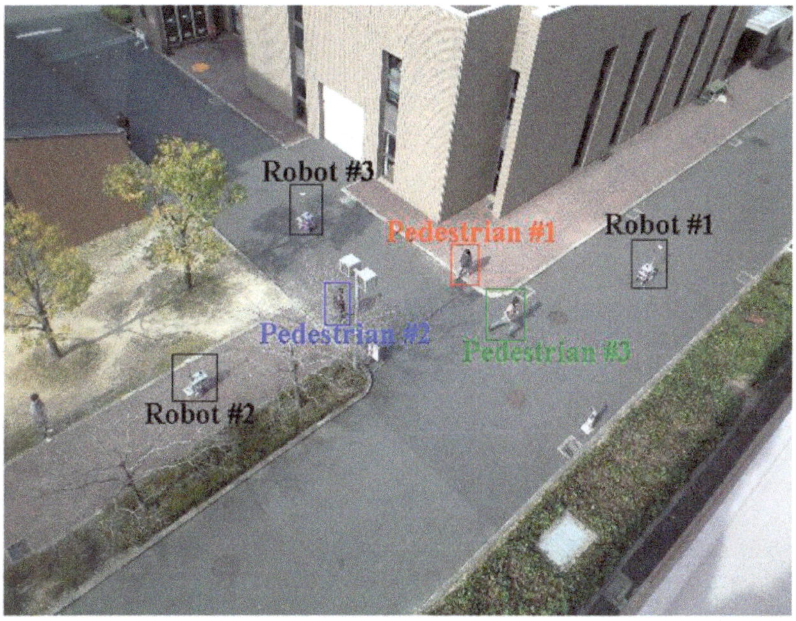

Figure 9: View of experimental environments.

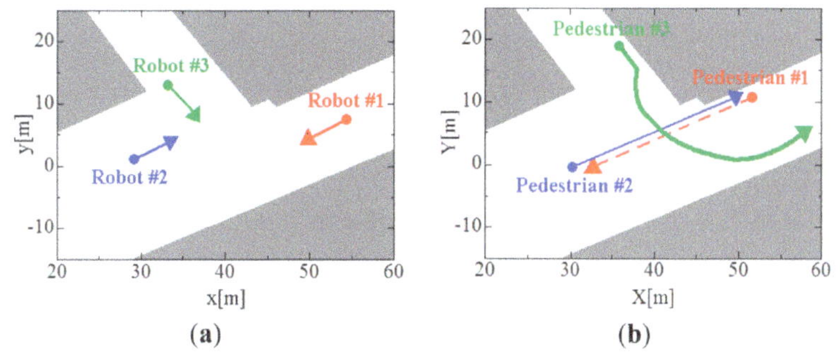

Figure 10: Movement path of robots and pedestrians. (**a**) Robot path. (**b**) Pedestrian path.

Figure 11 shows the results of pedestrian tracking only by individual tracking; figures (a), (b) and (c) show the tracks of three pedestrians estimated by robot #1, #2, and #3, respectively. Each robot partially tracks pedestrians because the pedestrians exist inside and outside the sensing area of the LRS.

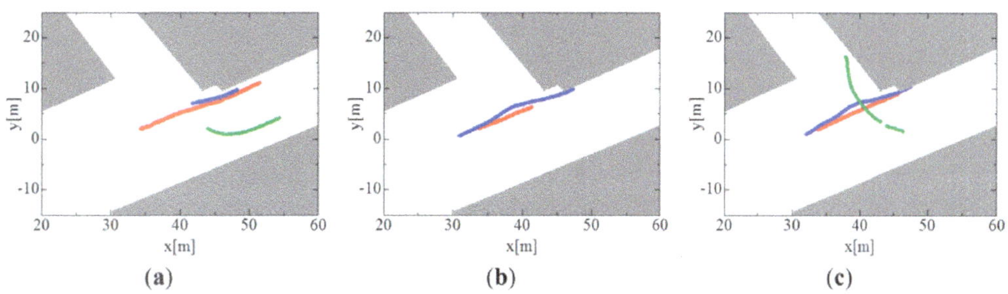

Figure 11: Pedestrian tracks estimated by individual tracking. (**a**) Robot #1. (**b**) Robot #2. (**c**) Robot #3. Red, blue and green lines indicate paths of pedestrians #1, #2 and #3, respectively.

Figure 12 shows the tracks of three pedestrians estimated by individual and cooperative tracking; because the three robots share the tracking data with each other, all three robots can track the three pedestrians for an extended period.

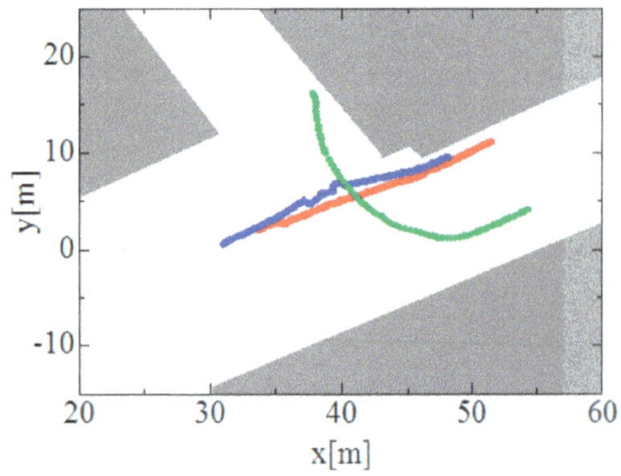

Figure 12: Pedestrian tracks estimated by individual and cooperative tracking. Red, blue and green lines indicate paths of pedestrians #1, #2 and #3, respectively.

Figure 13 shows the duration of pedestrian tracking; figures (a), (b) and (c) show the times during which robots #1, #2 and #3, respectively, track pedestrians using individual tracking. Figure 14 shows the duration of pedestrian tracking by the individual and cooperative tracking. From these results, cooperative tracking provides a better tracking performance than individual tracking; for example, cooperative tracking detects pedestrian #3 who runs into the road 34 scan (3.4 s) faster than individual tracking. The faster the pedestrians can be detected, the safer becomes robot's navigation.

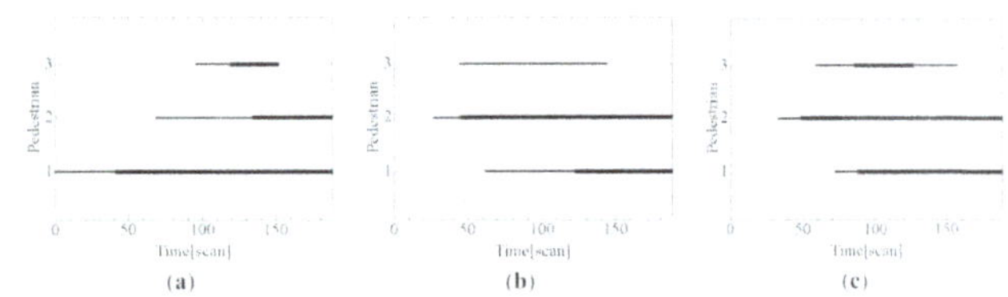

Figure 13: Duration of individual tracking. (**a**) Robot #1. (**b**) Robot #2. (**c**) Robot #3. The thin line indicates the time during which the pedestrian exits the sensing area of the robot. The bold line indicates the time during which the robot tracks pedestrians using the individual tracking.

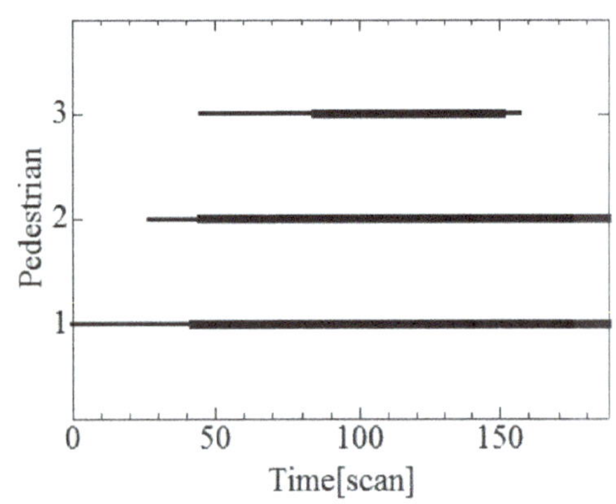

Figure 14: Duration of individual and cooperative tracking. The thin line indicates the time during which the pedestrian exits the sensing area of each robot. The bold line indicates the time during which the robot tracks pedestrians using the individual and cooperative tracking.

CONCLUSIONS

This paper presents a laser-based pedestrian tracking method using multiple mobile robots. Pedestrians were tracked by each robot using Kalman filter and GNN based data association. The tracking data obtained by each robot was broadcast to others robots and was combined by the CI method. Our method shares the pedestrian tracking data with all robots, and thus, collectively they can always recognize pedestrians that may be invisible to individual robots. The method was validated by simulation and experiment. Our tracking system worked effectively in a decentralized manner without any central server.

In the experiment, three pedestrians were tracked in a sparse environment. We will next conduct pedestrian tracking experiments in crowded environments. To achieve cooperative tracking, the robots must always identify their own postures with a high degree of accuracy in a common coordinate frame, for which, in this paper, we applied two localization methods: RTK-GPS-based and relative-scan matching-based. However, in outdoor environments such as areas surrounded by high buildings and roadside trees, it is difficult for robots to obtain posture information accurately by GPS due to GPS multipath and diffraction problems and so on. To cope with this problem, we will embed a simultaneously localization and mapping (SLAM) method into our tracking system; SLAM-GPS fusion based localization will always maintain a high degree of positioning accuracy, and therefore, it will enhance the robustness of our cooperative pedestrian tracking system in GPS-denied environments such as urban cities.

ACKNOWLEDGMENTS

This study was partially supported by Scientific Grant #23560305, Japan Society for the Promotion of Science (JSPS).

APPENDIX

Relative-Scan Matching

We determine the posture of robot #2 ${}^2z_1 = ({}^2x_1, {}^2y_1, {}^2\psi_1)^T$ relative to that of robot #1 based on laser-scan matching method. The matching is based on point-to-point scan matching by the iterative closest point (ICP) algorithm [29].

From the laser scan images taken by robots #1 and #2, we compute the relative posture ${}^2z_1$ using the weighted least-squares method; the cost function is given as:

$$J = \sum_{i=1}^{541} w_i \{\boldsymbol{q}_j - (\boldsymbol{R}\boldsymbol{p}_i + \boldsymbol{T})\}^2 \tag{A1}$$

where $\boldsymbol{R} = \begin{pmatrix} \cos^2\psi_1 & -\sin^2\psi_1 \\ \sin^2\psi_1 & \cos^2\psi_1 \end{pmatrix}$ and $\boldsymbol{T} = \begin{pmatrix} {}^2x_1 \\ {}^2y_1 \end{pmatrix}$.

Here, $\boldsymbol{p}_i = (p_{xi}, p_{yi})^T$, where i = 1, 2, ..., 541, denotes the distance sample (scan image) from robot #1; $\boldsymbol{q}_j = (q_{xj}, q_{yj})^T$, where j= 1, 2, ... 541, denotes the distance sample from robot #2, as shown in Figure A1.

Each sample $\boldsymbol{p}_i$ corresponds to the minimum distance sample $\boldsymbol{q}_j$ of all samples in the scan by robot #2; w_i denotes the weight. $\boldsymbol{R}$ and $\boldsymbol{T}$ denote the rotational matrix and translational vector, respectively.

To reduce the effects of correspondence errors in the distance samples in the two laser images, we define weight w_i according to the errors between correspondence points $w_i = (1+d_{ij}/C)^{-1}$, where d_{ij} denotes the distance error between the two laser images and C is a constant. In our experiment described in Section 5, C is set at 0.1.

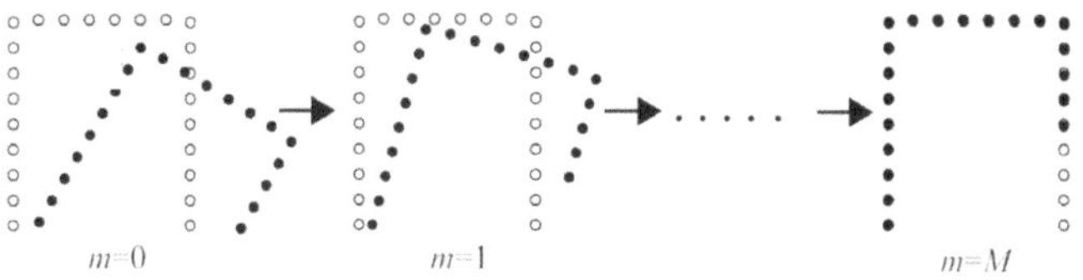

Figure 15: Relative-scan matching. Solid and open circles indicate scan images taken by robots #1 and #2, respectively.

From Equation (A1), the iterative least-squares method is used to update the relative posture z2(m−1)1as follows:

$$^{2}z_1^{(m)} = {}^{2}z_1^{(m-1)} + (\boldsymbol{H}^T\boldsymbol{W}\boldsymbol{H})^{-1}\boldsymbol{H}^T\boldsymbol{W}(\boldsymbol{q}-\boldsymbol{p}) \tag{A2}$$

where $\boldsymbol{p} = (\boldsymbol{p}_1^T, \boldsymbol{p}_2^T, \ldots, \boldsymbol{p}_{541}^T)^T, \boldsymbol{q} = (\boldsymbol{q}_1^T, \boldsymbol{q}_2^T, \ldots, \boldsymbol{q}_{541}^T)^T, \boldsymbol{W} = \mathrm{diag}(w_1, w_2, \ldots, w_{541})$ and $\boldsymbol{H} = \partial\boldsymbol{p}/\partial^2 z_1$.
The convergent value of $^{2}z_1^{(m)}$ gives the relative posture $^{2}z_1$.

REFERENCES

1. People Detection and Tracking. Proceedings of the IEEE International Conference on Robotics and Automation Workshop, Kobe, Japan, 12–17 May 2009.
2. Jia, Z.; Balasuriya, A.; Challa, S. Autonomous Vehicles Navigation with Visual Target Tracking Technical Approaches. Algorithms 2008, 1, 153–182.
3. Arras, K.O.; Mozos, O.M. Special issue on people detection and tracking. Int. J. Soc. Robot. 2010, 2, 1–107.
4. Ogawa, T.; Sakai, H.; Suzuki, Y.; Takagi, K.; Morikawa, K. Pedestrian Detection and Tracking using In-Vehicle Lidar for Automotive Application. Proceedings of the IEEE Intelligent Vehicles Symposium, Baden-Baden, Germany, 5–9 June 2011; pp. 734–739.
5. Scholer, F.; Behley, J.; Steinhage, V.; Schulz, D.; Cremers, A.B. Person Tracking in Three-Dimensional Laser Range Data with Explicit Occlusion Adaption. Proceedings of the 2011IEEE International Conference on Robotics and Automation, Shanghai, China, 9–13 May 2011; pp. 1297–1303.
6. Hashimoto, M.; Ogata, S.; Oba, F.; Murayama, T. A Laser Based Multi-Target Tracking for Mobile Robot. Intell. Auton. Syst. 2006, 9, 135–144.
7. Sato, S.; Hashimoto, M.; Takita, M.; Takagi, K.; Ogawa, T. Multilayer Lidar-Based Pedestrian Tracking in Urban Environments. Proceedings of the IEEE Intelligent Vehicles Symposium, San Diego, CA, USA, 21–24 June 2010; pp. 849–854.
8. Hashimoto, M.; Matsui, Y.; Takahashi, K. Moving-Object Tracking with In-Vehicle Multi-Laser Range Sensors. J. Robot. Mechatron. 2008, 20, 367–377.
9. Dias, M.B.; Zlot, R.; Kalra, N.; Stentz, A. Market-Based Multirobot Coordination: A Survey and Analysis. Proc. IEEE2006, 94, 1257–1270.
10. Michael, N.; Fink, J.; Kumar, V. Experimental Testbed for Large Multirobot Teams. IEEE Robot. Autom. Mag. 2010, 15, 53–61.
11. Jung, B.; Sukhatme, G.S. Cooperative Multi-Robot Target Tracking. Proceedings of the 8th International Symposium on Distributed Autonomous Robotic Systems, Minneapolis/St. Paul, MN, USA, 12–14 July 2006; pp. 81–90.
12. Li, Y.; Liu, Y.; Zhang, H.; Wang, H.; Cai, X.; Zhou, D. Distributed Target Tracking with Energy Consideration Using Mobile Sensor Networks. Proceeding of 2008 IEEE/RSJ International Conference on Intelligent Robots and Systems, Nice, France, 22–26 September 2008; pp. 3280–3285.
13. La, H.M.; Sheng, W. Adaptive Flocking Control for Dynamic Target Tracking in Mobile Sensor Networks. Proceedings of 2009 IEEE/RSJ International Conference on Intelligent Robots and Systems, St. Louis, MO, USA, 11–15 October 2009; pp. 4843–4848.
14. Hashimoto, M.; Konda, T.; Bai, Z.; Takahashi, K. Laser-Based Tracking of Randomly Moving People in Crowded Environments. Proceedings of the IEEE International Conference on Automation and Logistics, Macau, China, 16–20 August 2010; pp. 31–36.
15. Hashimoto, M.; Bai, Z.; Konda, T.; Takahashi, K. Identification and Tracking Using Laser and Vision of People Maneuvering in Crowded Environments. Proceedings of IEEE International Conference on Systems, Man and Cybernetics, Istanbul, Turkey, 10–13 October 2010; pp. 3145–3151.
16. Fod, A.; Howard, A.; Mataric, M.J. A Laser-Based People Tracker. Proceedings of the 2002 IEEE International Conference on Robotics and Automation, Washington, DC, USA, 11–15 May 2002; pp. 3024–3029.
17. Nakamura, K.; Zhao, H.; Shibasaki, R.; Sakamoto, K.; Ooga, T.; Suzukawa, N. Tracking Pedestrians by Using Multiple Laser Range Scanners. Proceedings of the ISPRS Congress, Istanbul, Turkey, 12–23 July 2004; 35, pp. 1260–1265.
18. Noguchi, H.; Mori, T.; Matsumoto, T.; Shimosaka, M.; Sato, T. Multiple-Person Tracking by Multiple Cameras and Laser Range Scanners in Indoor Environments. J. Robot. Mechatron. 2010, 22, 221–229.
19. Bellotto, N.; Hu, H. Multisensor-Based Human Detection and Tracking for Mobile Service Robots. IEEE Trans. Syst. Man Cybernetics Part B 2009, 39, 167–181.
20. Jung, B.; Sukhatme, G.S. Real-time Motion Tracking from a Mobile Robot. Int. J. Soc. Robot. 2010, 2, 63–78.
21. Julier, S.; Uhlmann, J.K. General Decentralized Data Fusion with Covariance Intersection (CI). Handbook of Data Fusion; Hall, D.L., Llinas, J., Eds.; CRC Press: New York, NY, USA, 2001; pp. 12:1–12:25.
22. Tsokas, N.A.; Kyriakopoulos, K.J. Multi-Robot Multiple Hypothesis Tracking for Pedestrian Tracking. Auton. Robot.2012, 32, 63–79.
23. Chou, C.T.; Li, J.Y.; Chang, M.F.; Fu, L.C. Multi-Robot Cooperation Based Human Tracking System Using Laser Range Finder. Proceedings of the 2011 IEEE International Conference on Robotics and Automation, Shanghai, China, 9–13 May 2011; pp. 532–537.
24. Bar-Shalom, Y.; Fortmann, T.E. State Estimation for

Linear Systems. In Tracking and Data Association; Academic Press, Inc.: San Diego, CA, USA, 1988; pp. 52–122.

25. Konstantinova, P.; Udvarev, A.; Semerdjiev, T. A Study of a Target Tracking Algorithm Using Global Nearest Neighbor Approach. Proceedings of the 4th International Conference on Systems and Technologies, Ruse, Bulgaria, 18–19 June 2003; pp. 290–295.

26. Kuhn, H.W. The Hungarian Method for the Assignment Problem. Nav. Res. Logist. Q 1955, 2, 83–98.

27. Blackman, S.S. Multiple Hypothesis Tracking for Multiple Target Tracking. IEEE Aerosp. Elect. Syst. Mag. 2004, 19, 6–18.

28. Tsokas, N.A.; Kyriakopoulos, K.J. Multi-Robot Multiple Hypothesis Tracking for Pedestrian Tracking with Detection Uncertainty. Proceedings of the 2011 IEEE International Conference on Mediterranean Conference on Control and Automation, Corfu, Greece, 20–23 June 2011; pp. 315–320.

29. Lu, F.; Milios, E. Robot Pose Estimation in Unknown Environments by Matching 2D Range Scans. J. Intell. Robot. Syst.1997, 20, 249–275.

CHAPTER 15 GENERATION MECHANISM AND PREDICTION MODEL FOR LOW FREQUENCY NOISE INDUCED BY ENERGY DISSIPATING SUBMERGED JETS DURING FLOOD DISCHARGE FROM A HIGH DAM

Jijian Lian, Wenjiao Zhang, Qizhong Guo and Fang Liu

Keywords: LFN, high dam flood discharge, energy dissipation by submerged jets, vorticity fluctuation, prediction model

INTRODUCTION

Noise pollution is a serious environmental problem to human beings, which has further been shown to be associated with reduced quality of life and wellbeing [1,2]. At present, most of the academic literature is related to high frequency noise sources from traffic, industry and human activities [3,4,5], while low frequency noise (LFN) is also considered annoying for humans [6]. The LFN found in living environments is mainly emitted from many artificial sources such as road vehicles, aircraft, and air movement machinery including wind turbines, compressors, and ventilation [7,8], and it is claimed that exposure to LFN has a negative impact on humans' physiological and psychological health. The physiological problems include headaches, hormone changes, dizziness or vertigo, tinnitus and the sensation of aural pain or pressure, and the psychological impact can cause sleep disturbance, dysphoria, difficulty concentrating, irritability and fatigue [9,10].

In the field of hydraulic engineering in China, the majority of high dams were constructed in recent decades with the features of high water head, large flow capacity, a deep narrow valley and large flood discharge power. Issues such as energy dissipation and scour protection, vibration control of a hydraulic structure, vapor atomization protection, aeration for cavitation protection, and security warnings associated with high dam flood discharge have attracted significant attention and led to abundant technological achievements [11,12,13,14]. Some environmental problems from LFN have also recently been found around some hydropower stations during the flood discharge period. For instance, when the Xiangjiaba Hydropower Station (Zhaotong, at the border of Sichuan and Yunnan provinces, China) began to release flow through the dam orifices at a high water level on 10 October 2012, the roller shutter doors of some shops and the windows and doors of residential buildings experienced sustained vibration in downstream Shuifu County. The dominant frequencies of the LFN observed on-site were approximately 0–2 Hz. In a recent study [15], the impact of ground vibration induced by the flood discharge of Xiangjiaba on on-site roller shutter doors was eliminated by means of a vibration response analysis, and it was deduced that the doors shaking resulted from the resonant interaction between the flow-induced LFN and the door structures. Moreover, the windows of buildings located at the left abutment of the Ertan Hydropower Station (Panzhihua, Sichuan Province, China) oscillated noticeably when the flood was released. The windows and doors of residential buildings in a downstream village about 700–1500 m from the Huangjinping Hydropower Station (Kangding, Sichuan Province, China) experienced sustained vibrations

Citation: Jijian Lian, Wenjiao Zhang, Qizhong Guo and Fang Liu, Generation Mechanism and Prediction Model for Low Frequency Noise Induced by Energy Dissipating Submerged Jets during Flood Discharge from a High Dam, doi:10.3390/ijerph13060594.

when the flood was released. The LFN observed during the flood discharge period of both the Jinping Hydropower Station (Liangshan Yi Autonomous Prefecture, Sichuan Province, China) and Xiluodu Hydropower Station (Zhaotong, at the border of Sichuan and Yunnan provinces, China) had sound pressure levels (SPL) approximately 20–50 dB higher than the background noise, and the dominant LFN frequencies were between 0.5 Hz and 1.5 Hz. It is found that the inaudible LFN observed around these hydropower stations causes audible secondary noise from surrounding buildings. In Japan, there are some criteria specific to the vibration and rattle attributable to the effects of LFN from stationary sources in worksites, shops and neighborhood residences [16]. The lowest noise frequency listed in the criteria is 5 Hz with its reference SPL limit of 70 dB. For the flow-induced LFN, the LFN's effects are extended to a lower frequency of about 0–2 Hz.

Japanese scholars have conducted some research on flow-induced LFN. As Japan is a densely populated country and many hydropower stations are close to residents' living quarters, the problems of LFN were noted earlier in Japan. Most of the research focused on the waterfalls formed by the discharge flow through the weir or dam's floodgate. The oscillation of such waterfalls can cause LFN, which has an adverse impact on surrounding buildings and inhabitants [17,18,19]. In addition, since LFN is a low frequency wave, it has a low attenuation rate in air [17,20,21], which is difficult to control, and thus can spread over long distances, even tens of kilometers, to resonate with buildings. Among the previous studies, Nakamura [17] reported the phenomenon of LFN induced by flood discharge as early as 1978. Takebayashi [18] and Nakagawa [19] carried out systematic research on the mechanisms of self-excited oscillation of cavity-waterfall systems and the characteristics of the induced LFN based on site observations. Some control measures, such as placing deflectors, were put forward and applied effectively in the research. Ochiai et al. [22] studied various influencing factors of the SPL of the noise from windows and doors induced by LFN by a series of model tests. Saitoh et al. [23] studied the sound sources of the hydraulic jump through hydraulic model experiments on the hydraulic jump and cylinder nozzle, and they found that the noise energy at a high frequency of 500–600 Hz induced by the hydraulic jump was derived from bubble cloud oscillations. These research results indicate that LFN induced by flood discharge can cause environmental problems within large areas, and the generation of LFN closely correlates with the discharge flow regime. The problems of LFN can be controlled and reduced by adjusting the flow regime. The previous studies mainly focus on LFN induced by a relatively continuous waterfall of some water projects with low water head, small flow rate and a simple flow regime. However, for water projects that dissipate energy through multi-horizontal submerged jets, LNF energy can still be observed on site although a waterfall does not exist during flood discharge. Therefore, there must be some differences in the generation mechanisms of LFN between energy dissipation by the submerged jet of high dam flood discharge and the oscillation of the waterfall. An extensive literature search has not found any research reports on LFN induced by energy dissipation through the submerged jets of a high dam. The objective of this study was to identify the mechanisms and key influencing factors of LFN induced by energy dissipation through the submerged jets of a high dam. First, in light of the prototype observation results of LFN, the on-site spatial distribution and propagation laws of the induced LFN along the downstream are analyzed, and the correlation between LFN and the discharge flow regime is discussed. Next, the vortex sound model and numerical turbulent flow model for the flood discharge and energy dissipation of the high dam are presented as the theoretical bases. Then, the flow field's distribution characteristics and effective regions of acoustic source for the LFN are studied and identified, according to the numerical simulation results of the gas-liquid turbulent flow model of flood discharge. The mathematical prediction model of the LFN intensity induced by the submerged jets is established based on the vortex sound model and turbulent flow model, and the prediction model is verified by prototype observation results. Finally, the findings are summarized, and the conclusions of this study are drawn.

MATERIALS AND METHODS

Prototype Observation

Prototype

The Xiangjiaba Hydropower Station is the third largest hydropower station in China and the sixth largest in the world. It is the last cascade on the Jinsha River and is adjacent to Shuifu County of Yunnan Province downstream, with the nearest distance being about 0.5 km [24]. The dam is 162 m in height, with a reservoir capacity of 5×10^9 m^3 and planned flood discharge rate of 41,200 m^3/s. The total installed capacity of the station is 6400 MW. The station consists of water-retaining structures, flood-releasing and sediment-flushing structures, a water diversion and power generation system, navigation structures, and irrigation water intakes. Because of the large scale of the flood discharge and energy dissipation and the requirement of the flood discharge and sediment flushing of the hydropower station, the water release structure was chosen as a gravity spillway dam, composed of 12 crest overflowing orifices and 10 mid-discharge orifices, as shown in Figure 1. However, after comparing a range of orifice layout schemes through a hydraulic model test, it was determined that the water release structure had to dissipate energy by the submerged jets and that the 12 crest overflowing orifices and the 10 mid-discharge orifices ought to be alternately arranged for the sake of balancing the unit width discharge, facilitating flood discharge and sediment flushing, decreasing the sluice gate's size and reducing the influence of the discharge atomization on the surrounding town and enterprise [25]. There are two symmetrical energy dissipation areas separated by the middle guide wall.

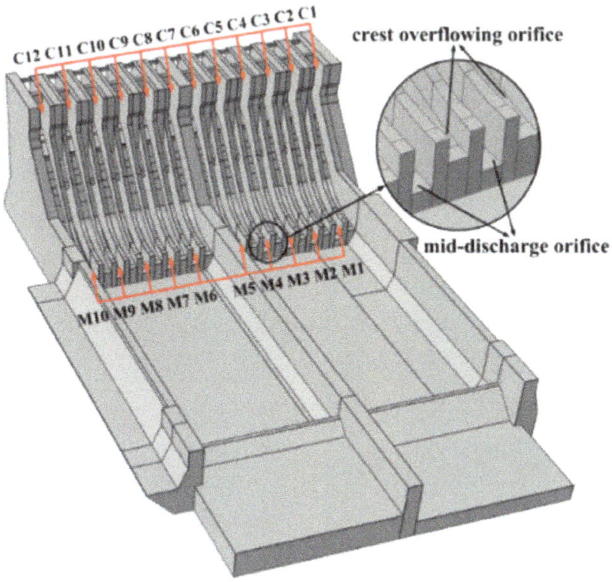

Figure 1: Layout of the water release structures.

Observation System and Conditions

Since 2012, Tianjin University has organized prototype observations of LFN many times during the flood period of Xiangjiaba. The observation equipment of LFN on site consists of an infrasound microphone, a multi-channel digitizer and a computer. The infrasound microphone, which was fixed above the ground a little over 2 m, has a frequency range of 0.1–500 Hz and a sensitivity of 107 mv/Pa. The multi-channel digitizer connects the microphone with the computer, which supplies power to the microphone and saves data in high speed. The digitizer has a built-in GPS, so the equipment can work well in the open air. With regard to the uncertainties in the observed data from the self-noise of the microphone and digitizer, the errors from the digitizer are small sufficiently to be neglected, and the maximum error from the microphone is below 2%, which corresponds to ±1 dB.

The observation range covered the dam area, construction area and urban area of Shuifu, placing 25 observation points in total (T1–T25 in Figure 2). The observation work was conducted at least twice at the same point under the same discharge conditions. The observation time period was set as 3–6 min. The observation frequency was set as 1000 Hz. Due to the daily, traffic and industrial noise in the county [26,27], especially with respect to the LFN generated by the surrounding living environments [8,28,29], the observation work was carried out in the evening to avoid signal mixing or distortion and ensure, as much as possible, the validity of the flow-induced LFN data. Table 1 shows the details of the observed discharge conditions of Xiangjiaba.

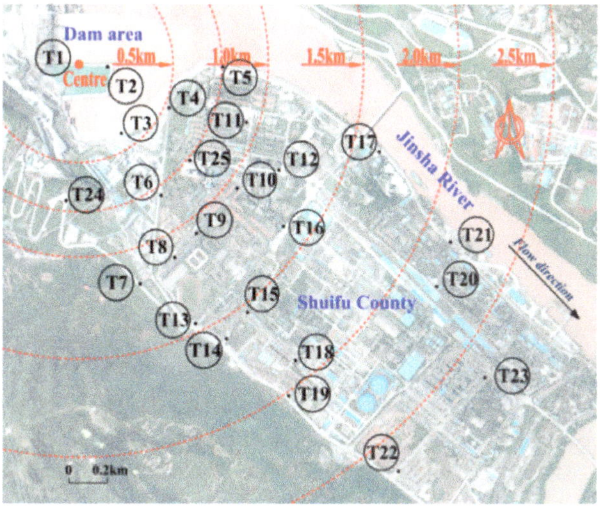

Figure 2: LFN observation point arrangement.

Table 1: Observed discharge conditions *

No.	Q (m^3/s)	Water Level Elevation (m)		Crest Overflowing Orifice		Mid-Discharge Orifice	
		Upstream	Downstream	Number	Opening (m)	Number	Opening (m)
1	0	/	/	/	/	/	/
2	1150	353.43	267.71	/	/	M6, M8, M10	3.3
3	2760	379.42	269.81	C8–C11	1.5	M6–M10	1.5
4	3340	379.34	270.52	C8–C11	1.5	M1–M10	1.5
5	4470	379.42	271.27	C2–C5, C8–C11	1.5	M1–M10	1.5
6	5140	379.30	271.87	C2–C5, C8–C11	1.7	M1–M10	1.9
7	6370	379.19	272.77	C2–C5, C8–C11	3.0	M1–M10	2.0
8	6600	374.71	275.89	C1–C6 C7–C12	3.0 2.0	M1–M10	1.2

* Q is the total discharge through the dam orifices. The discharge orifice's number is named from the left bank to the right bank.

Prototype Observation Data

Figure 3 compares the time history and power spectral density (PSD) curves of the LFN observed at T8 between the no-releasing discharge condition (Condition 1) and the small discharge condition (Condition 2) in Table 1. In this study, all of the PSD for the LFN is estimated using the autoregressive (AR) model with the Burg algorithm, which is one of the most frequently used parametric methods for processing various signals [30,31], and the order of the AR model is determined by the final prediction error (FPE) criterion. As shown, during the flood release period, the LFN intensity increases greatly, and the LFN energy presented in the PSD has an apparent peak frequency below 20 Hz. It is obvious that LFN is induced in the process of flood discharge.

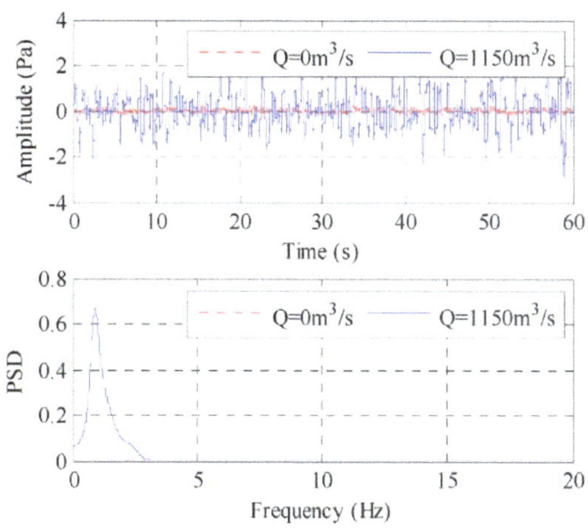

Figure 3: Comparisons of the time history and PSD curves of LFN.

Spatial Distribution and Propagation Patterns

The spatial distribution patterns of the induced LFN along the downstream area were obtained through statistical analysis. The LFN amplitude decreases gradually with increasing distance to the dam area under all observed discharge conditions. Figure 4 shows the contour map of the LFN amplitude observed under Condition 8. In this condition, the LFN amplitude achieved the maximum value, approximately 9.3 Pa (113.3 dB), at the dam area. To evaluate the human hearing and perception sensitivity for the LFN, the G-weighting (dB(G)) [32] is proposed as an appropriate metric for noise limits for LFN [33,34]. The maximum G-weighting SPL for the observed LFN was calculated as approximately 78.6 dB(G), well below the average 95 dB(G) hearing threshold [35] and the LFN limit of 85 dB(G) from Australian and Danish recommendations [36,37]. Therefore, it is noted that, under the discharge conditions in Table 1, the LFN observed around Xiangjiaba has no direct negative effect on local residents' health in general.

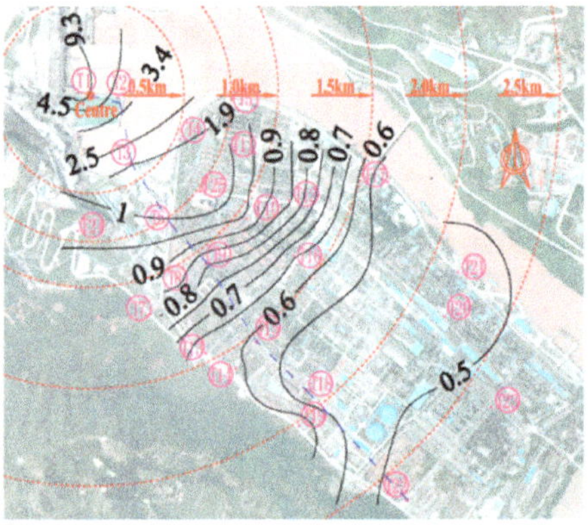

Figure 4: Distribution of the LFN amplitude. Q = 6600 m^3/s; the unit is Pa.

To analyze the attenuation rates of the flow-induced LFN in the process of propagation along the downstream city, the SPLs of 8 LFN observation points (dashed line in Figure 2) under all observed discharge conditions were calculated, as observed in Figure 5. Table 2 shows the SPL attenuation coefficients at different distances from the flood release and energy dissipation area. As a result, the SPLs near the dam area are greatly attenuated and have a maximum attenuation coefficient, 0.0347 dB/m, as a statistical average. The SPL attenuation coefficient gradually decreases with increasing distance to the dam area, and it decreases to 0.000459 dB/m for distances greater than 2 km.

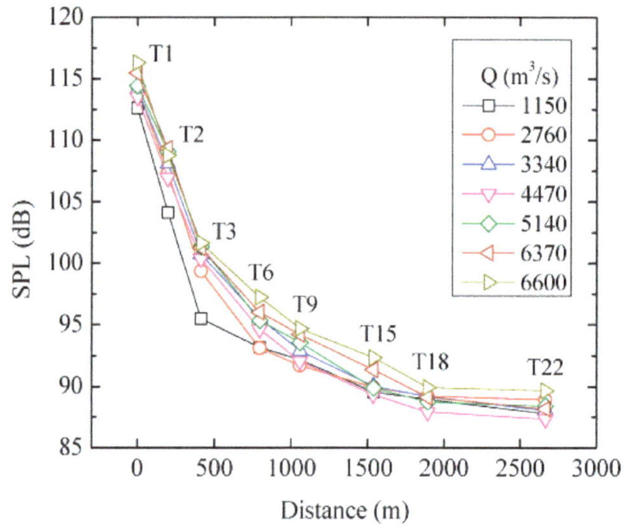

Figure 5: SPL distribution along the downstream city.

Table 2: SPL attenuation coefficients at different propagation ranges *

Range	Q (m^3/s)							Statistical Average
	1150	2760	3340	4470	5140	6370	6600	
Dam area	0.0434	0.0336	0.0322	0.0356	0.0274	0.0314	0.0386	0.0347
<0.5 km	0.0251	0.0220	0.0213	0.0202	0.0224	0.0234	0.0206	0.0221
0.5–1 km	0.00594	0.0108	0.0118	0.0124	0.0112	0.0102	0.0106	0.0104
1–2 km	0.00236	0.00190	0.00276	0.00319	0.00360	0.00406	0.00388	0.00311
>2 km	0.000324	0.000368	0.000397	0.000777	0.000416	0.000596	0.000332	0.000459

* The unit is dB/m.

Correlation between LFN and Discharge Flow Regime

Figure 6 shows the relationship between the RMS values of the LFN amplitude analyzed and the flow rate for several observation points in the downstream city area. In general, the LFN amplitude increases gradually with increasing Q, and the discharge operation conditions have some influence on the LFN amplitude. It is seen from Table 1 and Figure 6 that the joint discharge operation schemes of the crest overflowing orifices and mid-discharge orifices are advantageous for reducing the LFN amplitude. For instance, the LFN observed under Condition 4 (Q = 3340 m^3/s) has a higher amplitude than the LFN observed under Condition 5 (Q = 4470 m^3/s). The LFN amplitude observed under Condition 8 (Q = 6600 m^3/s) shows a significant decrease compared to the amplitude observed under Condition 7 (Q = 6370 m^3/s). As joint discharge operation schemes can avoid concentrated flow in the stilling basin, maintaining uniform and smooth flow in the stilling basin plays a significant role in reducing the LFN amplitude induced by flood discharge. Figure 7 shows the relationship between the analyzed dominant frequency of the LFN and the flow rate. As shown in the figure, the dominant frequencies of the LFN are concentrated between 0 Hz to 2 Hz and decrease gradually with increasing Q under the observed discharge conditions.

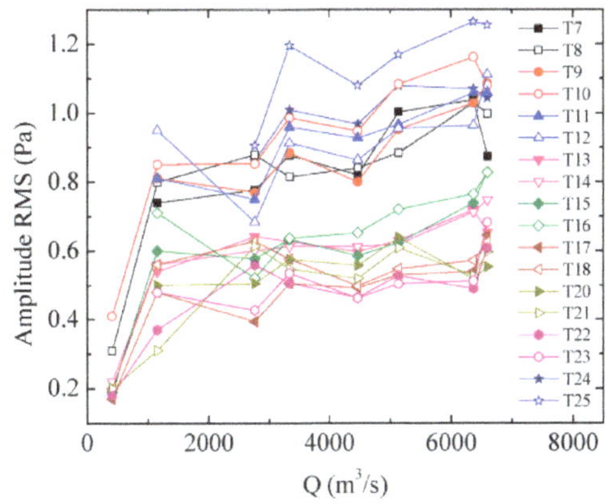

Figure 6: Correlation between the LFN intensity and Q.

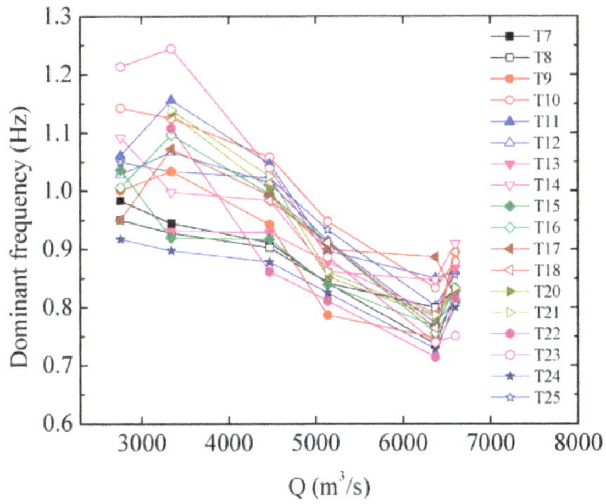

Figure 7: Correlation between the LFN dominant frequency and Q.

Theoretical Model of Vortex Sound

In acoustic research, Powell [38] and Howe [39,40] et al. studied the basic issues of the internal mechanism of fluid vocalization and the interaction between acoustic wave and turbulent flow in terms of vortex dynamic theory since the 1960s, and they established the vortex sound theory. Their research results indicated that the vortex is a significant acoustic source and that vortex sound theory is feasible and effective for identifying the generation mechanism of flow-induced noise at a low Mach number. In the turbulent flow field of flood discharge and energy dissipation from a high dam, various vortices always exist. In a stilling basin, the large amount of energy carried by high-velocity submerged jets is dissipated through strong turbulent mixing. Numerous low-frequency large coherent structures are present in the strong shear layers of the submerged jets, and they have evident vortical structures, high regularity and repeatability. The acoustic energy's formation and transition coming from the interaction of the vortical structures, potential flow and solid boundaries cannot be neglected. Therefore, the vortex sound theory is applied to determine the mechanism for LFN induced by energy dissipation through submerged jets in this study. Assuming that the flow is incompressible and isentropic and based on the equations of continuity and motion, the vortex sound equation [38] can be deduced and simplified as:

$$\nabla^2 p - \frac{1}{c^2}\frac{\partial^2 p}{\partial t^2} = -\nabla \cdot \rho\,(\omega \times \boldsymbol{u}), \tag{1}$$

where p is the acoustic pressure in the far field; c is the acoustic velocity; ρ is the density; $\boldsymbol{\omega}$ is the vorticity vector; and $\boldsymbol{u}$ is the flow velocity vector. The equation shows that the vortex sound equation is a typical nonhomogeneous wave equation, with the differential expression of the acoustic wave's propagation process in a nonhomogeneous fluid on the left and the acoustic source term of the vortex on the right. Equation (1) indicates that the flow-induced acoustic pressure is directly related to the sizes, variations and motion of the vortices. To predict the acoustic pressure, the general solution of Equation (1) for the far field is written as follows using the Green function method and the Helmholtz vortex equation, according to Möhring [41]:

$$p(x,t) = \rho\frac{\partial}{\partial t'}\int G\left(x,t;y,t'\right)\cdot\omega\left(y,t'\right)\mathrm{d}^3y\,dt', \tag{2}$$

where $\boldsymbol{x}$ is the displacement vector of the predicted point in the far field and $\boldsymbol{y}$ is the displacement vector of the acoustic source point in the flow region. The Green function must satisfy ∇yG(x,t;y,t′)=∇y×G(x,t;y,t′).

The integrand on the right side of Equation (2) only contains the vorticity variable $\boldsymbol{\omega}$, which identifies that the region with time-varying vorticity is the effective region of the acoustic source. Furthermore, because the flow velocity is not contained in the integrand, the acoustic pressure of the far field can be directly calculated by the effective vorticity fluctuation data. As $\boldsymbol{G}$ is a symmetric function, according to Powell's deduction, $\boldsymbol{G}$ can be confirmed as:

$$\boldsymbol{G}\left(x,t;y,t'\right) = \frac{1}{12\pi c^2 x^3}\delta''\left(t-t'-\frac{x}{c}\right)(\boldsymbol{x}\cdot\boldsymbol{y})\,\boldsymbol{x}\times\boldsymbol{y}, \tag{3}$$

where δ is the Dirac delta function. Then, Equation (2) leads to:

$$p(x,t) = \frac{\rho}{12\pi c^2 x^3}\frac{\partial^3}{\partial t^3}\int(\boldsymbol{x}\cdot\boldsymbol{y})\,\boldsymbol{x}\times\boldsymbol{y}\cdot\omega\mathrm{d}^3y, \tag{4}$$

where the vorticity values depend on the time t′ = t − x/c. As the attenuation of acoustic pressure during propagation is not included above, a sonar equation is obtained as:

$$SL - TL = 20\log_{10}\frac{P_e}{P_{ref}} - \left(\int_L \alpha\mathrm{d}L\right)_y = 20\log_{10}\frac{P'}{P_{ref}}, \tag{5}$$

where SL is the RMS value of the SPL without regard to attenuation; TL is the SPL attenuation value; P_e is the RMS value of acoustic pressure calculated by Equation (4); P_{ref} is the reference value of the acoustic pressure, set as 2×10^{-5} Pa; α is the SPL attenuation coefficient; and P′ is the virtual value of the acoustic pressure finally achieved with regard to attenuation. The mathematical prediction model of LFN is generated using Equations (4) and (5), and the LFN intensity can be predicted by substituting the results of numerical simulation of the near-field turbulent flow into the prediction model.

Numerical Turbulent Flow Model

Gas-Liquid Turbulent Flow Model

A large number of numerical simulation studies have shown that the k-ε turbulent flow model is a reasonable method for simulating the hydrodynamic characteristics of turbulent flow [42,43]. However, the RNG k-ε turbulent flow model, first developed by Yakhot and Orszag [44], can better simulate flow with high strain rate and large streamline curvature compared to the traditional k-ε turbulent flow model. The VOF method, proposed by Hirt and Nichols [45] in 1975 based on the MAC method, is also an effective way to calculate the complex free water surface [46]. Therefore, the RNG k-ε turbulent flow model and the VOF method are employed in this study to analyze the hydraulic parameters in the flow field of flood discharge and energy dissipation of a high dam. The major governing equations are listed below:

Basic equations of turbulent flow:

$$\begin{cases} \frac{\partial \rho}{\partial t} + \frac{\partial \rho u_i}{\partial x_i} = 0 \\ \frac{\partial u_i}{\partial t} + \frac{\partial u_i u_j}{\partial x_j} = -\frac{1}{\rho}\frac{\partial p}{\partial x_i} + \frac{1}{\rho}\frac{\partial}{\partial x_j}\left[\left(\mu + C_\mu \frac{k^2}{\varepsilon}\right)\left(\frac{\partial u_i}{\partial x_j} + \frac{\partial u_j}{\partial x_i}\right)\right] \end{cases}, \tag{6}$$

k equation:

$$\frac{\partial (\rho k)}{\partial t} + \frac{\partial (\rho k u_i)}{\partial x_i} = \frac{\partial}{\partial x_j}\left[\alpha_k \mu_{eff} \frac{\partial k}{\partial x_j}\right] + G_k + \rho\varepsilon, \tag{7}$$

ε equation:

$$\frac{\partial (\rho \varepsilon)}{\partial t} + \frac{\partial (\rho \varepsilon u_i)}{\partial x_i} = \frac{\partial}{\partial x_j}\left[\alpha_\varepsilon \mu_{eff} \frac{\partial \varepsilon}{\partial x_j}\right] + \frac{C_{1\varepsilon}^{*}\varepsilon}{k} G_k - C_{2\varepsilon}\rho \frac{\varepsilon^2}{k}, \tag{8}$$

Gas-liquid VOF equation:

$$\begin{cases} \alpha_a = 1 - \alpha_w \\ \frac{\partial \alpha_w}{\partial t} + u_i \frac{\partial \alpha_w}{\partial x_i} = 0 \end{cases}, \tag{9}$$

where $\mu_{eff} = \mu + \rho C_\mu \frac{k^2}{\varepsilon}$, $C_{1\varepsilon}^{*} = C_{1\varepsilon} - \frac{\eta(1-\eta/\eta_0)}{1+\beta\eta^3}$, $\eta = \left(2E_{ij} \cdot E_{ij}\right)^{1/2} \frac{k}{\varepsilon}$, $E_{ij} = \frac{1}{2}\left(\frac{\partial u_i}{\partial x_j} + \frac{\partial u_j}{\partial x_i}\right)$, $C_\mu = 0.0845$, $\alpha_k = \alpha_\varepsilon = 1.39$, $C_{1\varepsilon} = 1.42$, $C_{2\varepsilon} = 1.68$, $\eta_0 = 4.377$, $\beta = 0.012$, and α_w and α_a are the volume fractions of water and air in a unit grid, respectively.

Simulation Domain and Boundary Conditions

The numerical turbulent flow model established is for a 1:1 scale simplified spillway dam of Xiangjiaba, as shown in Figure 8, where X is along the flow direction. The calculation conditions of the numerical simulation are listed in Table 1. The simulation domain includes the water body extended 80-m into the reservoir; the spillway dam segments with the crest overflowing orifices and mid-discharge orifices, which are 132 m long; the 228 m long stilling basin; and the 40 m long tail-weir. Because the stilling basins of Xiangjiaba are symmetrical on two sides of the middle guide wall, the simulations are done on just a half portion of the discharge structures which is on one side of the middle guide wall in the transverse Z direction to reduce the numeration workload. The model is meshed by block-structured grids. For the main flood discharge and energy dissipation regions, such as the spillway dam area and the front of the stilling basins, the grid size is set as 1 m long × 1 m wide × 1 m high. In the remaining regions, the grid size is set as 2 m long × 1 m wide × 1 m high. Mesh refinement is applied on the overflow dam face, and the mesh refinement method adaptive to the variations of the pressure gradient and free water surface is employed during the calculation. The total number of grid units is approximately 3.52 million. Furthermore, based on the RNG k-ε turbulent flow model and the VOF method, the methods of finite volume, PISO and wall function are employed in the process of calculation.

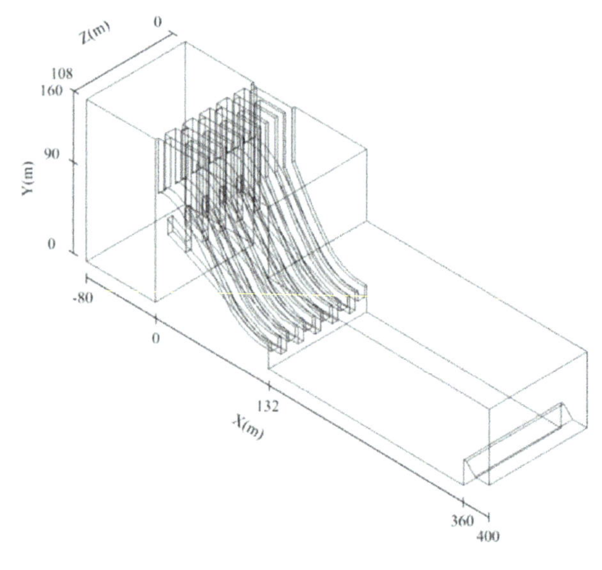

Figure 8: Turbulent flow model and simulation domain.

When setting the boundary conditions of the turbulent flow model, the upstream inlet and downstream outlet are set as the pressure inlet and pressure outlet, respectively, and the water levels of upstream and downstream are set equal to the measured values shown in Table 1. The top boundary of the model is all set as the air pressure inlet with the atmospheric pressure. The other boundary is set as a no-slip wall.

RESULTS AND DISCUSSION

From solving Powell's equation in the above section, it is suggested that the region with the time-varying vorticity is the effective region of acoustic source, and the mathematical prediction model of LFN is developed accordingly. The numerical turbulent flow model is also built as a feed to the LFN prediction model. In this section, the numerical calculation results from the turbulent flow model and the vortex sound model are analyzed to verify the theoretical analysis results. The flow field and vorticity fluctuation characteristics of the energy dissipation area are analyzed to identify the main regions of acoustic source for the LFN. The prediction model developed is used to calculate the LFN intensity on site.

Verification and Validation of Turbulent Flow Model

To evaluate the numerical uncertainty and discretization errors of the simulation results, the Grid Convergence Index (GCI) analysis based on the Richardson extrapolation method, which was outlined by Celik et al. [47,48], is adopted herein. Two additional refined grids are set to do the computations, which represent a relevant contribution to the numerical uncertainty. The three different grids consist of 3.52 (N_1), 4.81 (N_2) and 6.41 (N_3) million cells respectively, which are called as the coarse, medium and fine grids. The GCI of the coarse and fine grids are calculated as follows:

$$\begin{cases} GCI_{coarse}^{21} = \frac{1.25e_{21}r_{21}^{P}}{r_{21}^{P}-1} \\ GCI_{fine}^{32} = \frac{1.25e_{32}}{r_{32}^{P}-1} \end{cases}, \quad (10)$$

where $P = \frac{1}{\ln(r_{21})}\left|\ln\left|(f_3-f_2)/(f_3-f_2)\right| + q(P)\right|$, $e_{21} = \left|\frac{f_1-f_2}{f_1}\right|$, and $e_{32} = \left|\frac{f_2-f_3}{f_2}\right|$.

Here P is the order of accuracy; e is the relative error; r is the grid refinement factor; and f_1, f_2 and f_3 denote the solutions at the coarse, medium, and fine levels, respectively. As Roach [49] recommended a minimum 10% change in point of the computational cost and time consumption, the grid refinement factor is set to r = 1.11. The variables, f_1, f_2 and f_3, are considered as the flow velocity magnitudes near the stilling basin floor for different grid sizes.

Table 3 summarizes the numerical uncertainty assessment results for the turbulent flow model under Condition 8. All of the solutions for the velocity magnitudes show good mesh convergence behavior with errors of less than 2.5%, even to the coarse grid.

Table 3: Numerical uncertainty assessment for the turbulent flow model *

Z (m)	X (m)	Velocity Magnitude (m/s) f_1	f_2	f_3	r_{21}	r_{32}	p	e_{21}	e_{32}	GCI_{coarse}^{21} (%)	GCI_{fine}^{32} (%)
44	155	2.628	2.610	2.599	1.11	1.11	4.28	0.0067	0.0043	2.33	0.96
	170	2.516	2.498	2.491	1.11	1.11	8.79	0.0072	0.0029	1.49	0.24
	185	2.182	2.171	2.176	1.11	1.11	7.73	0.0053	0.0024	1.19	0.24
54	155	2.773	2.763	2.768	1.11	1.11	5.79	0.0035	0.0019	0.95	0.29
	170	2.509	2.511	2.509	1.11	1.11	2.51	0.0009	0.0007	0.49	0.29
	185	2.411	2.400	2.395	1.11	1.11	8.56	0.0049	0.0020	1.04	0.18

* The solutions for the velocity magnitudes with Z = 44 m and Z = 54 m are at the midlines of the crest overflowing orifice and the mid-discharge orifice, respectively.

To further validate the turbulent numerical simulation model, the data among the physical model experiment, hydraulic prototype observation and numerical simulation results were compared. The measured and simulated results of the water surface profile of Condition 8 in the stilling basin are shown in Figure 9. The observed and simulated results of the time-average pressure distribution and flow velocity near the stilling basin floor of Condition 8 are shown in Figure 10 and Figure 11, respectively.

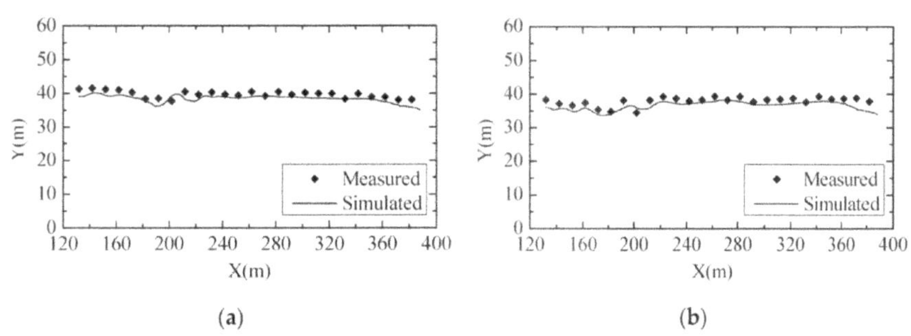

Figure 9: Comparisons of the water surface profile in the stilling basin between the measured (from the physical model) and simulated results: (**a**) Midline of the crest overflowing orifice; (**b**) Midline of the mid-discharge orifice.

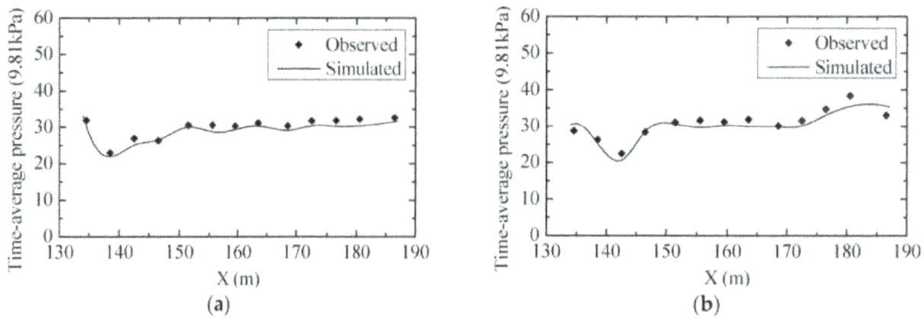

Figure 10: Comparisons of the time-average pressure distribution on the stilling basin floor between the observed (from the prototype) and simulated results: (**a**) Midline of the crest overflowing orifice; (**b**) Midline of the mid-discharge orifice.

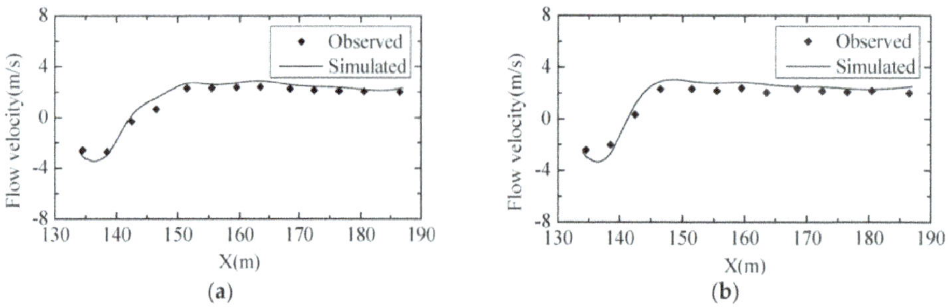

Figure 11: Comparisons of the flow velocity near the stilling basin floor between the observed and simulated results: (**a**) Midline of the crest overflowing orifice; (**b**) Midline of the mid-discharge orifice.

The figures show that the simulation results of the water surface profile, pressure distribution and flow velocity near the floor are in good agreement with those from the experiments and observations, and the other conditions' validation results are all consistent. As a result, it is feasible to analyze the hydraulic characteristics of the flow field by the numerical methods chosen in this study.

Analyses of Numerical Modeling Results

Flow Velocity Distribution

Because the distribution patterns of the flow field obtained by numerical calculation under every condition are similar, only the velocity contours on the sections along midlines of the crest overflowing orifice and mid-discharge orifice of Condition 8 are shown here in Figure 12. In Figure 12, the flow velocity begins to increase after the water discharges from the reservoir through the orifices, and the maximum velocity reaches 39.67 m/s near the end of the overflow dam surface. Then, the high velocity flow enters the stilling basin and forms submerged jets, where the water level rises locally. The multi-horizontal submerged jets maintain a certain distance in space from each other, and a steady three-dimensional flow structure similar to a submerged hydraulic jump is developed with strong turbulence and shearing action. As a result, the enormous energy carried by the high velocity flow is dissipated. Nevertheless, the kinetic energy of the high-velocity jets cannot vanish until the jets nearly reach the middle of the stilling basin.

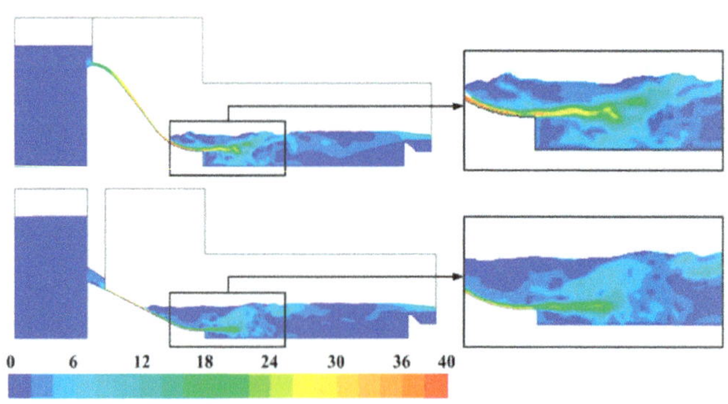

Figure 12: Flow velocity contours of the midline sections of the crest overflowing orifice and mid-discharge orifice. The unit is m/s.

Figure 13 shows the instantaneous velocity vector distribution. Plenty of vortices of different sizes and strengths form and develop continuously around the high-velocity submerged jets in the zone of the three-dimensional flow structure (submerged hydraulic jump) as a result of the large flow velocity gradient.

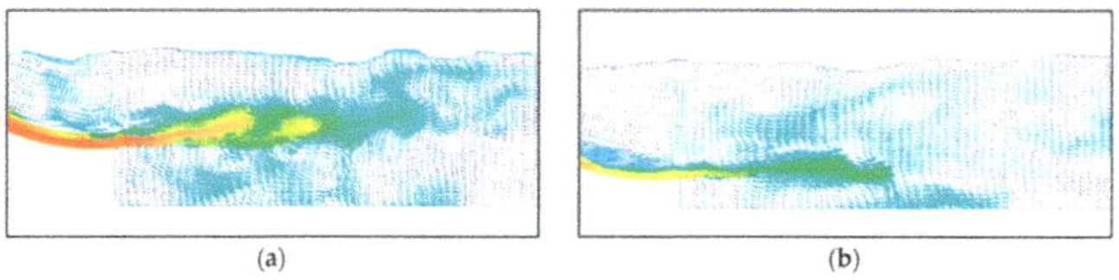

Figure 13: Flow velocity vector distribution in the energy dissipation area: (a) Midline of the crest overflowing orifice; (b) Midline of the mid-discharge orifice.

Vortical Structures' Characteristics

When analyzing the characteristics of the vorticity field, the Eulerian method, such as the Q-criterion developed by Hunt et al. [50], is frequently utilized to identify and visualize the vortical structures, and the patterns of the vortex motion can be found in an instantaneous flow field [51]. In addition, the Q-criterion method relies on the derivatives of the velocity and is derived from the invariants of the velocity gradient tensor. The velocity gradient tensor A for incompressible flow can be divided into two parts, S and W, respectively, for the symmetric and antisymmetric parts:

$$A_{ij} = S_{ij} + W_{ij} = \frac{1}{2}\left(\frac{\partial u_i}{\partial x_j} + \frac{\partial u_j}{\partial x_i}\right) + \frac{1}{2}\left(\frac{\partial u_j}{\partial x_i} - \frac{\partial u_i}{\partial x_j}\right), \quad (11)$$

where S_{ij} is the strain rate tensor representing the irrotational motion, and W_{ij} is the rotation tensor representing the rotational motion. Q is the second invariant of A_{ij} defined as:

$$Q = \frac{1}{2}\left(W_{ij}W_{ij} - S_{ij}S_{ij}\right), \quad (12)$$

As a consequence, if the Q value in some regions is positive, the rotational motion plays a dominating role in those regions. Figure 14 displays the visualization results of the vortical structures in the flow field of energy dissipation by the Q-criterion. To display the interaction of the vortices more clearly in the process of energy dissipation, the contours of the Qvalue ($Q > 0$) of the midline sections of the crest overflowing orifice and mid-discharge orifice under Condition 8 are presented in incremental time steps in Figure 15.

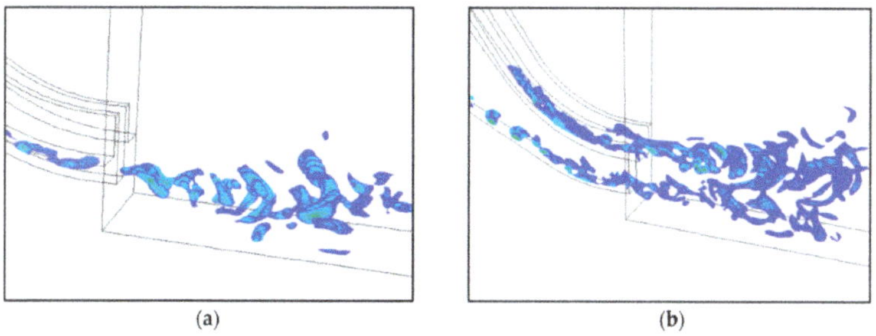

Figure 14: Instantaneous vortical structures in the energy dissipation area by Q = 10: (a) Condition 2; (b) Condition 8.

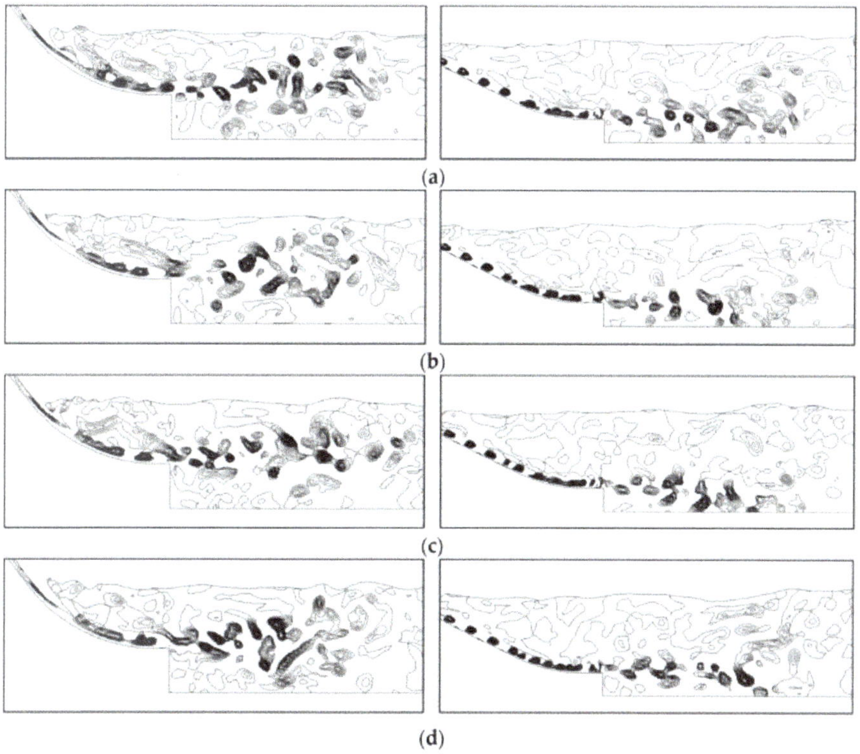

Figure 15: Contours of the Q value ($Q > 0$) of the midline sections of the crest overflowing orifice and mid-discharge orifice in incremental time steps: (a) $t_0 = 180$ s; (b) $t_0 + 1$ s; (c) $t_0 + 2$ s; (d) $t_0 + 3$ s.

At different discharge moments, the shear layers grow gradually along the overflow dam surface with increasing flow velocity, where the vortical structures that are tubular in the horizontal axial direction (Z direction) begin to appear. At the end of the overflow dam surface and the beginning of the submerged jets, the vortical structures in the shear layers turn to axial breakup.The breakup becomes more evident when the submerged jets enter the stilling basin, and much more vortical structures of different sizes and strengths are developed. It is seen that the three-dimensional characteristics of the flow regime in the stilling basin are distinct, and the vortical structures in the flow field are unsteady, movable and intermittent under different discharge conditions. All of the visualization results show that the vortical structures are mainly located at the beginning of the stilling basin, where the shearing motion of the flow is violent.

Correlation Analysis of Vorticity Fluctuating Characteristics and Acoustic Source

As the radiated acoustic pressure in the far field is directly related to the vorticity fluctuation characteristics and vortex motions in the flow region according to previous theoretical deductions, the dynamic characteristics of the vorticity in the energy dissipation area are derived by numerical calculation in this section to verify the deductions and estimate the location of the effective regions of acoustic source for the LFN. Figure 16 shows the vorticity contours of the midline sections of the crest overflowing orifice and mid-discharge orifice of Condition 8. It is clearly observed that the distributions of the vorticity and velocity fields are approximately similar. The vorticity intensity is large in two regions, one located at the free surface of the overflow water, and the other one located at the strong shear layers around the submerged jets. Before the jets get into the stilling basin, the Z-component of the vorticity, representing the quantity of horizontal vortices in the shear layers, is maximum. When the horizontal vortices flow into the stilling basin, vertical vortices begin to develop, and the components of the vorticity in every direction tend toward uniformity. The distributions of the vorticity field correspond to the identification results of the vortical structures in Figure 14 and Figure 15.

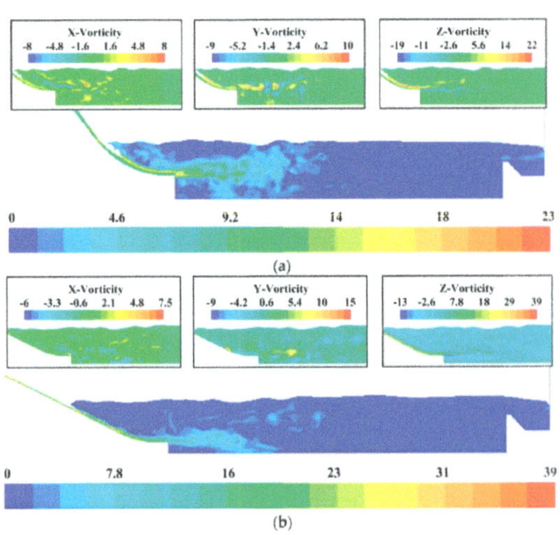

Figure 16: Vorticity contours in the energy dissipation area: (**a**) Midline of the crest overflowing orifice; (**b**) Midline of the mid-discharge orifice. The unit is s^{-1}.

To analyze the vorticity fluctuation characteristics in the flow field, several monitoring points have been set on the midline sections of the crest overflowing orifice and mid-discharge orifice in the numerical models, as shown in Figure 17. Specifically, S1 and M1 are set at the free surface of the overflow water; S2 and M2 are set under the high-velocity overflow water; and the rest of the measuring points are set in the energy dissipation region of the submerged jets.

Figure 17: Vorticity monitoring points on the midline sections of the crest overflowing orifice and mid-discharge orifice.

The time-averaged values, RMS values and dominant frequencies of the vorticity fluctuation of every monitoring point under all conditions were computed according to the transient monitoring data obtained through the numerical calculation. The results for Condition 8 are listed in Table 4. It is shown that, for the monitoring points located at the free surface of overflow water (S1, M1), the time-averaged values of the vorticity fluctuation are larger, whereas the RMS values are smaller, and the dominant frequency does not exist. The points located at the strong shear layers around the high-velocity submerged jets (S4, S5, S7, M4, M6, and M8) have the vorticity fluctuation with larger RMS values and dominant frequencies of 0–2 Hz.

Table 4: Analyses of the vorticity monitoring data

Midline Section of Crest Overflowing Orifice				Midline Section of Mid-Discharge Orifice			
Monitoring Point	Time-Averaged Value (s^{-1})	RMS Value (s^{-1})	Dominant Frequency (Hz)	Monitoring Point	Time-Averaged Value (s^{-1})	RMS Value (s^{-1})	Dominant Frequency (Hz)
S1	15.143	0.663	/	M1	30.021	1.656	/
S2	3.022	0.031	/	M2	7.986	0.394	/
S3	1.828	0.812	/	M3	0.654	0.243	/
S4	**18.737**	**1.149**	**1.000**	**M4**	**20.610**	**2.530**	**1.060**
S5	**11.686**	**1.599**	**0.960**	M5	1.630	0.665	/
S6	1.518	0.680	/	**M6**	**10.589**	**1.162**	**0.875**
S7	**8.948**	**1.839**	**1.020**	M7	1.791	0.967	/
S8	2.036	1.056	/	**M8**	**6.045**	**1.812**	**0.425**

Contrastive analyses were conducted between the monitoring data of the vorticity fluctuation and the prototype observation data of the LFN under the same condition. Figure 18 compares the time-history curves and autocorrelation PSD between monitoring point S5 and prototype observation point T8 under Condition 8. As a consequence, the vorticity fluctuation data from the strong shear layers (S4, S5, S7, M4, M6, and M8) have approximately the same periodic pattern with the prototype observation data of LFN.

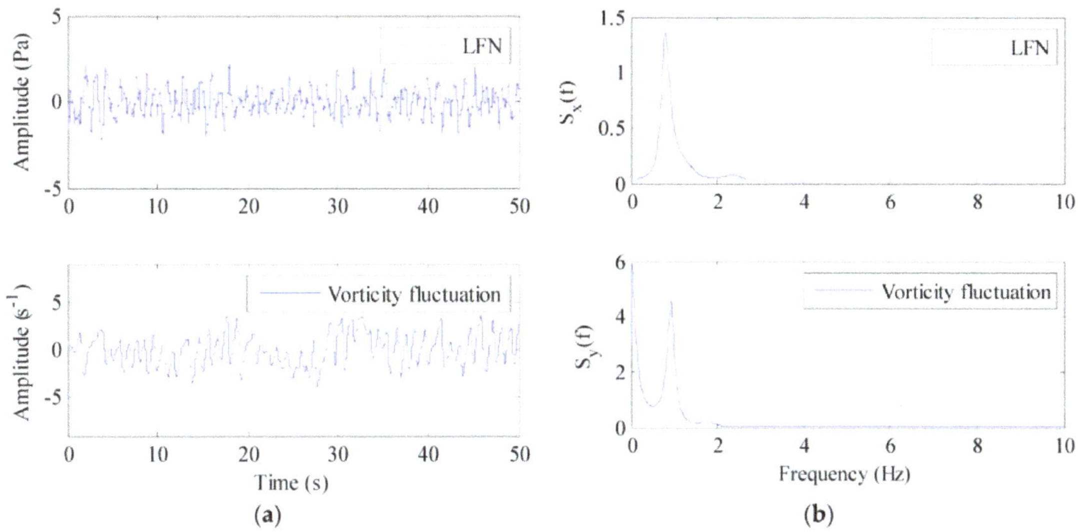

Figure 18: Comparisons between the vorticity monitoring point S5 and the prototype observation point T8: (**a**) Time-history curve; (**b**) Auto-correlation PSD.

Additionally, for clearly quantifying the correlation between the vorticity fluctuation and LFN, the spectral correlation coefficient ρ_F [52] is defined as:

$$\rho_F = \frac{\sum_{k=1}^{n} F_x(k)\, F_y(k)}{\sqrt{\sum_{k=1}^{n} F_x^2(k) \sum_{k=1}^{n} F_y^2(k)}}, \qquad (13)$$

where F_x and F_y represent the module of the Fourier spectrum of the vorticity fluctuation and the observed LFN, respectively. Figure 19 displays the statistical results of the spectral correlation coefficients between every monitoring point of the vorticity fluctuation and the prototype observation point T8 under all conditions. It is seen that the values of ρ_F at the monitoring points in the strong shear layers (S4, S5, S7, M4, M6, and M8) are relatively larger, which can be up to 0.762. The average values of ρ_F of the monitoring points in the strong shear layers of the crest overflowing orifice and the mid-discharge orifice are 0.681 and 0.601, respectively.

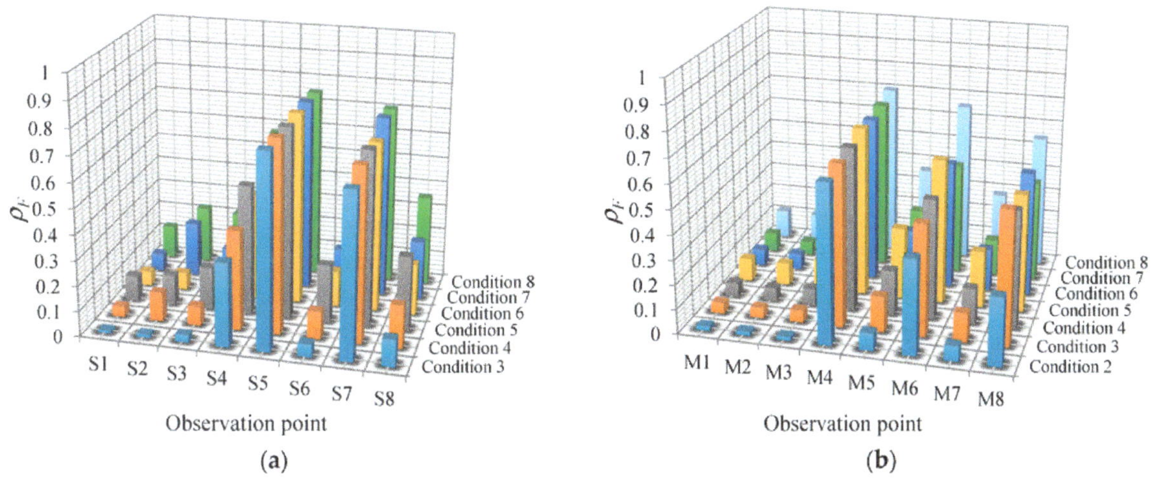

Figure 19: Statistical distribution of the spectral correlation coefficients between the vorticity monitoring points and the prototype observation point: (**a**) Midline of the crest overflowing orifice; (**b**) Midline of the mid-discharge orifice.

The analysis results of the vorticity fluctuating characteristics above verify the close correlation between the vorticity fluctuation in the strong shear layers around the high-velocity submerged jets and the flow-induced LFN propagated to the far field. In addition, combined with the identification results of the vortical structures above, it is concluded that, during the energy dissipation process of the submerged jets, a large number of turbulent low-frequency large-scale vortical structures come into existence in the strong shear layers around the submerged jets because of the large velocity gradient, and the LFN is generated in the intensive interactions of the vortical structures. Therefore, the strong shear layers around the submerged jets are judged as the main regions of acoustic source for the LFN.

Verification of the Mathematical Prediction Model of LFN

According to the mathematical prediction model of LFN (Equations (4) and (5)) built theoretically in the previous section, the numerical simulation results of the main acoustic source regions in the flow field are put into the prediction model to calculate the LFN intensity of prototype observation point T1, which is located at the energy dissipation region under all conditions. The comparisons between the prediction results and prototype observation results are listed in Table 5.

Table 5: Comparisons of results from the prediction model and prototype observation of LFN *

Condition	Prototype Observation Results			Prediction Model Results		
	LFN (Pa)	SPL (dB)	Frequency (Hz)	LFN (Pa)	SPL (dB)	Frequency (Hz)
2	8.57	112.64	1.117	8.84	112.91	1.150
3	9.49	113.53	1.022	9.99	113.97	1.125
4	10.72	114.59	0.978	11.54	115.23	0.933
5	10.38	114.30	0.899	10.62	114.50	0.895
6	10.54	114.44	0.833	11.08	114.87	0.975
7	11.87	115.47	0.894	12.93	116.21	1.261
8	13.08	116.31	1.178	13.46	116.56	1.250

* The values of the LFN intensity and SPL in the table are the RMS values.

The comparison results indicate that the LFN amplitudes, dominant frequencies and variation laws predicted theoretically are approximately consistent with those of the prototype observation. The predicted amplitudes gradually increase with increasing flow rate, and the dominant frequencies of the LFN predicted are concentrated between 0 Hz and 2 Hz. Depending on the prediction results of T1 and the SPL attenuation rate in Table 2, the LFN amplitudes of several observation points (T2, T3, T6, T9, T15, T18, and T22) downstream were estimated. Figure 20 compares the normalized PSD of T1 and the LFN amplitudes along the downstream between the observed and predicted LFN data under all conditions. It is shown that the variation tendencies of the LFN amplitudes of the observed and predicted LFN are nearly the same, while the predicted LFN amplitudes of some points are a little larger than the observed data.

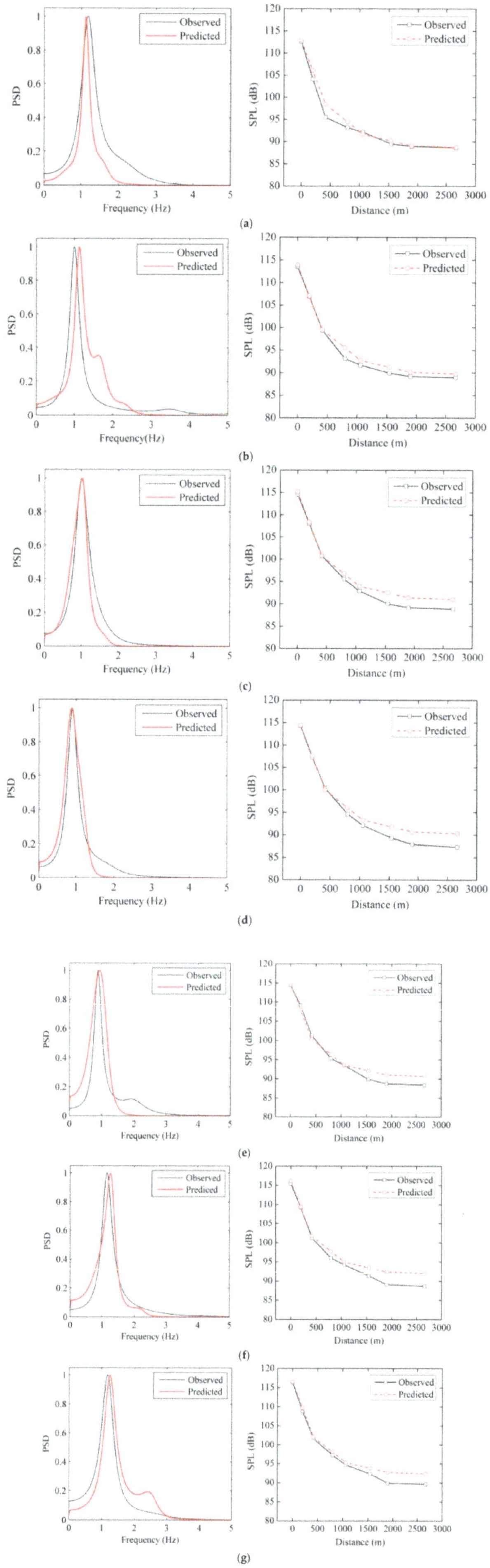

Figure 20: Comparisons of the normalized PSD of T1 and the LFN amplitudes along the downstream between the observed

and predicted data: (**a**) Condition 2; (**b**) Condition 3; (**c**) Condition 4; (**d**) Condition 5; (**e**) Condition 6; (**f**) Condition 7; (**g**) Condition 8.

In order to evaluate the proposed prediction model, the power spectral entropy, derived from Shannon entropy [53] as a common method [54,55], is introduced here to quantify and compare the uncertainties and complexities of PSD estimation between the observed and predicted LFN data. The power spectral entropy is defined as [54,55]:

$$H_F = -\sum_{f=0.1}^{f=20} p(f)\log(p(f)), \tag{14}$$

where $p(f) = s(f) \Big/ \sum_{f=0.1}^{f=20} s(f)$, $s(f)$ is the spectral component, the base of the logarithm is two, and the unit of H_F is bits. In current work, because the infrasound microphone' minimum measuring threshold is 0.1 Hz and the on-site LFN energy is generally below 20 Hz, the power spectral entropy is estimated based on the PSD within 0.1–20 Hz in Equation (14). In essence, the greater the spectral entropy, the more irregular the PSD distribution. The power spectral entropy between the observed and predicted LFN data of T1 under all conditions are shown in Figure 21. It is seen that the power spectral entropy of the observed and predicted LFN data are very close, and the power spectral entropy of the observed LFN are always a little larger than those of the predicted LFN. The average power spectral entropy of the observed and predicted LFN data are 2.632 and 2.540, respectively, and the difference in the average values is approximately 0.0923. A comparison of the results reveals that the PSD between the observed and predicted LFN data has uncertainties and complexities of nearly the same degree.

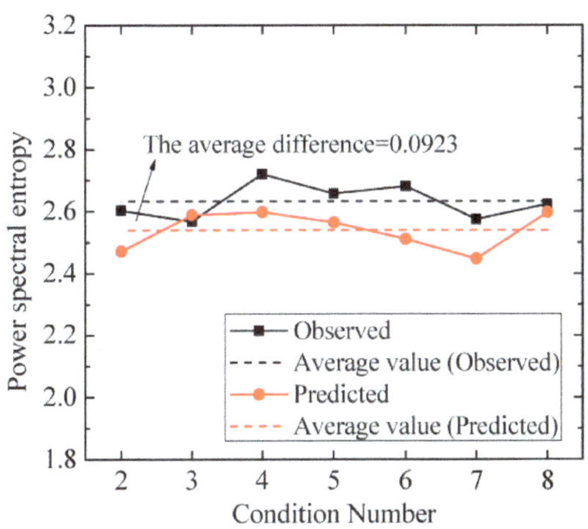

Figure 21: Comparisons of the power spectral entropy between the observed and predicted LFN data.

Moreover, the spectral correlation coefficients ρ_F between the observed and predicted LFN data of T1 were calculated through Equation (13). As observed in Figure 22, the values of ρ_F under all conditions are relatively large, at approximately 0.645–0.717. Thus, according to the analyses of the uncertainties of PSD and the spectral correlation coefficients, it can be seen that the mathematical prediction model is feasible and reasonable for preliminary estimation of the LFN intensity induced by energy dissipation through the submerged jets of a high dam.

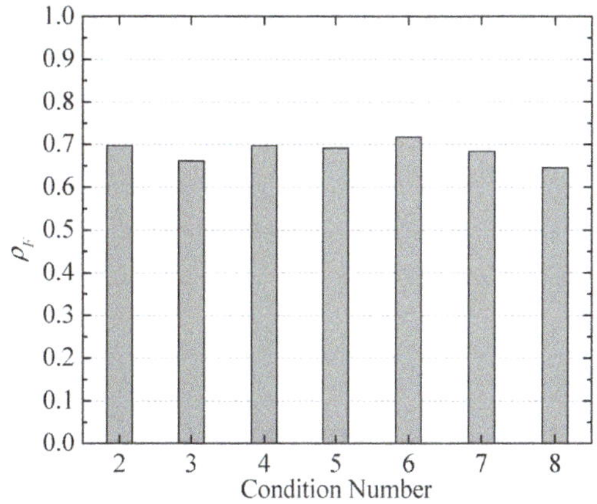

Figure 22: Spectral correlation coefficients between the observed and predicted LFN data.

CONCLUSIONS

In this study, extensive prototype observations and analyses of low frequency noise (LFN) induced by the process of energy dissipation through the submerged jets of a high dam are carried out. It is found that the LFN amplitude reaches the maximum value near the dam area, which is below the hearing threshold and the LFN limits. In general, the observed LFN has no directly negative effect on local residents' health. The sound pressure level (SPL) attenuation coefficient of LFN decreases gradually with increasing distance to the dam area, reaching maximum in the vapor atomization area near the overflow dam. The flow-induced LFN intensity has a close relationship with the discharge operation schemes and flow regimes in the stilling basin, and maintaining uniform and smooth flow patterns in the stilling basin using the joint discharge operation scheme has the benefit of reducing the LFN intensity.

The results from numerical simulation of the flow field characteristics of energy dissipation by the submerged jets indicate that the vortical structures are mainly located at the beginning of the stilling basin where the shearing motion of the flow is violent. The results from correlation analysis of the vorticity fluctuation characteristics and acoustic source indicate that the vorticity fluctuation from the strong shear layers around the high-velocity submerged jets have larger RMS values, where the vorticity fluctuation data are highly correlated to the on-site LFN data. Besides, the strong shear layers are the main regions of acoustic source for the LFN. The mathematical prediction model of the LFN intensity for energy dissipation by the submerged jets is established by combining the vortex sound theory and turbulent flow model, and the model is verified by the prototype observations. The intensity and dominant frequency of the predicted and observed LFN data show satisfactory agreement.

This study for the first time provides the reference data, theoretical foundation and prediction method for addressing the environmental problem of LFN induced by energy dissipating submerged jets during flood discharge from a high dam. In further studies, the specific and scientific discharge operation schemes for reducing the LFN intensity around the station should be built using the current models, which will provide an operation guide and improvements for the station in engineering practices. The theoretical and numerical methods adopted in this study should be applied to other similar engineering projects on LFN issues for further verification of the universality of the prediction model and optimization of the model parameters. Moreover, further problems, such as the propagation and attenuation patterns of LFN and the effects of the water surface wave on the LFN energy, should be studied systematically.

ACKNOWLEDGMENTS

This work was supported by the National Natural Science Foundation of China under Grant No. 51379140 and 51579173.

AUTHOR CONTRIBUTIONS

Jijian Lian conceived the study and contributed to the design of the study. Wenjiao Zhang completed the modeling and numerical analysis works and prepared the manuscript. Qizhong Guo conducted the analysis and contributed to presenting the methods. Fang Liu and Wenjiao Zhang contributed to the prototype observation data collection and analyses. All authors reviewed and approved the final manuscript.

REFERENCES

1. Öhrström, E.; Skånberg, A.; Svensson, H.; Gidlöf-Gunnarsson, A. Effects of road traffic noise and the benefit of access to quietness. J. Sound Vib. 2006, 295, 40–59.
2. Ragettli, M.S.; Goudreau, S.; Plante, C.; Perron, S.; Fournier, M.; Smargiassi, A. Annoyance from road traffic, trains, airplanes and from total environmental noise levels. Int. J. Environ. Res. Public Health 2016, 13.
3. Basner, M.; Griefahn, B.; Berg, M.V.D. Aircraft noise effects on sleep: Mechanisms, mitigation and research needs. Noise Health 2010, 12, 95–109.
4. Marquis-Favre, C.; Morel, J. A simulated environment experiment on annoyance due to combined road traffic and industrial noises. Int. J. Environ. Res. Public Health 2015, 12, 8413–8433.
5. Baloye, D.O.; Palamuleni, L.G. A comparative land use-based analysis of noise pollution levels in selected urban centers of Nigeria. Int. J. Environ. Res. Public Health 2015, 12, 12225–12246.
6. Leventhall, H.G. Low frequency noise and annoyance. Noise Health 2004, 6, 59–72.
7. Zajamsek, B.; Hansen, K.L.; Doolan, C.J. Characterisation of wind farm infrasound and low-frequency noise. J. Sound Vib. 2016, 370, 176–190.
8. Alves, J.A.; Silva, L.T.; Remoaldo, P.C.C. The influence of low-frequency noise pollution on the quality of life and place in sustainable cities: A case study from Northern Portugal. Sustainability 2015, 7, 13920–13946.
9. Alimohammadi, I.; Sandrock, S.; Gohari, M.R. The effects of low frequency noise on mental performance and annoyance. Environ. Monit. Assess. 2013, 185, 7043–7051.
10. Salt, A.N.; Hullar, T.E. Responses of the ear to low frequency sounds, infrasound and wind turbines. Hear. Res. 2010, 268, 12–21.
11. Mu, J.; Ma, C.; Zhao, J.Q.; Lian, J.J. Optimal operation rules of Three-gorge and Gezhouba cascade hydropower stations in flood season. Energy. Convers. Manag. 2015, 96, 159–174.
12. Zhang, Y.; Lian, J.J.; Liu, F. An improved filtering method based on EEMD and wavelet-threshold for modal parameter identification of hydraulic structure. Mech. Syst. Signal Process. 2016, 68–69, 316–329.
13. He, L.J.; Lian, J.J.; Ma, B. Intelligent damage identification method for large structures based on strain modal parameters. J. Vib. Control 2014, 20, 1783–1795.
14. Lian, J.J.; Li, C.Y.; Liu, F.; Wu, S.Q. A prediction method of flood discharge atomization for high dams. J. Hydraul. Res.2014, 52, 274–282.
15. Lian, J.J.; Guo, J.S.; Liu, F. Analysis and research of the shutter doors vibration induced by low frequency noise of flood discharge. J. Hydraul. Eng. 2015, 46, 1207–1212. (In Chinese)
16. Kamigawara, K.; Yue, J.; Saito, T.; Hirano, T. Publication of "Handbook to deal with low frequency noise (2004)". J. Low Freq. Noise Vib. Act. 2006, 25, 153–156.
17. Nakamura, R. Problems of the low frequency air vibration. J. Environ. Conserv. Eng. 1978, 7, 848–849. (In Japanese)
18. Takebayashi, S. Low frequency noise of dam and weir. J. INCE Jpn. 1999, 23, 324–328. (In Japanese)
19. Nakagawa, H.; Takebayasi, S.; Mukuno, S. A study on low frequency noise of water fall. In Proceedings of the International Symposium on Hydraulics for High Dams, Beijing, China, 15–18 November 1988; pp. 1205–1217.
20. Wise, S.; Leventhall, G. Active noise control as a solution to low frequency noise problems. J. Low Freq. Noise Vib. Act.2010, 29, 129–137.
21. Groot-Hedlin, C.D.D.; Hedlin, M.A.H. Infrasound detection of the Chelyabinsk meteor at the USArray. Earth Planet. Sci. Lett. 2014, 402, 337–345.
22. Ochiai, H.; Taya, K. On the Threshold level of rattling fittings due to low frequency sound pressure level. J. INCE Jpn.2002, 26, 120–128. (In Japanese)
23. Saitoh, H.; Maeda, N.; Kyotoh, H.; Nishimura, H. Sound generated by hydraulic jump. J. Hydraul. Eng. 2003, 47, 775–780. (In Japanese)
24. Lian, J.J.; Zhang, Y.; Liu, F.; Zhao, Q.H. Analysis of the ground vibration induced by high dam flood discharge using the cross wavelet transform method. J. Renew. Sustain. Energy 2015, 7, 043146.
25. Zhang, Y.T. Design on water release structure of Xiangjiaba Hydropower Station. China Three Gorges Constr. 2008, 5, 28–31. (In Chinese)
26. Nejadkoorki, E.; Yousefi, E.; Naseri, F. Analysing street traffic noise pollution in the city of Yazd. Iran J. Environ. Health Sci. Eng. 2010, 7, 53–62.
27. McTague, M.F.; Galusha, D.; Dixon-Ernst, C.; Kirsche, S.R.; Slade, M.D.; Cullen, M.R.; Rabinowitz, P.M. Impact of daily noise exposure monitoring on occupational noise exposures in manufacturing workers. Int. J. Audiol. 2013, 52, S3–S8.

28. Ziaran, S. The assessment and evaluation of low-frequency noise near the region of infrasound. Noise Health 2014, 16, 10–17.

29. Camargo, H.E.; Ravetta, P.A.; Burdisso, R.A.; Smith, A.K. Application of phased array technology for identification of low frequency noise sources. J. Low Freq. Noise Vib. Act. 2009, 28, 237–244.

30. Liu, X.K.; Zhang, C.L.; Ji, Z.; Ma, Y.; Shang, X.; Zhang, Q.; Zheng, W.; Li, X.; Gao, J.; Wang, R. Multiple characteristics analysis of Alzheimer's electroencephalogram by power spectral density and Lempel-Ziv complexity. Cogn. Neurodyn.2016, 10, 121–133.

31. Wang, Z.Q.; Zhou, J.H.; Zhong, M.C.; Li, D.; Li, Y.M. Calibration of optical tweezers based on an autoregressive model. Opt. Express 2014, 22, 16956–16964.

32. International Organization for Standardization. Acoustics-Frequency-Weighting Characteristic for Infrasound Measurements; ISO 7196; International Organization for Standardization: Geneva, Switzerland, 1995.

33. Berger, R.G.; Ashtiani, P.; Ollson, C.A.; Whitfield Aslund, M.; McCallum, L.C.; Leventhall, G.; Knopper, L.D. Health-based audible noise guidelines account for infrasound and low-frequency noise produced by wind turbines. Front. Public Health 2015, 3, 31.

34. Salt, A.N.; Kaltenbach, J.A. Infrasound from wind turbines could affect humans. Bull. Sci. Technol. Soc. 2011, 31, 296–302.

35. Watanabe, T.; Møller, H. Low frequency hearing thresholds in pressure field and in free field. J. Low Freq. Noise Vib. Act. 1990, 9, 106–115.

36. Roberts, C. Ecoaccess guideline for the assessment of low frequency noise. In Proceedings of the ACOUSTICS, Gold Coast, Australia, 3–5 November 2004.

37. Jakobsen, J. Danish guidelines on environmental low frequency noise, infrasound and vibration. J. Low Freq. Noise Vib. Act. 2001, 20, 141–148.

38. Powell, A. Theory of vortex sound. J. Acoust. Soc. Am. 1964, 36, 177–195.

39. Howe, M.S. Contribution to the theories of aerodynamic sound with application to excess jet noise and to theory of the flute. J. Fluid Mech. 1975, 71, 625–673.

40. Howe, M.S. On the absorption of sound by turbulence and other hydrodynamic flow. PMM-J. Appl. Math. Mech. 1984, 84, 187–209.

41. Möhring, W. On vortex sound at low Mach number. J. Fluid Mech. 1978, 85, 685–691.

42. Karim, O.A.K.; Mali, K.H.M. Prediction of flow patterns in local scour holes caused by turbulent water jets. J. Hydraul. Res. 2000, 38, 279–287.

43. Deng, J.; Xu, W.L.; Zhang, J.M.; Qu, J.X.; Yang, Y.Q. A new type of plunge pool—Multi-horizontal submerged jets. Sci. China Technol. Sci. 2008, 51, 2128–2141.

44. Yakhot, V.; Orszag, S.A. Development of turbulent models for shear flows by a double expansion technique. Phys. Fluids 1992, 4, 1510–1520.

45. Hirt, C.W.; Nichols, B.D. Volume of Fluid (VOF) for the dynamics of free boundarier. J. Comput. Phys. 1981, 39, 201–225.

46. Ramamurthy, A.S.; Qu, J.; Vo, D. Numerical and experimental study of dividing open-channel flows. J. Hydraul. Eng.2007, 133, 1135–1144.

47. Celik, I.B.; Ghia, U.; Roache, P.J.; Freitas, C.J.; Coleman, H.; Raad, P.E. Procedure for estimation and reporting of uncertainty due to discretization in CFD applications. J. Fluids Eng. ASME 2008, 130, 078001.

48. Park, S.; Park, S.; Rhee, S.H. Influence of blade deformation and yawed inflow on performance of a horizontal axis tidal stream turbine. Renew. Energy 2016, 92, 321–332.

49. Roache, P.J. Perspective a method for uniform reporting of grid refinement studies. J. Fluid. Eng. 1994, 116, 405–413.

50. Hunt, J.C.R.; Wray, A.; Moin, P. Eddies, Stream and Convergence Zones in Turbulent Flows; Technical Report; Center for Turbulence Research: Stanford, CA, USA, 1988; pp. 193–208.

51. Wan, Z.H.; Zhou, L.; Sun, D.J. A study on large coherent structures and noise emission in a turbulent round jet. Sci. China Phys. Mech. 2014, 57, 1552–1562.

52. Cheng, W.; Zhang, Z.S.; He, Z.J.; Wang, Y.X.; Chen, L. Spectral correlation technique and its application in analyzing vibration signals' transfer characteristics of submarine structures. J. Vib. Shock 2010, 29, 8–10. (In Chinese)

53. Shannon, C.E. A mathematical theory of communication. Bell Syst. Tech. J. 1984, 27, 597–610, 623–656.

54. Zhang, J.J.; Zhang, T. Parameter-induced stochastic resonance based on spectral entropy and its application to weak signal detection. Rev. Sci. Instrum. 2015, 86, 025005.

55. Zhang, R.; Xu, P.; Chen, R.; Li, F.; Guo, L.; Li, P.; Zhang, T.; Yao, D. Predicting inter-session performance of SMR-based brain-computer interface using the spectral entropy of resting-state EEG. Brain Topogr. 2015, 28, 680–690.

CHAPTER 16 MAXIMUM STRESS ESTIMATION MODEL FOR MULTI-SPAN WALER BEAMS WITH DEFLECTIONS AT THE SUPPORTS USING AVERAGE STRAINS

Sung Woo Park, Byung Kwan Oh and Hyo Seon Park

Keywords: Average strain, vibrating wire strain gauge, maximum stress, structural health monitoring

INTRODUCTION

Safety monitoring of multi-span beam structures under distributed loads has been an issue of interest since structural health monitoring (SHM) has been introduced [1,2,3,4,5,6,7,8]. If the maximum stress in a beam structure exceeds the specified strength or allowable stress due to unexpected loads, then it can be considered that the safety of the beam structure has reached a limit state. Therefore, structural responses, including the maximum stress in beam structures must be monitored in the long term to assess the safety of a structure [9,10,11,12,13].

To estimate the maximum strain or stress of multi-span beams, mathematical models based on measured average strains have been developed and applied to monitoring the safety of beam structures [14]. In the mathematical models, the maximum strain of a multi-span beam is obtained by defining the relationship between the average strain measured from sensors and the estimated maximum strain. It is shown that the maximum strain of a multi-span beam can be accurately estimated by using average strains obtained from long gauge fiber optic sensors (LGFOS) or vibrating wire strain gauges (VWSG) [15,16,17,18]. However, for the case of beam structures subjected to deflections at the supports, the mathematical models of multi-span beam without consideration of the effect of deflections at supports on the magnitude of the maximum stress cannot be used in the assessment of the safety of beam structures. A typical example of a beam structure subjected to deflections at its supports is the waler beam in an anchored retaining wall system shown in Figure 1. The waler beam is used to transfer retained loads due to lateral pressures from the retained soil evenly between the ground anchors [19].

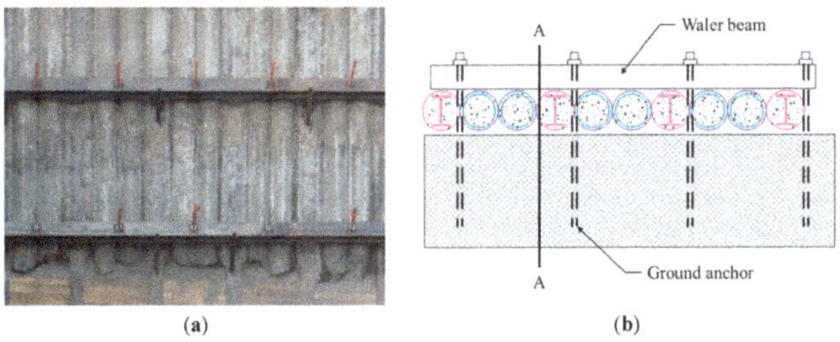

Figure 1: A multi-span waler beam supported by ground anchors in a retaining wall system: (**a**) Front view; (**b**) Side view.

As shown in Figure 2, the ground anchors in a retaining wall can be modelled as supports for the waler beam. Since the ground anchors are subjected to the axial tension due to lateral soil pressure, axial displacement along the axis of the ground anchor occurs and then the waler beam supported by the ground anchors is subjected to differential support deflections.

Citation: Sung Woo Park, Byung Kwan Oh and Hyo Seon Park, Maximum Stress Estimation Model for Multi-Span Waler Beams with Deflections at the Supports Using Average Strains, doi:10.3390/s150407728.

Excessive structural deformation in waler beams due to lateral pressure and differential support deflections is a main cause of damage or collapse of retaining wall systems [20]. The safety of the retaining wall system can be assessed by monitoring the safety of the waler beam subjected to deflections at the supports. Thus, it is necessary to develop a generalized mathematical model for estimation of the maximum stress in a multi-span beam subjected to support deflections.

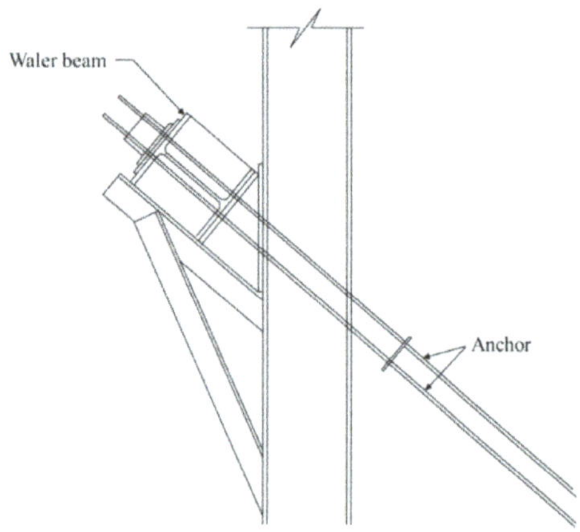

Figure 2: Section A-A of a waler beam supported by ground anchors in a retaining wall system.

Therefore, in this paper, an analytical model to estimate the maximum stress for a multi-span beam with deflections of supports under distributed loads is presented. For the estimation, the relationship between the maximum stress and the average strain of each span is derived and the average strain is measured at the middle of each span. In addition to estimation of the maximum stress, using average strains from VWSGs, estimation models for deflections at supports, support reactions, and the magnitudes of distributed loads are presented. For the safety monitoring of multi-span beam structures in retaining wall systems, the model is applied to estimation of the maximum stress and distribution of moments along the length of a beam.

AVERAGE STRAIN OF WALER BEAMS MEASURED BY VWSGS

In Figure 3, a typical waler beam supported by $n + 1$ ground anchors in a retaining wall system is modeled a linear elastic n-span continuous beam with $n + 1$ supports. l_i and w_i are the length and the magnitude of distributed lateral load of the ith span, respectively. k_i and δ_i are the stiffness of the anchor and the deflection at the ith supports, respectively. It is assumed that the deflection at each support has a different value.

For a linear elastic multi-span beam as shown in Figure 3, the relationship between the bending stress, $\sigma(x_i)$, and strain, $\varepsilon(x_i)$, as a function of the position of x_i in the ith span along the axis of the beam can be written as:

$$\sigma(x_i) = \frac{M(x_i)}{Z} = \varepsilon(x_i)E \tag{1}$$

where $M(x_i)$ is the bending moment, Z is the elastic section modulus and E is the modulus of elasticity of the beam [21]. Using the predetermined value for E, the stress in Equation (1) can be estimated by measuring the strain $\varepsilon(x_i)$. For the measurement of strains, VWSGs are the most frequently used sensors in the construction field of building structures since operating principle is simple and installation cost is relatively low compared to other sensors [16,17]. In addition, VWSGs have excellent endurance and are free from electromagnetic interference (EMI) so that VWSGs suitable for long-term monitoring in the field [22,23,24].

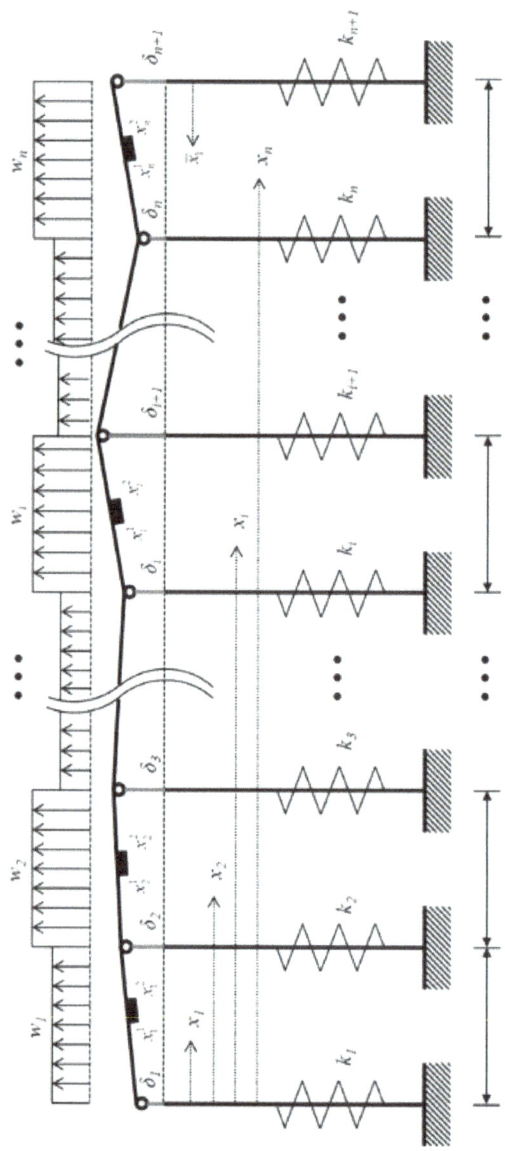

Figure 3: n-Span continuous beam supported by $n + 1$ ground anchors.

Figure 4: Typical vibrating wire strain gauge attached to the surface of a steel section for measurement of average strains.

As shown in Figure 4, a VWSG consists of three components of two mounting blocks which are attached to the surface of a structural member to be measured, a vibrating wire whose frequency changes due to tension or compression between the two mounting blocks, and plucking and pickup coil, which excited the vibrating wire and measure its resonance frequencies. The change of length of a vibrating wire which is fixed at the two mounting blocks causes the variation of the natural frequency of the wire. For this reason, a VWSG measures the average strain over the gauge length between the two mounting blocks. As shown in Figure 3, if the gauge length is the distance between the mounting blocks and the locations of the mounting blocks for the VWSG in the *i*th span of the waler beam are set to x_i^1 and x_i^2, then the average strain for the *i*th span of the waler beam, $\varepsilon_{avg,i}$, can be given as:

$$\varepsilon_{avg,i} = \frac{\int_{x_i^1}^{x_i^2} \varepsilon(x_i)dx}{x_i^2 - x_i^1} \tag{2}$$

Then, the relationship between the bending moment at the *i*th span, $M(x_i)$ in Equation (1) and the measured average strain $\varepsilon_{avg,i}$ in Equation (2) can be expressed by:

$$\varepsilon_{avg,i} = \frac{\int_{x_i^1}^{x_i^2} M(x_i)dx}{EZ(x_i^2 - x_i^1)} \tag{3}$$

ESTIMATION MODEL FOR MAXIMUM STRESS

Based on the three moment theorem, the relationship among the bending moments, M_{i-1}, M_i and M_{i+1}, at three consecutive supports of the multi-span beam in Figure 3 is given by Equation (A1) in Appendix A. The beam is assumed to have constant moment of inertia and elastic modulus. For the continuous beam with $n + 1$ supports, $n - 1$ three moment equations can be formulated using Equation (A1) in Appendix A and expressed in matrix form as:

$$[A]\{M\} = [B]\{w\} + [C]\{\delta\} \tag{4}$$

where [A] is a square matrix of dimensions (n−1)×(n−1), [B] is a matrix of dimensions (n−1)×n, and [C] is a matrix of dimensions (n−1)×(n+1), given by Equations (B1)–(B3) in Appendix B. Also, $\{M\}$, $\{w\}$ and $\{\delta\}$ are the bending moment vector of dimension (n−1) ×1, the magnitude of distributed lateral load vector of dimension n×1, and the deflection at support vector of dimension (n+1)×1, respectively. Then, the equations in Equation (4) can be solved simultaneously for the moments at supports. Having found the moments at $n + 1$ supports, the vector of dimension (n+1)×1 for the reactions at the supports of the continuous beam, {R}, can be found by:

$$\{R\} = [G]\{w\} + [C']\{M\} \tag{5}$$

where [G] is a matrix of dimensions (n−1)×n and $[C']$ is a matrix of dimensions (n+1)×(n−1), given by Equations (B4) and (B5) in Appendix B. As shown in Figure 3, the reaction at the *ith* support can also be found by multiplying the stiffness *ki* by the deflection *δi* of the support. Then, the reaction vector in Equation (5) can be expressed by substituting the moment vector $\{M\}$ in Equation (4) into Equation (5).

$$\{R\} = [K]\{\delta\} = [G]\{w\} + [C']\left[[A^{-1}B]\{w\} + [A^{-1}C]\{\delta\}\right] \tag{6}$$

Where $[K]$ is the stiffness matrix for the $n + 1$ supports in Figure 3. From Equation (6), the relationship between the vectors $\{w\}$ and $\{\delta\}$ is found by:

$$[G + C'A^{-1}B]\{w\} = [K - C'A^{-1}C]\{\delta\} \tag{7}$$

Using the support reactions $\{R\}$ and distributed lateral loads $\{w\}$, the bending moment of each span of the beam $\{\bar{M}(x)\}$ can be given by:

$$\{\bar{M}(x)\} = [P]\{R\} - [Q]\{w\} \tag{8}$$

where [P] is a matrix of dimensions (n+1)×(n+1) and [Q] is a matrix of dimensions (n+1)×1, given by Equations (B6) and (B7) in Appendix B.

By considering the gauge length of a VWSG in Equation (3), the relationship in Equation (8) for the *i*th span can be evaluated by numerical integration over the gauge length of $x_i^2 - x_i^1$ as below:

$$\left\{\int_{x_i^1}^{x_i^2} \bar{M}(x)dx\right\} = [\bar{P}]\{R\} - [\bar{Q}]\{w\} \tag{9}$$

where $[\bar{P}]$ and $[\bar{Q}]$ are integrals of [P] and [Q] in Equations (B6) and (B7) in Appendix B over the gauge length, respectively. The integral of the moment over the interval of the gauge length in Equation (9) can be expressed in term of the average strain in Equation (3):

$$\left\{\int_{x_i^1}^{x_i^2} \bar{M}(x)dx\right\} = \{\varepsilon_{avg,i}\} EZ(x_i^2 - x_i^1) = [\bar{P}]\{R\} - [\bar{Q}]\{w\} \tag{10}$$

Then, by substituting {R} in Equation (6) and {w} in Equation (7) into Equation (10), Equation (10) can be given by:

$$EZ(x_i^2 - x_i^1)\{\varepsilon_{avg,i}\} = [\bar{P}][K]\{\delta\} - [\bar{Q}]\left[[G + C'A^{-1}B]^T[G + C'A^{-1}B]\right]^{-1}[G + C'A^{-1}B]^T[K - C'A^{-1}C]\{\delta\} \tag{11}$$

From Equation (11), deflections at supports of the waler beam {δ} can be found from the measurement of average strains from VWSGs $\{\varepsilon_{avg,i}\}$ as below:

$$\{\delta\} = EZ(x_i^2 - x_i^1)\left[[\bar{P}][K] + [\bar{Q}]\left[[\bar{Y}]^T[\bar{Y}]\right]^{-1}[\bar{Y}]^T[\bar{X}]\right]^{-1}\{\varepsilon_{avg,i}\} \tag{12}$$

where $[\bar{X}] = [K - C'A^{-1}C]$ and $[\bar{Y}] = [G + C'A^{-1}B]$. Thus, by substituting the deflections determined from the average strains in Equation (12) into Equations (6) and (7), reaction at each support {R} and the magnitude of the distributed lateral load at each span {w} can be identified. Consequently, the bending moment of each span can be determined by substituting {R} and {δ} into Equation (9). Finally, based on the relationship in Equation (1), the maximum stress of an *n*-span waler beam can be determined from the average strains measured by VWSGs.

TEST OF THE MODEL ON MULTI-SPAN BEAMS

Four-Span Beam

The setup for simulation of the four-span test beam is shown in Figure 5. The section of beam is H-300 × 300 × 10 × 15, with a depth of 300 mm, a flange width of 300 mm, a web thickness of 10 mm, and a flange thickness of 15 mm. The material for the beam section is SS400 grade steel with a modulus of elasticity of 205 GPa and yield strength of 235.3 MPa. The test model is designed to simulate the collapse of the waler beam in a retaining wall at a construction site in 2009. The waler beam in the retaining wall was designed to carry a uniformly distributed load of 206.5 kN/m. The waler beam was supported by ground anchors of prestressed concrete (PC) strands of Φ12.7 mm × 4 at a horizontal spacing of 1.6 m. The axial stiffness of the anchor was given by 10,117.6 kN/m. With these properties, the four-span beam in the construction site can be modeled as a conceptual model in Figure 5c.

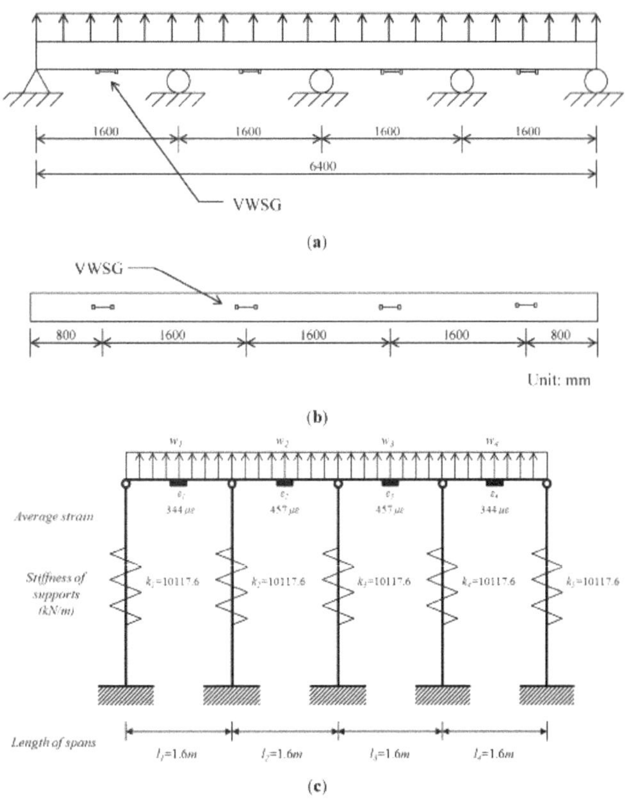

Figure 5: Experimental setup for the test of 4-span beam: (**a**) Side view; (**b**) Bottom view; (**c**) Conceptual model.

To test the conceptual model, it is assumed that the vibrating wire strain gauges with the gauge length of 15 cm are installed to measure the average strains in Figure 5b. As an input, the measured average strains are assumed to be $\varepsilon_1 = \varepsilon_4$ = 344 με and $\varepsilon_2 = \varepsilon_3$ = 457 με. For the given values for average strains obtained from the numerical analysis with the properties of the model in Figure 5c, the magnitudes of deflections of supports, reactions, and distributed loads in Table 1 are estimated from the analytical model in Equations (6), (7) and (12), respectively.

Table 1: Estimated values for deflections, reactions, and distributed loads for the 4-span beam

Deflection (m)	Reaction (kN)	Distributed load (kN/m)
δ_1 = 0.0201	R_1 = 202.9	w_1 = 206.5
δ_2 = 0.0291	R_2 = 294.3	w_2 = 206.5
δ_3 = 0.0323	R_3 = 327.2	w_3 = 206.5
δ_4 = 0.0291	R_4 = 294.3	w_4 = 206.5
δ_5 = 0.0201	R_5 = 202.9	

The deformed shape based on estimated support deflections, support reactions, and applied loads of the test beam is shown in Figure 6. As can be seen in Figure 6, a maximum deflection of 3.23 cm and a maximum reaction of 327.2 kN were calculated at the third support.

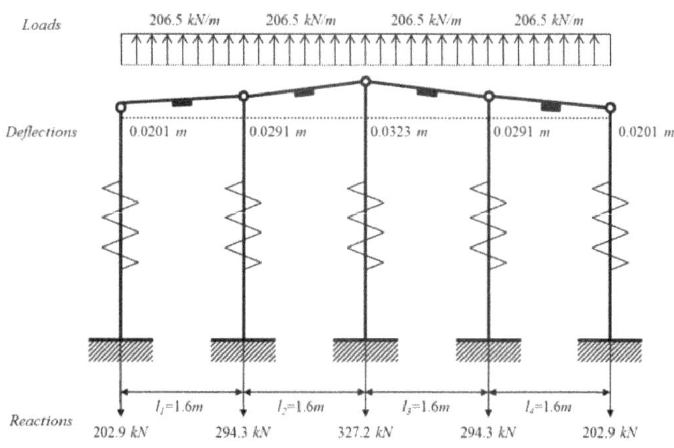

Figure 6: Deformed shape of 4-span beam subjected to the distributed lateral loads.

In Figure 7, the bending moment diagram with the consideration of support deflections are compared with the bending moment distribution without the consideration of deflections at the supports. As can be seen in Figure 7, maximum bending moments of 127.63 kNm were estimated at the points of 2.4 m and 3.93 m measured from the left end of the beam. Using the modulus of elasticity of 205 GPa in Equation (1), maximum stresses of 93.85 MPa were estimated at the same points of the four-span beam where the maximum moment occurs. Thus, for a given value of the average strains measured by VWSGs, it is confirmed that the reactions, distributed loads, bending moments, and maximum stress of the 4-span test beam can be estimated by the proposed analytical model.

As a practical application of the analytical model presented in the paper, the measured values of the average strain can be limited to a specific value to prevent the waler beam from reaching one of limit states for failure or collapse of waler beams. For the safety monitoring of the four-span waler beam, the average strains at the second and third spans, ε_2 and ε_3, may be limited to the value of 780 με to prevent the waler beam from reaching the allowable stress of 160 MPa for SS400 grade steel beam used in the construction site.

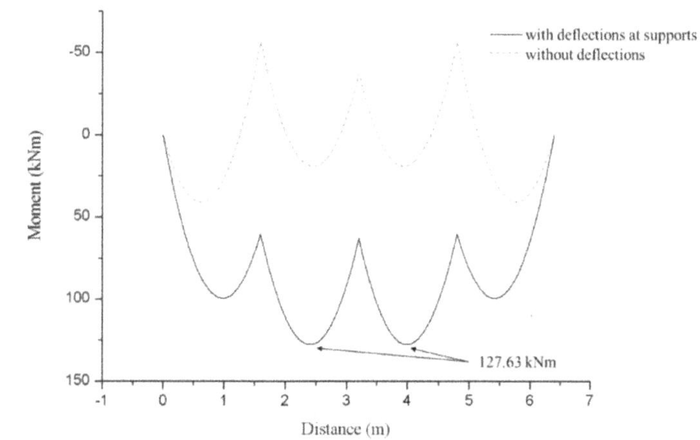

Figure 7: Bending moment diagram for a 4-span beam for estimation of maximum stress.

Eight-Span Beam

A schematic diagram for the eight-span test beam is shown in Figure 8. The material for the beam section of H-300 × 300 × 10 × 15 is SS400 grade steel with a modulus of elasticity of 205 GPa and yield strength of 235.3 MPa. To test the applicability of the estimation method, the test beam in Figure 8 is designed to have irregularities in geometry, mechanical properties of anchors, and the magnitudes of distributed lateral loads.

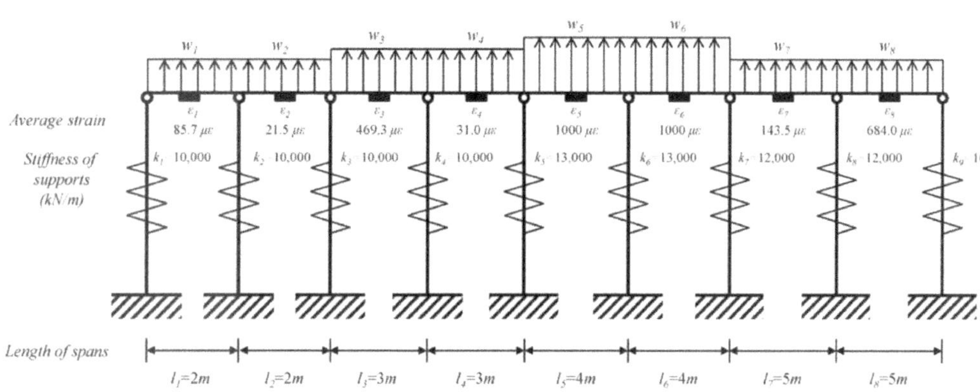

Figure 8: Schematic diagram for the 8-span beam with irregularities in geometry, mechanical properties, and magnitudes of lateral loads.

The waler beam was designed to carry three different magnitudes of distributed design loads of $w_1 = w_2 = w_7 = w_8 =$ 100 kN/m, $w_3 = w_4 = 200$ kN/m, and $w_5 = w_6 = 300$ kN/m. As input for the estimation model in Equation (12), the average strains from VWSGs are listed in Table 2. For the given values for average strains obtained from the numerical analysis with the properties of the model in Figure 8, the magnitudes of deflections of supports, reactions, and distributed loads in Table 3 are estimated from the analytical model in Equations (6), (7) and (12), respectively.

Table 2: Span length, stiffness of anchor, and average strain for an 8-span beam

Span Length (m)	Stiffness of Anchor (kN/m)	Average Strain (με)
$l_1 = 2.0$	$k_1 = 10{,}000.0$	$\varepsilon_1 = 85.7$
$l_2 = 2.0$	$k_2 = 10{,}000.0$	$\varepsilon_2 = 21.5$
$l_3 = 3.0$	$k_3 = 10{,}000.0$	$\varepsilon_3 = 469.3$
$l_4 = 3.0$	$k_4 = 10{,}000.0$	$\varepsilon_4 = 31.0$
$l_5 = 4.0$	$k_5 = 13{,}000.0$	$\varepsilon_5 = 1{,}000.0$
$l_6 = 4.0$	$k_6 = 13{,}000.0$	$\varepsilon_6 = 1{,}000.0$
$l_7 = 5.0$	$k_7 = 12{,}000.0$	$\varepsilon_7 = 143.5$
$l_8 = 5.0$	$k_8 = 12{,}000.0$	$\varepsilon_8 = 684.0$
	$k_9 = 10{,}000.0$	

Table 3: Estimated deflections, reactions, and magnitudes of distributed loads for an eight-span beam

Deflection (cm)	Reaction (kN)	Distributed Load (kN/m)
$\delta_1 = 3.01$	$R_1 = 75.3$	$w_1 = 100$
$\delta_2 = 8.88$	$R_2 = 222.0$	$w_2 = 100$
$\delta_3 = 14.83$	$R_3 = 378.8$	$w_3 = 200$
$\delta_4 = 23.23$	$R_4 = 586.6$	$w_4 = 200$
$\delta_5 = 31.85$	$R_5 = 995.5$	$w_5 = 300$
$\delta_6 = 42.92$	$R_6 = 1287.7$	$w_6 = 300$
$\delta_7 = 47.49$	$R_7 = 712.4$	$w_7 = 100$
$\delta_8 = 38.38$	$R_8 = 575.7$	$w_8 = 100$
$\delta_9 = 22.00$	$R_9 = 220.0$	

The deformed shape based on estimated support deflections, support reactions, and applied loads of the test beam are shown in Figure 9. As can be seen in Figure 9, maximum deflection of 47.5 cm and maximum reaction of 1287.7 kN were calculated at the 7th and 6th supports, respectively. In Figure 10, the bending moment diagram with the consideration of support deflections are compared with the bending moment distribution without the consideration of deflections at the supports. As can be seen in Figure 10, a maximum negative bending moment of −304.20 kNm was estimated at the 5th support. Using the modulus of elasticity of 205 GPa, a maximum stress of −223.67 MPa was estimated at the same points where the maximum moment occurs, therefore, it is found that the magnitude of the estimated maximum stress exceeded the allowable stress of 160 MPa for a SS400 grade steel beam.

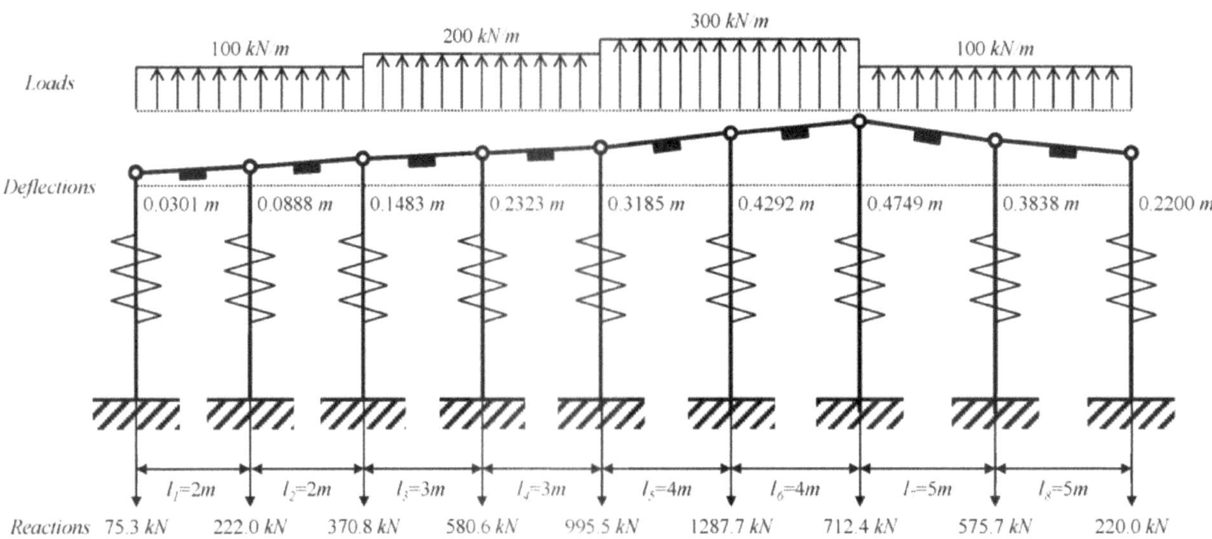

Figure 9: Deformed shape of an 8-span beam subjected to distributed lateral loads.

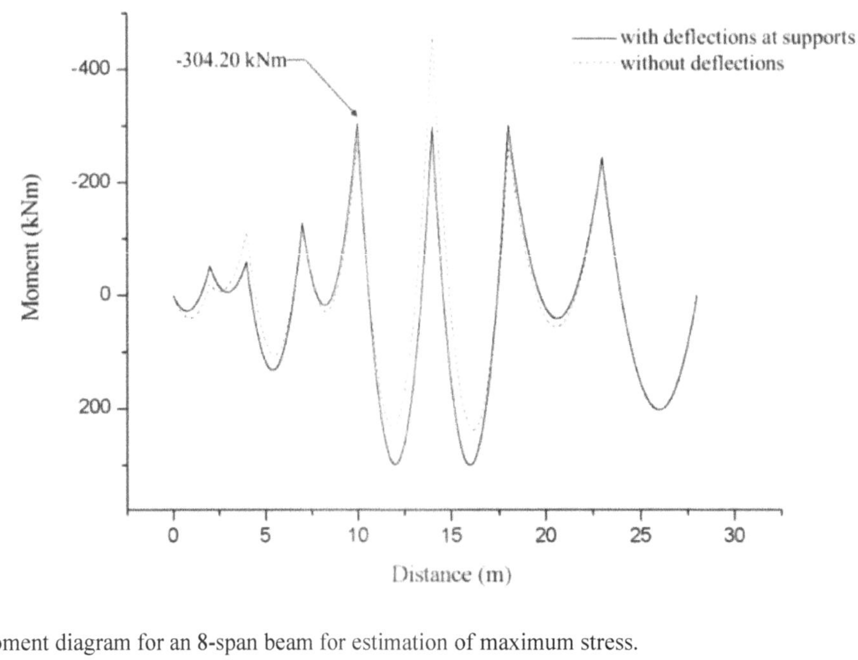

Figure 10: Bending moment diagram for an 8-span beam for estimation of maximum stress.

CONCLUSIONS

In this paper, an analytical model for estimation of the maximum stress of a multi-span beam with deflections of its supports is presented. The analytical model is derived by defining the relationship between the maximum stress and the average strain of each span of the multi-span beam. Based on the use of the average strains obtained from VWSGs as the input to the analytical model, the analytical model allows estimation of deflections at supports, support reactions, and the magnitudes of distributed loads. Using tests on two multi-span beams, the performance of the model is evaluated by estimating maximum stresses, support deflections, support reactions, and the magnitudes of distributed loads. From the simulation test, it can be concluded that safety monitoring of multi-span waler beams is made possible by comparing the given allowable stress of the beam with the estimated maximum stress with consideration of any deflections at the supports. Through periodic estimation of the maximum stress by the proposed method, the safety of retaining walls at construction sites will be improved.

ACKNOWLEDGMENTS

This work was supported by the National Research Foundation of Korea (NRF) Grant funded by the Korea government (MSIP) (No. 2011-0018360).

APPENDIX

A. Three Moment Equation

$$M_i l_i + 2M_{i+1}(l_i + l_{i+1}) + M_{i+2} l_{i+1} = -\frac{1}{4}\left(w_i l_i^3 + w_{i+1} l_{i+1}^3\right) + 6EI\left(-\frac{1}{l_i}(\delta_i) + \left(\frac{1}{l_i} + \frac{1}{l_{i+1}}\right)(\delta_{i+1}) - \frac{1}{l_{i+1}}(\delta_{i+2})\right) \tag{A1}$$

where Mi, li, wi and δi are the bending moment, length, distributed lateral load and deflection at the *i*th support of the beam, respectively. E and I are elastic modulus and moment of inertia, respectively.

B. Matrices in Section 3

$$[A] = \begin{pmatrix} 2(l_1+l_2) & l_2 & 0 & \cdots & \cdots & 0 \\ l_2 & 2(l_2+l_3) & l_3 & 0 & \cdots & \vdots \\ 0 & l_3 & 2(l_3+l_4) & l_4 & 0 & \vdots \\ \vdots & 0 & \ddots & \ddots & \ddots & 0 \\ \vdots & \vdots & 0 & l_{n-2} & 2(l_{n-2}+l_{n-1}) & l_{n-1} \\ 0 & \cdots & \cdots & 0 & l_{n-1} & 2(l_{n-1}+l_n) \end{pmatrix} \tag{B1}$$

$$[B] = -\frac{1}{4}\begin{pmatrix} l_1^3 & l_2^3 & 0 & \cdots & 0 \\ 0 & l_2^3 & l_3^3 & \ddots & \vdots \\ \vdots & \ddots & \ddots & \ddots & 0 \\ 0 & \cdots & 0 & l_{n-1}^3 & l_n^3 \end{pmatrix} \tag{B2}$$

$$[C] = 6EI\begin{pmatrix} -\frac{1}{l_1} & \frac{1}{l_1}+\frac{1}{l_2} & -\frac{1}{l_2} & 0 & \cdots & 0 \\ 0 & -\frac{1}{l_2} & \frac{1}{l_2}+\frac{1}{l_3} & -\frac{1}{l_3} & \ddots & \vdots \\ \vdots & \ddots & \ddots & \ddots & \ddots & 0 \\ 0 & \cdots & 0 & -\frac{1}{l_{n-1}} & \frac{1}{l_{n-1}}+\frac{1}{l_n} & -\frac{1}{l_n} \end{pmatrix} \tag{B3}$$

$$[G]=\frac{1}{2}\begin{pmatrix} l_1 & 0 & \cdots & \cdots & 0 \\ l_1 & l_2 & 0 & \cdots & \vdots \\ 0 & l_2 & l_3 & 0\cdots & \vdots \\ \vdots & 0 & \ddots & \ddots & 0 \\ \vdots & \vdots & 0 & l_{n-1} & l_n \\ 0 & \cdots & \cdots & 0 & l_n \end{pmatrix} \quad \text{(B4)}$$

$$[C']=-\frac{1}{6EI}[C]^T \quad \text{(B5)}$$

$$[P]=\begin{pmatrix} x_1 & 0 & \cdots & \cdots & 0 & 0 \\ x_2 & x_2-l_1 & 0 & \cdots & \ddots & \vdots \\ \vdots & \vdots & x_3-l_1-l_2 & 0 & \vdots & \vdots \\ \vdots & \vdots & \vdots & \ddots & 0 & \vdots \\ x_n & x_n-l_1 & x_n-l_1-l_2 & \cdots & x_n-l_1-\cdots-l_{n-1} & 0 \\ 0 & \cdots & \cdots & 0 & 0 & \bar{x}_0 \end{pmatrix} \quad \text{(B6)}$$

$$[Q]=\begin{pmatrix} \frac{1}{2}x_1^2 & 0 & & & & \\ l_1(x_2-\frac{l_1}{2}) & \frac{1}{2}x_2^2 & 0 & \cdots & \cdots & 0 \\ l_1(x_3-\frac{l_1}{2}) & l_2(x_3-l_1-\frac{l_2}{2}) & \frac{1}{2}x_3^2 & 0 & \ddots & \vdots \\ \vdots & \vdots & \vdots & \ddots & 0 & \vdots \\ l_1(x_n-\frac{l_1}{2}) & l_2(x_n-l_1-\frac{l_2}{2}) & \cdots & & l_{n-1}(x_n-l_1-\cdots l_{n-2}-\frac{l_{n-1}}{2}) & 0 \\ 0 & \cdots & \cdots & \cdots & 0 & \frac{1}{2}\bar{x}_0^2 \end{pmatrix} \quad \text{(B7)}$$

In Equations (B6) and (B7), x_i $(l_1+\cdots+l_{n-1}\le x_i<l_1+\cdots+l_n)$ is the distance from the left end of the beam, as shown in Figure 3, and $\bar{x}_0$ $(0\le\bar{x}_0<l_n)$ is the distance from the right side of the beam to the position in the last span.

REFERENCES

1. Jiang, X.; Adeli, H. Pseudospectra, music, and dynamic wavelet neural network for damage detection of highrise buildings. *Int. J. Numer. Meth. Eng.* 2007, *71*, 606–629.
2. Li, H.; Huang, Y.; Ou, J.; Bao, Y. Fractal dimension-based damage detection method for beams with a uniform cross-section. *Comput.-Aided Civ. Inf.* 2011, *26*, 190–206.
3. Rezaei, D.; Taheri, F. Damage identification in beams using empirical mode decomposition. *Struct. Health Monit.* 2011, *10*, 261–274.
4. Park, H.S.; Lee, H.M.; Adeli, H.; Lee, I. A new approach for health monitoring of structures: Terrestrial laser scanning. *Comput.-Aided Civ. Inf.* 2007, *22*, 19–30.
5. Mitchell, R.; Kim, Y.; El-Korchi, T. System identification of smart structures using a wavelet neuro-fuzzy model. *Smart Mater. Struct.* 2012, *21*, 115009.
6. Choi, S.W.; Lee, J.; Oh, B.H.; Park, H.S. Measurement Model for the Maximum Strain in Beam Structures Using Multiplexed Fiber Bragg Grating Sensors. *Int. J. Distrib. Sens. N* 2013. 894780.
7. Hong, K.; Lee, J.; Choi, S.W.; Kim, Y.; Park, H.S. A Strain-Based Load Identification Model for Beams in Building Structures. *Sensors* 2013, *13*, 9909–9920.
8. Ng, C.-T. Bayesian model updating approach for experimental identification of damage in beams using guided waves. *Struct. Health Monit.* 2014, *13*, 359–373.
9. Hampshire, T.A.; Adeli, H. Monitoring the behavior of steel structures using distributed optical fiber sensors. *J. Constr. Steel Res.* 2000, *53*, 267–281.
10. Park, H.S.; Jung, H.S.; Kwon, Y.H.; Seo, J.H. Mathematical models for assessment of the safety of steel beams based on average strains from long gage optic sensors. *Sens. Actuators A Phys.* 2006, *125*, 109–113.
11. Sari, M.; Bucher, A.E. Natural frequencies and critical loads of beams and columns with damaged boundaries using Chebyshev polynomials. *Int. J. Eng. Sci.* 2010, *48*, 862–873.
12. Wu, N.; Wang, Q. Experimental studies on damage detection of beam structures with wavelet transform. *Int. J. Eng. Sci.* 2011, *48*, 835–847.
13. Khorram, A.; Bakhtiari-Nejad, F.; Rezaeian, M.

Comparison studies between two wavelet based crack detection methods of a beam subjected to a moving load. *Int. J. Eng. Sci.* 2012, *51*, 204–215.

14. Park, H.S.; Jung, S.M.; Lee, H.M.; Kwon, Y.H.; Seo, J.H. Analytical models for assessment of the safety of multi-span steel beams based on average strains from long gage optic sensors. *Sens. Actuators A Phys.* 2007, *137*, 6–12.
15. Casas, J.R.; Cruz, J.S. Fiber optic sensors for bridge monitoring. *J. Bridge Eng. ASCE* 2003, *8*, 362–373.
16. Lee, H.M.; Kim, J.M.; Sho, K.; Park, H.S. A wireless vibrating wire sensor node for continuous structural health monitoring. *Smart Mater. Struct.* 2010, *19*, 055004.
17. Lee, H.M.; Choi, S.W.; Jung, D.; Park, H.S. Analytical Model for Estimation of Maximum Normal Stress in Steel Beam-Columns Based on Wireless Measurement of Average Strains from Vibrating Wire Strain Gages. *Comput.-Aided Civ. Inf.* 2013, *28*, 707–717.
18. Lee, H.M.; Park, H.S. Measurement of Maximum Strain of Steel Beam Structures Based on Average Strains from Vibrating Wire Strain Gages. *Exp. Tech.* 2013, *37*, 23–29.
19. Geotechnical Design Procedure for Flexible Wall Systems. Available online: https://www.dot.ny.gov/divisions/engineering/technical-services/technical-services-repository/GDP-11b.pdf (accessed on 28 January 2015).
20. Retaining Walls. Available online: http://www.dot.ca.gov/hq/esc/techpubs/manual/bridgemanuals/bridge-design-specifications/page/section5.pdf (accessed on 28 January 2015).
21. Gere, J.M.; Timoshenko, S.P. *Mechanics of Material*, 2nd ed.; PWS Engineering: Boston, MA, USA, 1984; pp. 212–220.
22. Park, H.S.; Lee, H.Y.; Choi, S.W.; Kim, Y.S. A Practical Monitoring System for the Structural Safety of Mega-Trusses Using Wireless Vibrating Wire Strain Gauges. *Sensors* 2013, *13*, 17346–17361.
23. Geokon. Available online: http://www.geokon.com/strain-gages/ (accessed on 28 January 2015).
24. RST. Available online: http://www.rstinstruments.com/Vibrating%20Wire%20Strain%20Gauges.html (accessed on 28 January 2015).

CHAPTER 17

A PRACTICAL MONITORING SYSTEM FOR THE STRUCTURAL SAFETY OF MEGA-TRUSSES USING WIRELESS VIBRATING WIRE STRAIN GAUGES

Hyo Seon Park, Hwan Young Lee, Se Woon Choi and Yousok Kim

Keywords: Vibrating wire strain gauge, structural health monitoring, wireless network, mega-truss, temperature stress, long-term monitoring, irregular building

INTRODUCTION

Visual inspection has traditionally played a critical role in quality management of the construction process and damage detection in structures subjected to various loadings [1,2]. However, because structures have become increasingly complex (e.g., high-rise and irregular designs), visual inspections are becoming increasingly time consuming and labor intensive and suffer from expensive and subjective evaluations; these aspects represent critical problems in the application of this method to real structures. Because visual inspection is restricted to post-event assessments, immediate damage detection and safety evaluations of structures are nearly impossible.

For these reasons, structural health monitoring (SHM) based on sensor technology has received considerable attention and has successfully replaced traditional visual inspection [3–6]. The SHM of buildings, which is based on a wired sensor network, was initially conducted for simple civil structures. The development of monitoring systems enabled a real-time response evaluation of a structure. However, the high installation cost of the cable that connects the sensor to the server and maintenance and management challenges remain unresolved issues.

For these reasons, Straser [7] attempted to solve the problems of existing wire-based monitoring systems by proving the effectiveness of the wireless sensor network (WSN) in an actual building. The WSN system (WSNS) significantly decreased the installation cost of a wired sensor network, and thus, the high-nodal density of sensor networsks was realized, which enabled local damage detection. Its applications and state-of-the-art design are well documented [7–12]. Different types of sensors have been employed in SHM to measure the responses of structures, such as acceleration [13–16], displacement [17–21], and strain [22–26].

In special circumstances, strain-type sensors are adopted to directly measure the strain of a structural element. The stress distribution estimated from a strain measurement can be utilized in a safety assessment of an element by comparing it with the yield stress of materials or the design strength of structural members. Several types of strain gauges are used to monitor structural responses, including electrical strain gauges (ESGs), fiber optic sensors (FOSs), [27–31], and vibrating wire strain gauges (VWSGs, [32–34]). Among these strain-type sensors, FOSs and VWSGs, which are immune to electromagnetic interference and provide superior endurance, are actively studied and employed in SHM. Despite the outstanding potential of FOSs in health monitoring applications [27], FOSs are extremely fragile, which contributes to a high

Citation: Hyo Seon Park, Hwan Young Lee, Se Woon Choi and Yousok Kim, A Practical Monitoring System for the Structural Safety of Mega-Trusses Using Wireless Vibrating Wire Strain Gauges, doi:10.3390/s131217346.

rate of installation failure in real structures. In addition, both the measurement system and sensors are relatively expensive compared with VWSGs. Other advantages of VWSGs include the ease and low cost of installation and the use of a smaller amount of data logger bandwidth due to the simple measurement principle of VWSGs. They are appropriate for long-term monitoring because vibrating wires exhibit minimal deterioration over time.

The accuracy and reliability of VWSGs as sensors for measuring strain have been verified from numerous laboratory experiments [35–38]. However, few studies have focused on its application in actual building construction, for which numerous obstacles hinder stable measurements. Although some researchers have conducted real-time monitoring using VWSGs in high-rise buildings during construction [39], the development of WSNS for buildings under construction remains a unique and challenging task.

As a practical monitoring technique, a WSNS based on VWSGs is proposed to obtain reliable data that can enhance structural safety and construction precision through long-term real-time monitoring. The proposed WSNS realized the functions required by wireless network systems through the application of a power-saving wireless VWSG sensor node, which offers three major built-in functions: data collection, data processing, and data transmission. With the exception of the signal cable that connects the sensor to the sensor node, the proposed system enables the entire network to be wireless. Therefore, an automatic system in which the manager can examine data collected from VWSGs in real time was realized.

The structural system of the subject building consists of two mega-trusses, which were installed above four mega-columns to form a three-dimensional irregular shape on a maximum scale. These mega-structures were designed to support the floor and roof of a large space. The mega-truss, whose length extends 35 m, is the cantilever structure and one of the main members that ultimately supports the other members; thus, precise construction is required during construction.

Because stress changes in the main structural elements are anticipated during subsequent phases of construction, a total of 48 VWSGs, 14 sensor nodes, and seven master nodes were implemented to monitor the strain variation in the mega-truss of the irregular large-scale building under construction. Based on strain data collected over a long-term monitoring for 16 months, a quantitative evaluation of the construction process was performed to determine the aspects that exhibit the greatest influence on member behavior and their effects. The numerical model was also verified by comparing the measurement results with the simulated results obtained from the preliminary analysis. In addition, variations in temperature during construction were obtained from temperature sensors, which were installed on major elements and used to investigate the effects of temperature stress on the structural elements. The feasibility of the proposed long-term monitoring system, which is based on wireless VWSGs, to evaluate the structural safety of an irregular building under construction was investigated.

WSNS USING VWSGS

A WSNS based on VWSGs consists of two nodes: the sensor node, which processes and transmits the raw data obtained from the VWSGs to the other network equipment (master node), and the master node, which receives all data from the sensor node and transmits the data to the monitoring server. As illustrated in Figure 1, the entire network of the WSNS in this study was wirelessly automated for the convenience of measurement and maintenance and was constructed so that the user can observe data in real time.

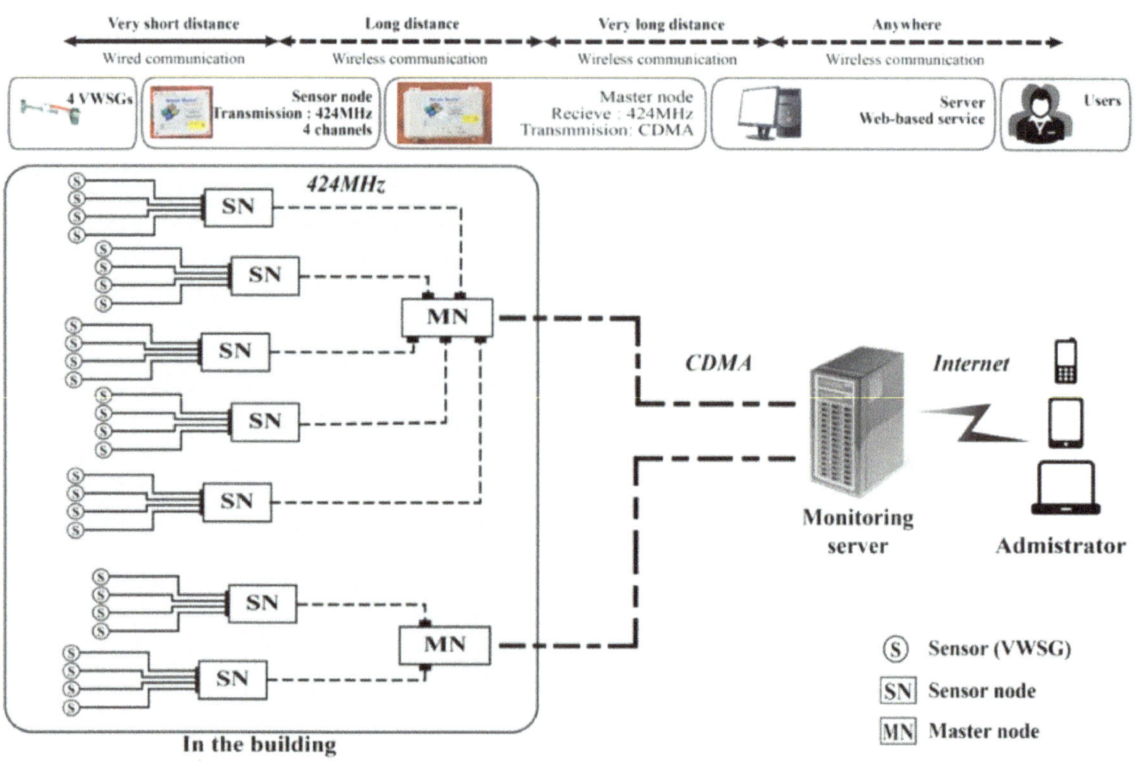

Figure 1: WSNS.

VWSG Sensor

The VWSG has a long transmission length compared with an electrical resistance strain gauge. It also has a lower exterior electromagnetic effect, is less affected by vibration and impact force, and enables semi-permanent measurements. Longitudinal deformation of the wire can occur due to tension and compression, and the natural frequency of the VWSG can vary accordingly. This output frequency measures the average strain for the length of the vibrating wire gauge connected between the mounting blocks (Figure 2). The strain ε can be simply calculated with the change in frequency as:

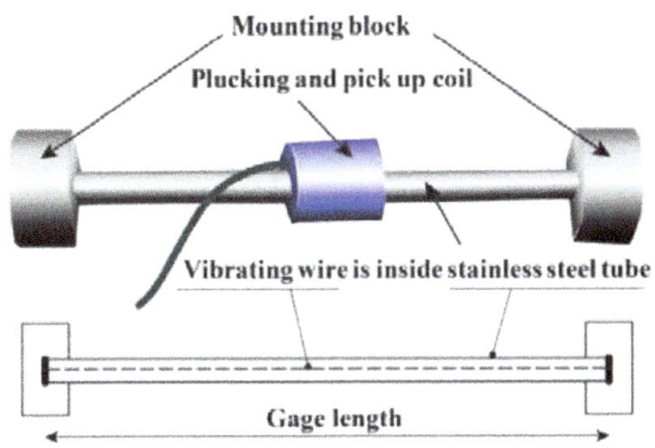

Figure 2: VWSG.

$$\epsilon = k\,(f_2^2 - f_1^2) \tag{1}$$

where k is the gauge factor determined by the properties of the vibrating wire and length and f_1 and f_2 denote the natural frequency prior to and after a change, respectively.

Wireless Sensor/Master Nodes

The VWSGs were grouped (a maximum of four) and connected through a signal cable to the sensor node located near the sensors. Each sensor node with a four-channel sensor interface can simultaneously receive and process data from four VWSGs (Figure 3a, [34]). The raw data from the VWSGs are processed by the data processor built into the sensor node. The processed data are transmitted to the master node through a short-range (Industrial, Scientific, and Medical band—ISM band) wireless communication module. Further information about the wireless sensor node is available in [34].

The master node shown in Figure 3b [40,41] receives processed data from the sensor nodes through short-range wireless communication (ISM band) and transmits these data to the monitoring server through long-range (Code Division Multiple Access—CDMA, [42]) communication. Four sensor nodes can be connected to each master node with a four-port interface, which can simultaneously receive data from 16 VWSGs via sensor nodes.

(a) (b)

Figure 3: Wireless nodes. (**a**) Sensor node; (**b**) Master node.

The communication between the sensor node and master node uses an ultra-high-frequency ISM band of 424 MHz. Compared with microwaves, there is less influence from interference and adequate diffraction of obstacles, thus, the ISM band is considered a suitable communication method for a construction site. A valid communication distance for short-range wireless communication with a sensor node is within 400 m with line of sight (LOS) and within 100 m with non-line of sight. The operating temperature range is approximately −30 to 85 °C.

The wireless communication method of CDMA between the master node and server involves spread-spectrum communication, thus, there is practically no loss of data during transmission and the method is suitable for long-term monitoring due to its low power consumption. Regardless of the communication distance, numerous managers can access the data at preferred times with a personal computer (PC), notebook, tablet PC, mobile phone, or other wireless communication in locations where Internet access is available.

Regarding power consumption, which has been a critical issue for WSNS, this research employed a power-saving circuit in sensor and master nodes, which activated the equipment to operating mode only when the network measurement equipment was in the process of measuring or transmitting. When not under these conditions, the system converted to sleep mode to minimize energy consumption. To apply such low-power technology, the measurement and transmission of all network equipment was synchronized. Only 170 μA of power was consumed during sleep mode, which enabled long-term use in locations that lacked a continuous supply of electricity.

APPLICATION OF THE WSNS

Target Structure: Mega-Truss in the D-Building

The subject building for the application of the WSNS based on VWSGs is the D-Building, which is located in Seoul, South Korea. The main structural frame, which is a reinforced-concrete (RC) and steel structure, consists of three underground floors and four stories, as shown in Figure 4. The steel structure of the mega-truss and mega-column comprise the main frames that form the three-dimensional curve of the streamline design. Zones A and B each contain one mega-truss, which is a large member with a depth of approximately 5–14 m. In zone A, two mega columns support the mega-truss, whose total length is 91 m; the length of the cantilever is 35 m. In zone B, two mega-columns support the mega-truss, whose total length is 34 m; the length of the cantilever is 14 m.

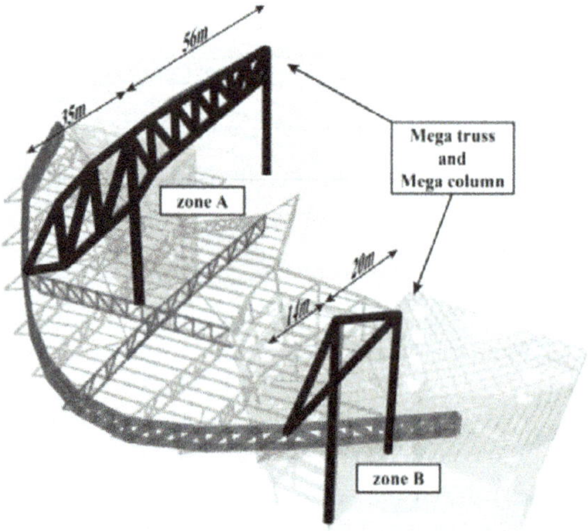

Figure 4: Structural frames.

The main gravity load that is directly conveyed to the mega-truss includes the floor load of the fourth floor and the finishing load of the roof level. The steel material consisted of SM570 with a yield strain and yield strength of 2,136 με and 355 MPa, respectively.

WSNS in the D-Building

A total of 48 VWSGs, 14 sensor nodes, and seven master nodes were applied to the mega-truss, on which long term monitoring for 16 months was conducted. As shown in Figure 5, the mega-truss and mega-column were divided into monitoring areas A1–A8 and B1–B6. Figure 6a shows A6 and A7 areas of the mega-truss during construction. The areas were divided based on adjacent sensors, which were connected to one sensor node through signal lines to narrow the wire range (signal line), as shown in Figure 6b. This process heightened the applicability of on-site measurements in the building during construction.

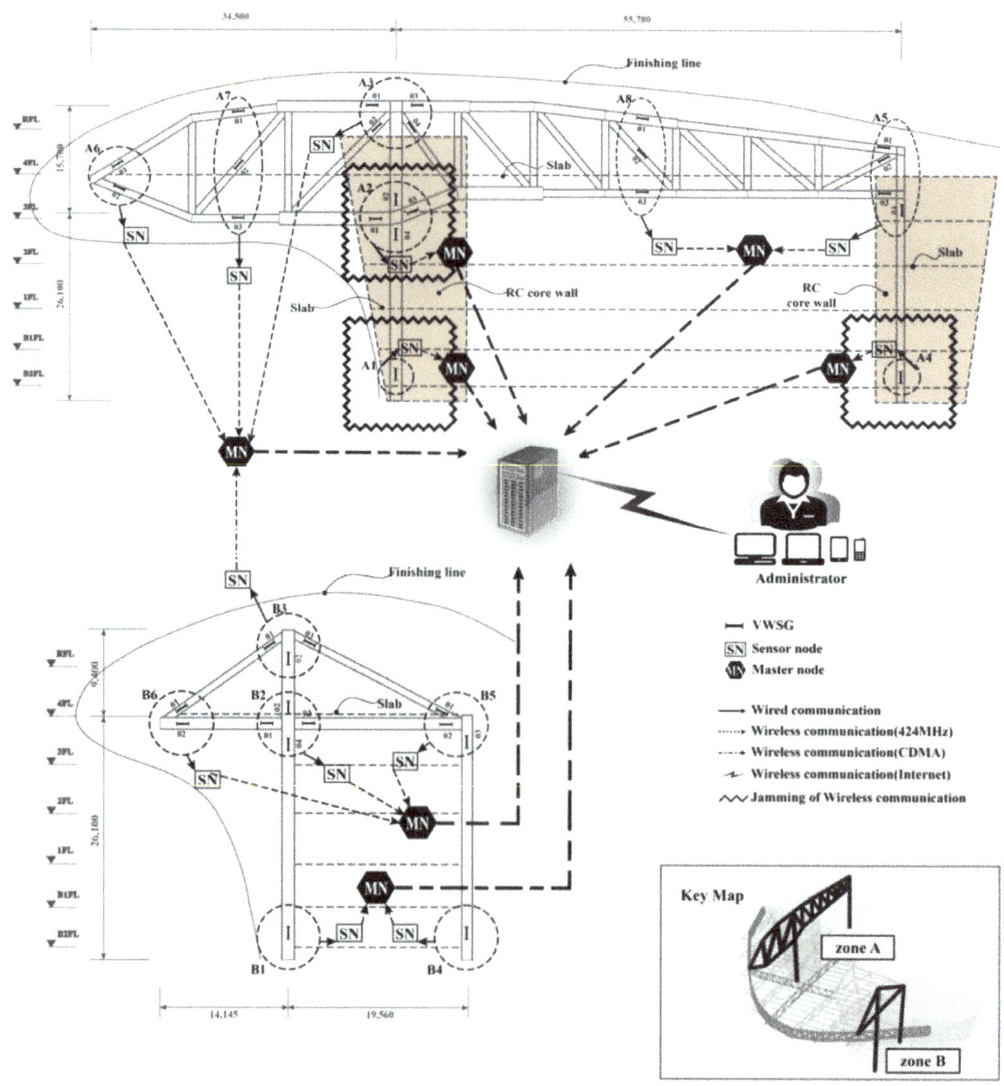

Figure 5: WSNS for the D-Building.

(a) (b)

Figure 6: Application site during construction. (**a**) Mega-truss (A6 and A7 areas) during construction; (**b**) VWSG and sensor node of zone A3 in the D-Building.

The raw data measured by the VWSGs are transmitted to the sensor node and pass through the data processing system built into the sensor node. The communication jamming (∿∿), which is shown in Figure 5, is caused by physical factors, such as the thick RC wall and slab without openings. Therefore, a separate master node was installed. Because other truss areas are exposed to the exterior, the master nodes were arranged based on the communication distance from the sensor node to other obstacles. The master node transmits all processed strain data to the monitoring server through the long-range wireless communication method of CDMA.

ANALYSIS OF STRAIN DATA FROM THE VWSG DURING CONSTRUCTION

Relationship between Construction Events and the Observed Strain Data

The variations in the measured strain data according to the various events of the construction schedule are evaluated in this section. The main construction events are listed in Table 1. The bents were installed to secure stability during the construction of the mega-truss, edge truss, and floor truss. As shown in Figure 7, there are a total of 10 bents on the mega-truss: two bents support the frame at the end of the cantilever of the mega-truss (#4 and #10), and a total of eight bents exist on the edge truss and floor truss. When the erection of the framework was nearly complete, the temporary bents were removed in the sequential order shown in Figure 7. Subsequently, the self-weight of the structure, which had been supported by the bents, was relocated to the mega-truss. Therefore, substantial vertical displacements and stress variations were expected to occur in the cantilever-type mega-truss and edge truss. Bent removal (Event 4 in Table 1) was the most significant event for the stability and precise construction of the building during the monitoring period in this study.

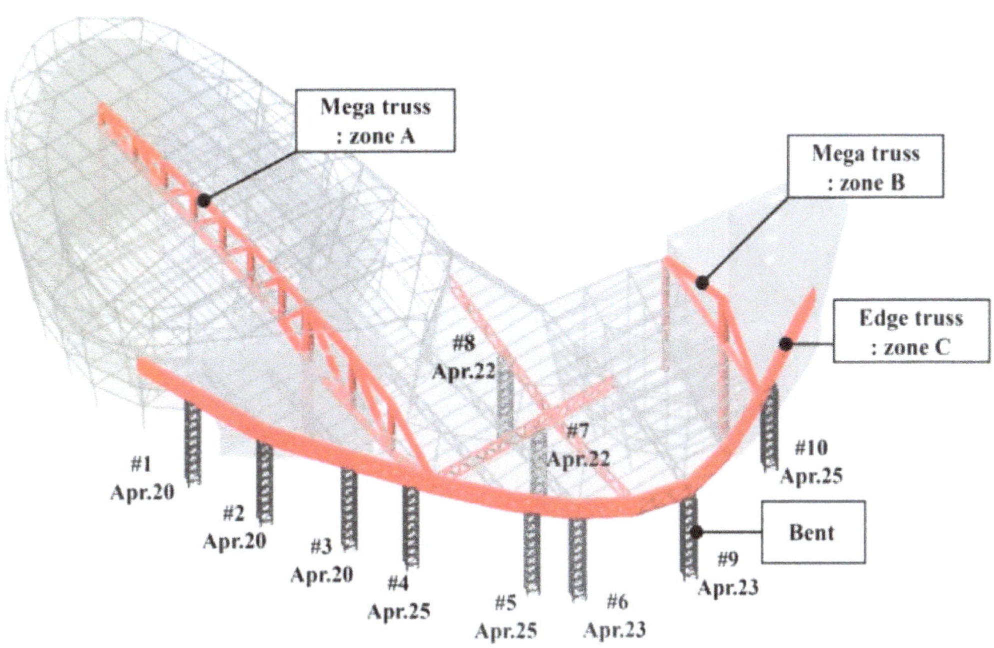

Figure 7: Removal schedule of temporary bents.

Table 1: Main construction schedule

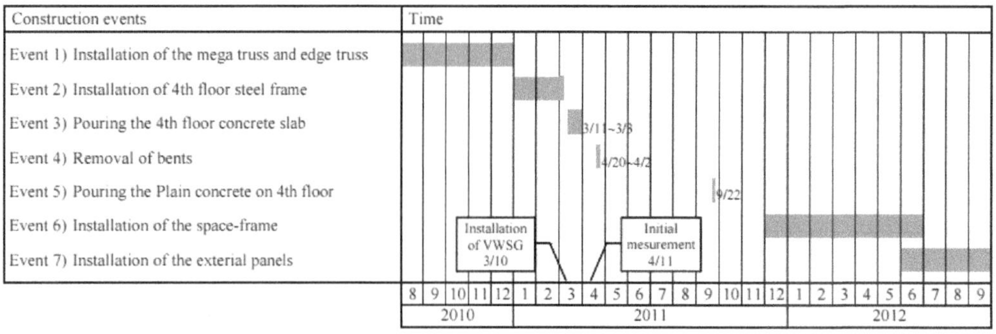

Figure 8 displays the variations in strain data measured by the VWSG that was installed in zone A6-01 according to the construction schedule (events 4–7). A substantial change occurred in the measured value from April 25, 2011, when all bents were removed. After event 4, minimal strain variations were observed until event 5 was initiated. Conversely, during event 5–7, the self-weight of the concrete, space-frame, and exterior panels on the fourth floor were applied, causing strain variations in the mega-trusses. However, their effect on strain variations was insignificant compared with the effect of the bent removal on strain variation.

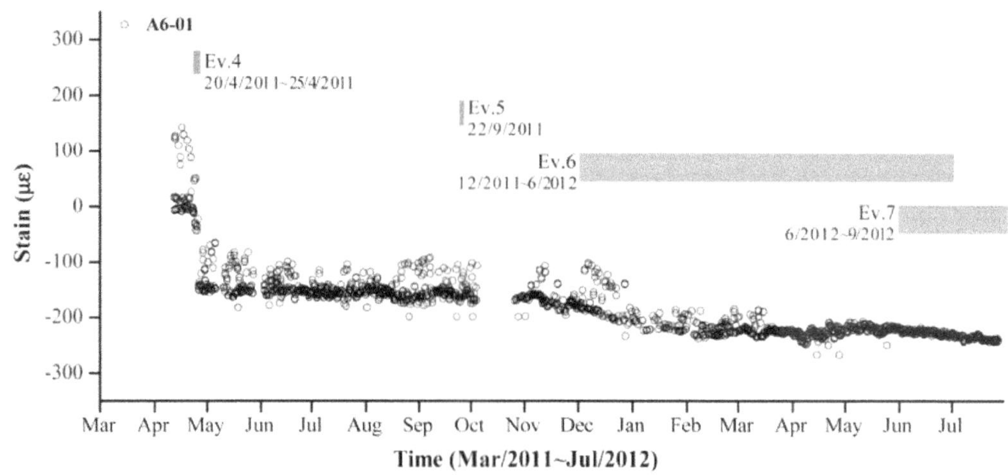

Figure 8: Construction events and strain of A6-01.

Figure 9 illustrates the safety grades according to the construction schedule. The safety grade signifies the ratio of the measured strain of the member to the yield strain of the member, which comprise the cumulative results until each event was finished. Although the variation in the measured data is in accordance with the construction events, the safety grade increases to the maximum value of approximately 15% at B2-04 after complete removal of the bents.

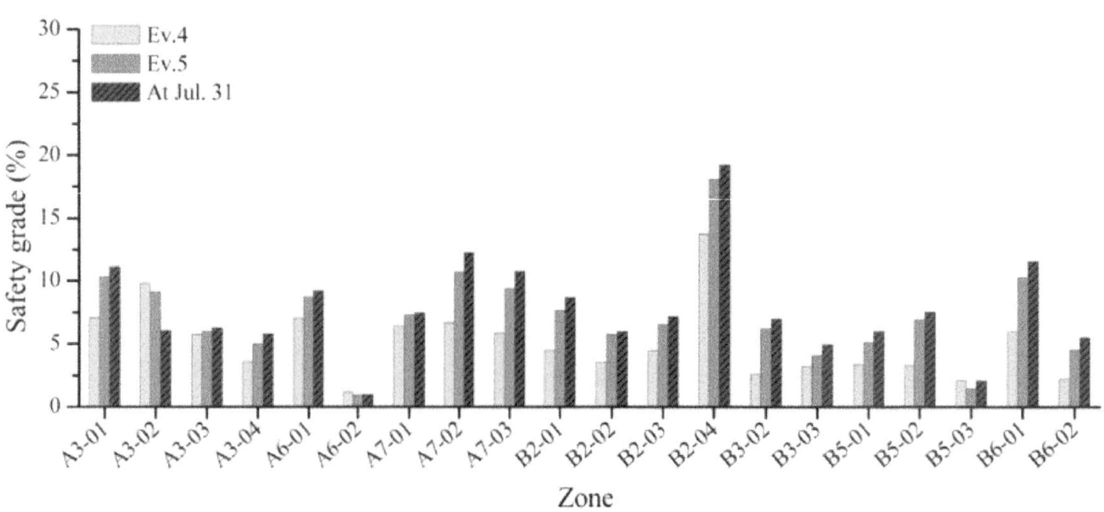

Figure 9: Structural safety during construction events.

Comparison of the Predicted and Measured Strain Responses

This section presents a comparison of the measured strain data for the mega-truss during the removal of the bents with the predicted results obtained from the preliminary analysis. The analysis program MIDAS/GEN Ver. 800 [43] was employed for the comparison. The mega-truss has a full-strength butt welding (FSBW) of approximately 2 m from the working point (WP); thus, a moment connection was applied to the modeling to reflect its structural integrity. Regarding the modeling, the mesh was divided into plate elements for the irregularity of the walls and the accuracy of the model. To correctly reflect the rigidity of the core and fourth slab, these elements were modeled with plate elements that were similar to the wall. The temporary bents were modeled according to the actual building data such that the vertical load was conveyed on the point where the bent and main structure connected. The effect of bent removal in the analysis model was considered by removing the points of the mega-truss and edge truss, which were supported by the bent, and modifying them to free ends. The main evaluating period for the measured data began on April 20, 2011, when bent removal commenced, and ended on April 25, 2011, when all temporary bents had been removed.

A comparison of the measured data for all members with the predicted values reveals the variations in stress prior to removal of the bents and after the complete removal of the last bent, as shown in Figure 10. A3-01 and A2-01 are horizontal members fixed to the mega-column (Figure 5). After the removal of bent #4 in Figure 7, the horizontal members became the fixed ends of the cantilever structure. Due to the deflection of the free end (point of bent removal), which is caused by self-weight, the tensile stress was generated in the upper member of A3-01 (−26 MPa) and compressive stress (+23.4 MPa) occurred in the lower member of A2-01. The largest changes in stress occurred at A6 and A7, in which a significant change in support occurred. The stress variation in zone B is smaller than that in zone A because the former has a smaller cantilever length (14.1 m *versus* 34.5 m).

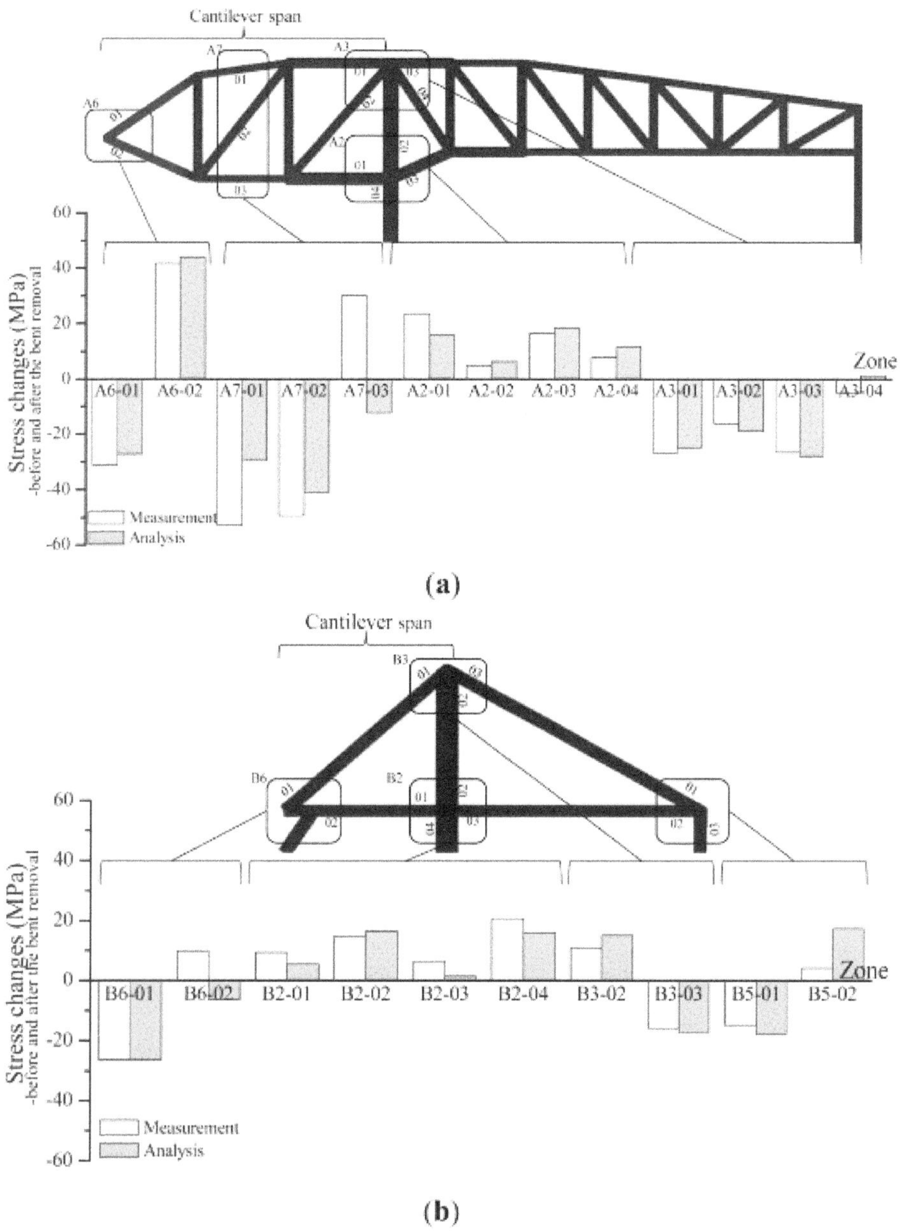

Figure 10: Stress changes before and after bent removal. (**a**) Zone A; (**b**) Zone B.

These results indicate that the preliminary analysis results accurately reflect the tendency of stress variation in the mega-truss members during bent removal, although a slight discrepancy exists between the calculated and measured strain data in quantitative terms. A more accurate analysis model may be constructed through an in-depth review of the measured results.

Thermal Stress Effect on Structural Responses

Figure 11 shows the variations in strain rates (solid line) and temperature (dotted line) which is the result of data collection three times a day (once every 8 h). Although the measurements were collected on a single day, there is a substantial disparity in their values. That is, the strain values are fluctuating with variation of temperature. It may be attributable to the spring temperatures in South Korea, which exhibit an expansive daily temperature range. The daily maximum temperature at the D-Building construction site was 20 °C.

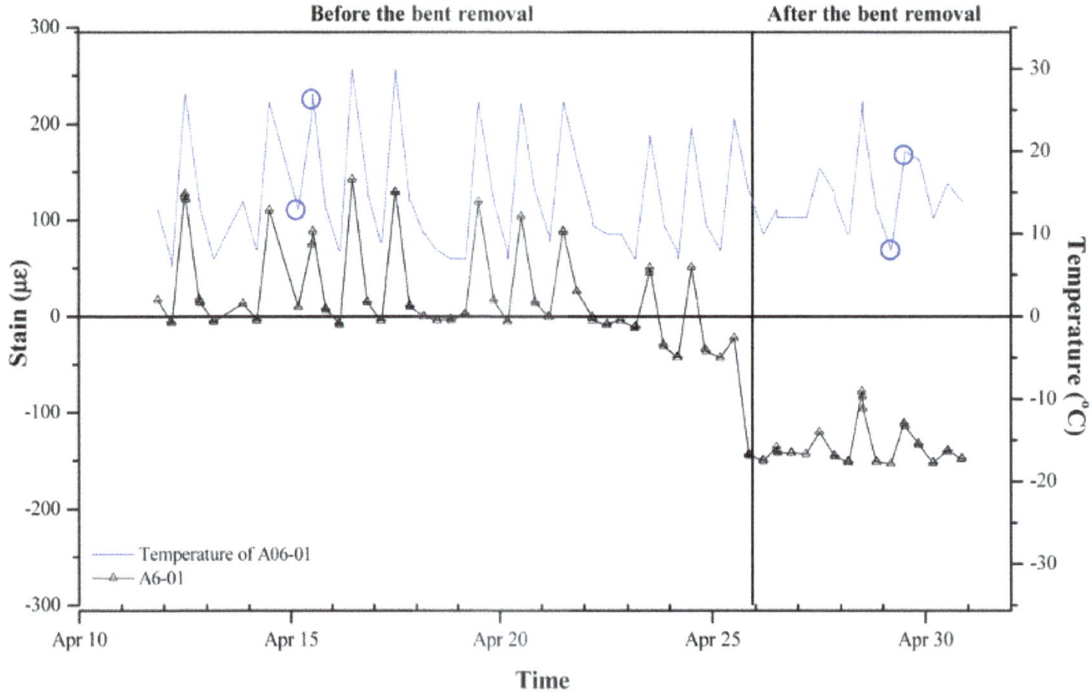

Figure 11: Relationship between temperature and strain (A6-01).

To extensively analyze the relationship of strain regarding temperature variation in this area, the temperature stress change for each member can be expressed as:

$$\Delta\sigma_T = R \cdot E \cdot \Delta T \quad (2)$$

where $\Delta\sigma_T$ is the stress change of the members according to the temperature change and R is the coefficient of thermal expansion of the member considering the confining effect of the frame. If the steel member is not confined, there is a linear expansion of 12×10^{-6}/°C for every temperature increase of 1 °C. E represents the Young's Modulus of the mega-truss, and ΔT is the temperature difference. When the member is within the elastic range, the following formula is valid according to Hook's Law:

$$\Delta\sigma = E \cdot \Delta\varepsilon \quad (3)$$

Therefore, when there is no load other than the temperature load, the stresses from Equations (2) and (3) are equal. For a member in the elastic range, R can be expressed as follows:

$$R = \frac{\Delta\varepsilon}{\Delta T} \quad (4)$$

Figure 12 shows the relationship of strain versus temperature change for the A6-01 sensor. Formula (4) represents the gradient in Figure 12. When the gradient is negative, tensile stress occurs in the member; when the gradient is positive, compressive stress occurs in the member.

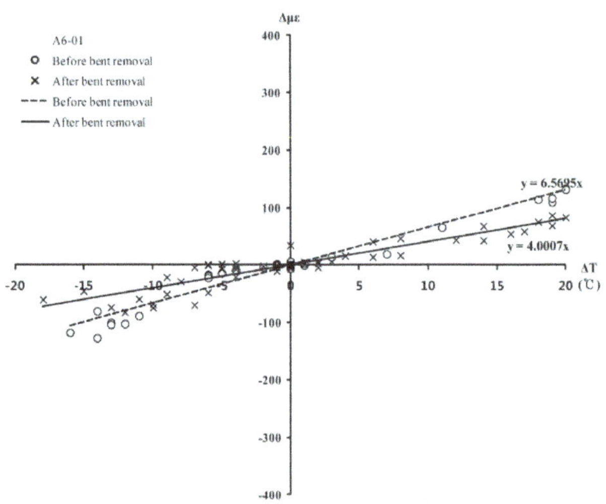

Figure 12: Change in thermal coefficient (R) of A6-01.

The change in the R value before and after bent removal was examined through linear approximation. The results indicate that the coefficient R was approximately 6.6×10^{-6} prior to bent removal but decreased to approximately 4.0×10^{-6} after bent removal. Because the R value is positive before and after bent removal, additional compression is applied with increasing temperature. The absolute value $|R|$ is smaller after bent removal; thus, the resulting change in stress due to temperature is smaller. For example, when there is a 10 °C increase in temperature, an additional compressive stress of 13.5 MPa is produced by the temperature prior to bent removal; however, after bent removal, an additional compressive stress of 8.2 MPa is produced. This difference may be due to the varying degrees of restraint from the change in supports, which resulted from bent removal. Considering that the maximum strain, which occurred during bent removal, was approximately 200 $\mu\varepsilon$, the strain caused by the variations in temperature is significant compared with the strain caused by bent removal.

CONCLUSIONS

This research developed a long-term (16-month) real-time WSNS using VWSGs, which were applied to an irregular building under construction. The data measured by the sensors (VWSGs) are transmitted to the sensor node, and the raw data are processed by the built-in ADC and data processor. The data processed in the sensor nodes are transmitted to the master node through a short-range wireless communication module (ISM), and the master node sends the received data to the monitoring server through long-range wireless communication (CDMA). Thus, the WSNS enables the user to review the measured data in real time. The structural response measurements that were conducted for the large-scale structural members (mega-truss), which support the large space of the structure, are described below.

By examining the stress variations in the mega-truss during load changes for each construction event, it was confirmed that the bent removal process exhibited the largest influence on the structural member. Bent removal, which was performed after the construction of the main structural members was completed, caused changes in the supporting condition of the mega-truss member and generated the largest stress change in the structural member during construction.

The variations in stress induced by the bent removal were examined through a numerical simulation. A comparison of the measured results produced slightly different results than the quantitative evaluation, but similar tendencies were also observed in the overall stress variations of the members. The accuracy of the analysis can be enhanced by updating the analysis model so that it accurately reflects the measured results, which is one of the main functions of a monitoring system based on sensors.

The comparison of the variations in stress and exterior temperature through temperature sensors indicated that the variations in stress due to temperature were comparable to the changes that resulted from the bent removal. Although temperature stress is temporary, it should be considered a significant factor when extensive variations in temperature occur during construction. The change in the constraint condition of the member, which was caused by the bent removal, was also confirmed to have a large influence on stress change due to temperature variation. The measurement results prove that the WSNS based on VWSGs proposed in this research can be applied as a practical monitoring method to ensure construction safety and the accurate construction of a building.

ACKNOWLEDGMENTS

This work was supported by the National Research Foundation of Korea(NRF) grant funded by the Korea government(MEST) (No. 2011-0018360).

REFERENCES

1. Doherty, J.E. Nondsestructive Evaluation. In *Handbook on Experimental Mechanics*, 2nd ed.; Kobayashi, A.S., Ed.; Society for Experimental Mechanics, Inc.: Bethel, CT, USA, 1987.
2. Arditi, D.; Gunaydin, H.M. Total quality management in the construction process. *Int. J. Proj. Manage.* 1997, *15*, 235–243.
3. Aktan, A.E.; Catbas, F.N.; Grimmelsman, K.A.; Tsikos, C.J. Issues in infrastructure health monitoring for management. *J. Eng. Mech. Asce.* 2000, *126*, 711–724.
4. Brownjohn, J.M.W. Structural health monitoring of civil infrastructure. *Philos. T. R. Soc. A* 2007, *365*, 589–622.
5. Liu, S.C.; Tomizuka, M. Vision and Strategy for Sensors and Smart Structures Technology Research. Proceedings of the 4th International Workshop on Structural Health Monitoring, Stanford, CA, USA, 15–17 September 2003; DEStech Publication, Inc.: Lancaster, PA, USA, 2003.
6. Fan, W.; Qiao, P. Vibration-based damage identification methods: A review and comparative study. *Struct. Health Monit.* 2011, *10*, 83–111.
7. Straser, E.G.; Kiremidjian, A.S. *A Modular, Wireless Damage Monitoring System for Structures*; Earthquake Engineering Center, Stanford University: Stanford, CA, USA, 1998.
8. Lynch, J.P.; Loh, K.J. A summary review of wireless sensors and sensor networks for structural health monitoring. *Shock Vib. Dig.* 2006, *38*, 91–128.
9. Krishnamurthy, V.; Fowler, K.; Sazonov, E. The effect of time synchronization of wireless sensors on the modal analysis of structures. *Smart Mater. Struct.* 2008, *17*, 055018.
10. Qiu, Z.; Wu, J.; Yuan, S. A wireless sensor network design and evaluation for large structural strain field monitoring. *Meas. Sci. Technol.* 2011, *22*, 075205.
11. Jang, S.; Sim, S.H.; Jo, H.; Spencer, B.F. Full-scale experimental validation of decentralized damage identification using wireless smart sensors. *Smart Mater. Struct.* 2012, *21*, 115019.
12. Park, H.S.; Shin, Y.; Choi, S.W.; Kim, Y. An integrative structural health monitoring system for the local/global responses of a large-scale irregular building under construction. *Sensors* 2013, *13*, 9085–9103.
13. DeWolf, J.T.; Lauzon, R.G.; Culmo, M.P. Monitoring

bridge performance. *Struct. Health Monit.* 2002, *1*, 129–138.

14. Park, K.T.; Kim, S.H.; Park, H.S.; Lee, K.W. The determination of bridge displacement using measured acceleration. *Eng. Struct.* 2005, *27*, 371–378.
15. Bao, C.; Hao, H.; Li, Z. Vibration-based structural health monitoring of offshore pipelines: Numerical and experimental study. *Struct. Control Health Monit.* 2013, *20*, 769–788.
16. Kim, J.T.; Ho, D.D.; Nguyen, K.D.; Hong, D.S.; Shin, S.W.; Yun, C.B.; Shinozuka, M. System identification of a cable-stayed bridge using vibration responses measured by a wireless sensor network. *Smart Struct. Syst.* 2013, *11*, 533–553.
17. Park, H.S.; Lee, H.M.; Adeli, H.; Lee, I. A new approach for health monitoring of structures: Terrestrial laser scanning. *Comput. Aided Civ. Inf.* 2007, *22*, 19–30.
18. Adewuyi, A.P.; Wu, Z.S.; Serker, N.H.M.K. Assessment of vibration-based damage identification methods using displacement and distributed strain measurements. *Struct. Health Monit.* 2009, *8*, 443–461.
19. Rodrigues, C.; Felix, C.; Figueiras, J. Fiber-optic-based displacement transducer to measure bridge deflections. *Struct. Health Monit.* 2011, *10*, 147–156.
20. Moschas, F.; Stiros, S. Measurement of the dynamic displacements and of the modal frequencies of a short-span pedestrian bridge using GPS and an accelerometer. *Eng. Struct.* 2011, *33*, 10–17.
21. Ye, X.W.; Ni, Y.Q.; Wai, T.T.; Wong, K.Y.; Zhang, X.M.; Xu, F. A vision-based system for dynamic displacement measurement of long-span bridges: algorithm and verification. *Smart Struct. Syst.* 2013, *12*, 363–379.
22. Majumder, M.; Gangopadhyay, T.K.; Chakraborty, A.K.; Dasgupta, K.; Bhattacharya, D.K. Fibre Bragg gratings in structural health monitoring—Present status and applications. *Sens. Actuators A: Phys.* 2008, *147*, 150–164.
23. Chacón, R.; Guzmán, F.; Mirambell, E.; Real, E.; Oñate, E. Wireless sensor networks for strain monitoring during steel bridges launching. *Struct. Health Monit.* 2009, *8*, 195–205.
24. Adewuyi, A.P.; Wu, Z.S. Modal macro-strain flexibility methods for damage localization in flexural structures using long-gauge FBG sensors. *Struct. Control .Health Monit.* 2010, *18*, 341–360.
25. Jo, H.; Park, J.W.; Spencer, B.F.; Jung, H.J. Develoment of high-sensitivity wireless strain sensor for structural health monitoring. *Smart Struct. Syst.* 2013, *11*, 477–496.
26. Kim, Y.S.; Sung, H.J.; Kim, H.W.; Kim, J.M. Monitoring of tension force and load transfer of groundanchor by using optical FBG sensors embedded tendon. *Smart Struct. Syst.* 2011, *7*, 303–317.
27. Li, H.N.; Li, D.S.; Song, G.B. Recent applications of fiber optic sensors to health monitoring in civil engineering. *Eng. Struct.* 2004, *26*, 1647–1657.
28. Park, H.S.; Jung, S.M.; Lee, H.M.; Kwon, Y.H.; Seo, J.H. Analytical models for assessment of the safety of multi-span steel beams based on average strains from long gage optic sensors. *Sens. Actuat. A Phys.* 2007, *137*, 6–12.
29. Park, H.S.; Jung, H.S.; Kwon, Y.H.; Seo, J.H. Mathematical models for assessment of the safety of steel beams based on average strains from long gage optic sensors. *Sens. Actuat. A: Phys.* 2006, *125*, 109–113.
30. Deng, L.; Cai, C.S. Applications of fiber optic sensors in civil engineering. *Struct. Eng. Mech.* 2007, *25*, 577–596.
31. Pang, C.; Yu, M.; Gupta, A.K.; Bryden, K.M. Investigation of smart multifunctional optical sensor platform and its application in optical sensor networks. *Smart Struct. Syst.* 2013, *12*, 23–29.
32. Coutts, D.R.; Wang, J.; Cai, J.G. Monitoring and analysis of results for two strutted deep excavations using vibrating wire strain gauges. *Tunn. Undergr. Sp. Tech.* 2001, *16*, 87–92.
33. Yu, F.; Gupta, N. An efficient model for improving performance of vibrating-wire instruments. *Measurement* 2005, *37*, 278–283.
34. Lee, H.M.; Kim, J.M.; Sho, K.; Park, H.S. A wireless vibrating wire sensor node for continuous structural health monitoring. *Smart Mater. Struct.* 2010, *19*, 055004.
35. Lee, H.M.; Park, H.S. Measurement of maximum strain of steel beam structures based on average strains from vibrating wire strain gages. *Exp. Tech.* 2013, *37*, 23–29.
36. Bourquin, F.; Joly, M. A magnet-based vibrating wire sensor: Design and simulation. *Smart Mater. Struct.* 2005, *14*, 247–256.
37. Neild, S.A.; Williams, M.S.; McFadden, P.D. Development of a vibrating wire strain gauge for measuring small strains in concrete beams. *Strain* 2005, *41*, 3–9.
38. Sreeshylam, P.; Ravisankar, K.; Parivallal, S.; Kesavan, K.; Sridhar, S. Condition monitoring of prestressed concrete structures using vibrating wire sensors. *Int. J. Comadem* 2008, *11*, 46–54.
39. Xia, Y.; Ni, Y.Q.; Zhang, P.; Liao, W.Y.; Ko, J.M. Stress development of a supertall structure during construction: Field monitoring and numerical analysis. *Comput. Aided Civ. Inf.* 2011, *26*, 542–559.
40. Choi, S.W.; Kwon, E.; Kim, Y.; Hong, K.; Par, H.S. A practical data recovery technique for long-term strain monitoring of mega columns during construction. *Sensors* 2013, *13*, 10931–10943.
41. Choi, S.W.; Kim, Y.; Kim, J.M.; Park, H.S. Field monitoring of column shortenings in a high-rise building during construction. *Sensors* 2013, *13*, 14321–14338.
42. Viterbi, A.J. *CDMA: Principles of Spread Spectrum Communication*; Addison-Wesley: Reading, MA, USA, 1995.

CHAPTER 18

OBTAINING THE THERMAL STRUCTURE OF LAKES FROM THE AIR

Michaella Chung, Carrick Detweiler, Michael Hamilton, James Higgins, John-Paul Ore and Sally Thompson

Keywords: Unmanned aerial systems (UAS), unmanned aerial vehicle (UAV), robotics, thermal structure, aquatic ecosystems

INTRODUCTION

The thermal structure and regime of aquatic ecosystems are primary physical determinants of habitat quality [1,2] and important physical forces driving macroscopic dynamics (e.g., stability and overturning of the water column, which affects biogeochemistry, oxygen demand and, ultimately, ecology [3,4,5,6,7,8]). While the bulk thermal properties of surface water bodies are often well represented by a vertically-resolved suite of measurements that captures the existence and migration of a thermocline, such measurements implicitly assume well-mixed conditions in the longitudinal and lateral directions. Increasingly, however, thermal heterogeneities in surface water bodies are recognized as drivers of mixing and providing critical habitat. For instance, lateral temperature variations in two alpine rivers in floodplain environments in Italy were shown to be as large as the diurnal variation at any given point [9]. Thermal gradients can drive mixing at stream confluences [10], and where water temperatures in tributaries are significantly lower than in main channels, such confluences can represent important thermal refugia [11]. Bank-side shading, deep pools and groundwater inputs [12] also create persistent thermal structure and cold-water refuges that are critical for fish habitat, particularly in streams impacted by anthropogenic changes, warm water inputs or climate warming [13,14,15]. In small lakes, lateral temperature gradients can be established by differential heating or cooling. Differential heating occurs when there are contrasts in lake depth, bottom sediments, shading or exposure to winds [16,17]. These gradients may be sufficient to establish convective currents within the lake or to contribute to changes in mixing layer deepening and subsequent overturning dynamics [17,18,19,20]. In large lakes, temperature differences may reflect changes in inflows, meso-scale climatic variations and upwelling [21]. It is clear that, across a wide range of surface water bodies, significant thermal gradients may arise both vertically and in the horizontal plane, with important implications for the physical mixing processes in the water bodies, as well as the water quality and habitat values that result. Empirically characterizing this heterogeneity in the thermal structure is non-trivial. At present, researchers have four broad options for measuring the spatio-temporal thermal structure of water bodies: manual sampling, *in situ* sensors and remote and mobile sensing.

Manual observations of temperature structure are spatially and temporally discrete. They require the presence of a researcher and, frequently, the use of a vehicle to transport the researcher to the measurement location or wading in shallow water. While the major challenges associated with manual sampling relate to workload, there is also the potential for the presence of a vehicle or researcher to disrupt the *in situ* thermal structure. This is particularly true in small water bodies or if dense observation grids are required in a limited time frame.

A wide range of continuously-reporting, *in situ* sensors is available and requires minimal maintenance and ingress to the water body [22], which makes them particularly suitable for long-term deployments. The major limitation associated with *in situ* measurements lies in capturing 3D spatial structure: doing so requires a dense grid of sensors, which can rapidly become costly, challenging to deploy and can disrupt recreational or commercial uses of the water body. Recent developments in thermal sensing using fiber optics provide greater spatial continuity in the measurements made. In these measurements,

Citation: Michaella Chung, Carrick Detweiler, Michael Hamilton, James Higgins, John-Paul Ore and Sally Thompson, Obtaining the Thermal Structure of Lakes from the Air, doi:10.3390/w7116467.

the exponential dependence of the scattering of light on fiber temperature allows travel times for light within a fiber optic cable to be used as a proxy for temperature, and local variations in temperature can then be inferred along the cable length [23]. However, this technique requires managing long cables, which, while well suited to longitudinal sampling along stream corridors, are poorly suited to the development of a 3D gridded observational array. In addition, cost remains a constraint for fiber optic cable-based temperature measurement.

Remotely sensing water temperatures through thermal infrared imaging is an appealing technique to capture the lateral structure of surface temperatures. Surface temperatures can be observed using manned helicopters [24], fixed-wing aircraft [25] or autonomous aerial vehicles [26] equipped with thermal cameras, and on much larger scales through satellite measurements. Temperatures are typically inferred from the relationship between the intensity of the emitted radiation of a water body, with radiant temperature corrected for emissivity to yield kinetic temperatures. The resulting datasets are robust, agreeing with *in situ* measurements within 0.5–1 °C [24,27,28]. Their obvious limitation, however, is that only surface observations can be made.

An attractive approach that combines rapid spatial coverage of remote sensing with the ability to probe multiple depths in the water column (*cf. in situ* or manual measurements) is to use mobile sensors that are moved to multiple measurement locations by autonomous vehicles. Autonomous surface vehicles (ASVs, *i.e.*, "robot boats") and autonomous underwater vehicles (AUVs, *i.e.*, "robot submarines") have growing applications in thermal mapping [29,30,31,32,33]. These systems are mobile, do not need a human operator and can be deployed for long periods of time, allowing researchers to generate high spatio-temporal resolution maps of the full 3D thermal structure of the water. The major challenges associated with these systems are that they can be difficult to deploy (for example, many require a boat ramp or dock to launch the vehicle), are costly and have limited obstacle avoidance capabilities that prevent operation in shallow areas near shore or where there is significant underwater vegetation. AUVs are further limited by the lack of GPS and radio communication underwater, which makes obtaining geo-referenced readings challenging and costly. Because of these limitations, ASVs and AUVs are typically only deployed in large bodies of water.

This paper proposes and evaluates an alternative automated temperature-sensing system based on the use of an unmanned aerial system (UAS) to quickly obtain 3D thermal maps of water bodies by lowering a temperature probe into the water at controlled depths. This builds on our prior work in obtaining water samples with a UAS [34]. UASs offer several advantages for thermal sensing in surface water bodies: they do not require shoreline access, can be quickly and easily deployed and redeployed, facilitating repeated sampling, and can rapidly move between measurement locations. In addition, unlike thermal infrared sensors, the UAS can resolve the full 3D thermal structure of the water. The proposed UAS is limited in the depths of the water column that can be feasibly sampled, making them most suitable for deployment in shallow water bodies. UASs are also subject to well-known and previously-reported limitations, related to vehicle restrictions, such as flight times and payloads, as well as regulatory constraints [35,36].

Here, we introduce the aerial sensor platform and the adjustments made to enable temperature measurements throughout a water column at multiple points. We explore preliminary results from initial field experiments, which were designed to directly compare the temperature measurements made from the UAS platform with measurements recorded by an *in situ*array while also considering the possibility of UAS platform-induced disturbances in the vertical thermal structure of the water column. We also identify key needs for improving future applications and explore the potential scalability of future UAS water sampling platforms for 3D mapping of temperature structure.

MATERIALS AND METHODS

Site Description

Temperature sensing experiments were undertaken at Big Lake in Blue Oak Ranch Reserve (BORR), a 1330-ha undeveloped ecological reserve managed as part of the University of California's Natural Reserve System. BORR is a locus for the development of ecological sensing technologies and was considered an ideal site to test the UAS platform due to its remote location and multiple surface water habitats. Big Lake is the largest of the perennial ponds at BORR. It has a surface area of 1.25 ha and a maximum depth of nearly 9 m.

Two vertical arrays of temperature sensors (6.5 m apart) were installed 10 m from the shoreline in 1.3 m water on the western boundary of the lake. Five HOBO Pendant temperature loggers were attached to the vertical array at depths of 0.25, 0.41, 0.56, 0.79 and 1.17 m. See Table 1 for sensor details.

Table 1: Sensor specifications

Sensor	Specifications
In situ temperature	Hobo Pendant Temperature/Light Data Logger Resolution: 0.14 °C at 25 °C Accuracy: ±0.53 °C Operating range: −20–50 °C
Pressure-temperature	Measurement Specialties MS5803-01BA sensor Pressure resolution: 0.012 mbar Pressure accuracy: ±2.5 mbar Temperature resolution: 0.01 °C Water contact detector

Vehicle and Sensors

Figure 1 shows the UAS system in flight at BORR. The UAS platform used in these experiments was an Ascending Technologies FireFly [37] Hexrotor with a 600-g payload and a flight time of approximately 20 min. The FireFly is equipped with a variety of sensors to measure position: a GPS, an inertial measurement unit and a pressure altimeter. The vehicle is controlled by input from two 802.15.4 XBee 2.4 GHz radios: one radio channel provides input from a remote control station on a laptop (allowing, for example, a pre-programmed flight path to be sent to the UAS) and another for manual control by a human pilot.

Figure 1: UAS in flight at Blue Oak Ranch Reserve.

To facilitate the use of the UAS for freshwater ecology, we augmented the base platform with several sensors and subsystems [34]. Most importantly, we added multiple downward-facing ultrasonic rangers (Maxbotix MB1240-EZ4: ±1-cm accuracy within 2 m) to improve altitude estimation over water and thus provide better flight elevation control. To enable sensing within the water column, we attached a custom-built pressure-temperature embedded sensor system to a 4-m tube, which terminates in a pump, allowing water samples to also be made with the same system. A basic conductivity sensor is also attached to the end of the tube, allowing a direct indication of whether it is submerged or in air. This is primarily a safety feature to protect the pump and to ensure that it is submerged before pumping. A schematic of the system is shown in Figure 2, and the sensor specifications are provided in Table 1. During flight, the tube is reeled into the UAS, using a thin thread that runs through clips. The clips are attached to the tube at 0.75-m intervals, and the tube is retracted by spooling the thread beneath the UAS, causing the tube to coil below the vehicle. A long length of tube can become entangled in the environment, and its swinging motion during flight can induce dangerous and unstable flight dynamics. Therefore, the spooling mechanism reduces the hazards associated with flying with an object attached to the UAS.

To make measurements, the UAS hovers above the water surface, and the tube is extended through the water column. Varying water depths can be sampled either by changing the length of the tube that is extended into the water column or by changing the UAS altitude with a fixed length of tube extended. For each flight, the data from the UAS's pressure-temperature system are transmitted from the UAS to a laptop ground station, where it is time stamped and written to a data log along with the UAS's GPS coordinates and altitude.

Figure 2: Detail of the pressure-temperature embedded sensor system contained at the end of the UAS tube. A U.S. quarter piece is shown for scale.

Flights for Temperature Sampling

To compare temperature measurements from the UAS and *in situ* sensors, we made four flights. The fully-unspooled tube and thermal sensor were lowered through the water column near (~0.5-m radius) one *in situ* thermal array, while maintaining a maximal distance from the other array. Temperatures were recorded at 4 Hz for 30 s at each of the five points through the water column, at 0.2-m intervals from 0.2 m deep from the bed of the lake. We waited 30 min between flights to allow the array to re-equilibrate to the background value. We selected 30 min because we previously confirmed that this duration was sufficient to achieve equilibrium following a disturbance. Flights were made in the late afternoon of 20 March 2015, with an average air temperature of 18 °C and a wind speed of 2.1 m/s. A comparison of these conditions with *in situ*temperature logging made during the preceding days indicated that the water column experienced mild diurnal temperature variances, with no evidence of over-turning or significant change in the surface mixed layer depth. Due to these stable, stratified conditions, we did not expect to observe exaggerated temperature dynamics in the lake during the experiments.

Data from the *in situ* array and the UAS were post-processed as described in Section 2.4 and compared for each flight.

To explore the feasibility of larger scale mapping with the UAS, we made two flights, using the spooling mechanism, to measure vertical profiles at three points along two parallel lines, as shown in Figure 3. The two parallel lines are separated by 10 m, with points in each line spaced 5 m apart. We flew ~4 min at each location, starting at a higher altitude to submerge the pressure-temperature sensor to a shallow depth of ~0.5 m, then descending by 0.5-m increments every 30 s. After eventually submerging and holding the sensor at a maximum depth of ~2.5 m at each location, we ascended high enough to completely clear the water, then flew to the next location.

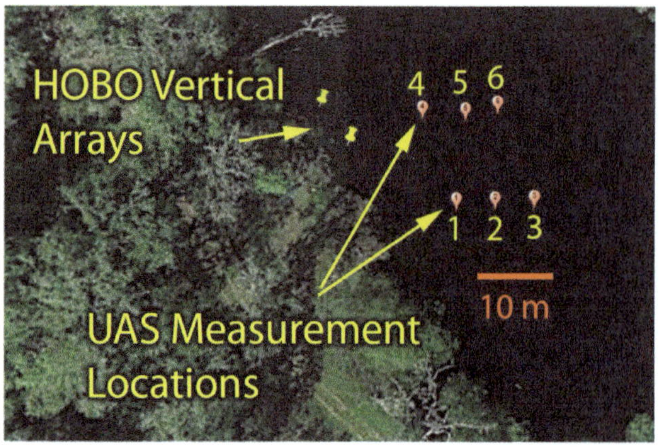

Figure 3: Locations of vertical profile measurements made by the UAS at Big Lake, Blue Oak Ranch Reserve, California, USA.

Post-Processing

Post-processing involved two steps: filtering UAS temperature data and calibrating the UAS and *in situ* sensors to a common temperature datum.

Filtering

We filtered the temperature data because unintentional altitude changes in the UAS during flight result in the temperature sensor moving periodically up and down 25 cm through the water column. This movement causes multiple temperature readings to be associated with a given target depth and is problematic because the thermal response time of the MS5803 sensor is long compared to the rate of vertical movements. On average, the sensor requires 90 ± 45 s to converge to 90% of the actual temperature (the temperature response time of the sensor system is slowed due to polyurethane waterproofing that protects the electronics). To avoid erroneous data associated with these long equilibration times, we filtered the temperature data based on the joint rates of the change of pressure sensor depth and temperature. Specifically, we excluded temperature readings when either temperature or depth changes quickly: $\left|\frac{\partial T}{\partial t}\right| \geq 0.04\ °Cs^{-1}$ or $\left|\frac{\partial m}{\partial t}\right| \geq 0.35\ \mathrm{ms}^{-1}$. These thresholds were determined by visually inspecting plots of the temperature and depth rates of change and observing that during periods of minimal change, nearly all values are bounded by the aforementioned thresholds. As shown in Figure 4, this filtering method retained 15% of the original readings and discarded most readings from the descent periods between the five measurement points in the water column.

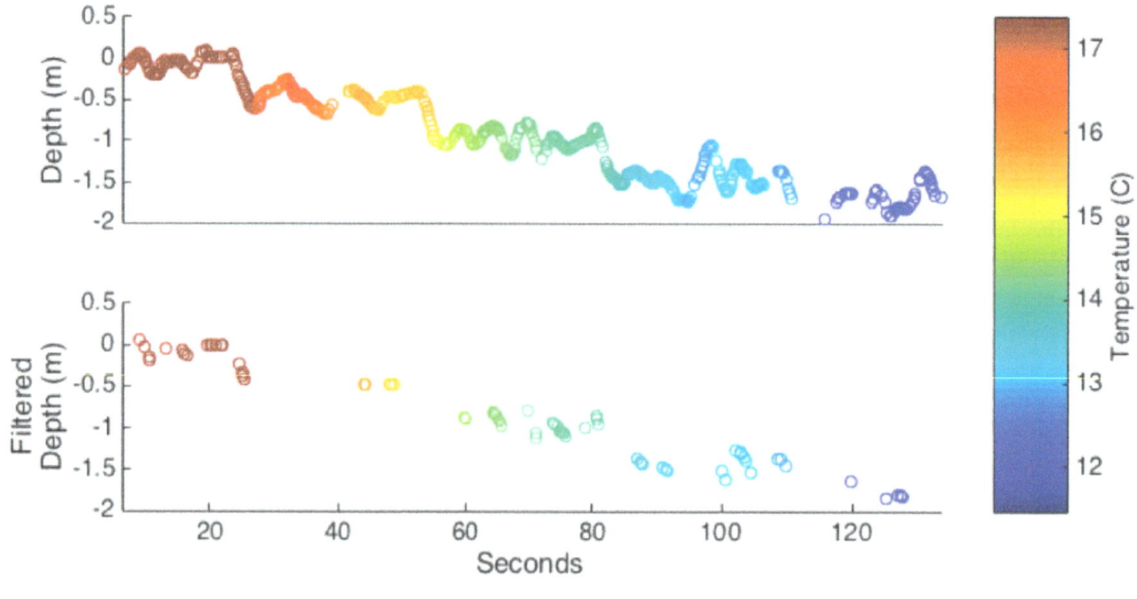

Figure 4: The filtering method removes data when temperature and depth change rapidly.

Calibration

Prior to comparing UAS and *in situ* temperature readings, we calibrated the sensors against each other in the laboratory using room temperature water (21.8 °C) and ice water (0.1 °C). The MS5803 (UAS) sensor reported higher temperatures than the HOBO (*in situ*) sensor, except during periods of rapid temperature change, when the two sensors reported comparable values. We assumed that temperature changes imposed during calibration would be large compared to temperature changes measured in the field and therefore adjusted the UAS temperature data to improve correspondence with the *in situ* sensor readings obtained during calibration. The UAS temperatures were offset by −0.3 °C.

RESULTS AND DISCUSSION

Comparison of UAS and In Situ Temperature Measurements

Over the four test flights, the UAS sensors reproduced the temperature variations in space associated with the water column's thermal gradient and the mean variations in time during the two-hour experimental window. However, the temperatures measured from the UAS were consistently higher than the *in situ* array-recorded temperatures, even following calibration. The magnitude of the temperature bias varied with depth in the water column and was highest in deeper areas of the water column. Figure 5 shows box plots of the bias associated with each temperature measurement made by the UAS over the four flights, binned by depth in the water column. The bias is on the order of 0.5 °C in the surface 0.5 m of the water column and increases to 1 °C at greater depths below the thermocline.

While it is encouraging that these deviations in temperature are comparable to those made by other methods (e.g., TIR [24]; fiber optics [23,38]), the deviations are much greater than previously seen in mobile sensing systems [30]. Furthermore, the depth dependence in the magnitude of the bias suggests that a physical mechanism may be responsible. We hypothesized that several mechanisms could explain the bias: thermal lag effects when the sensor was first introduced into the water column from the air or entrainment of cold water from a depth to surface layers as the sensor was raised from the lake bed to the surface. As a preliminary means to distinguish between these hypotheses, we explored whether the magnitude of the thermal bias at the surface varied between the first shallow readings and the final readings made during a flight.

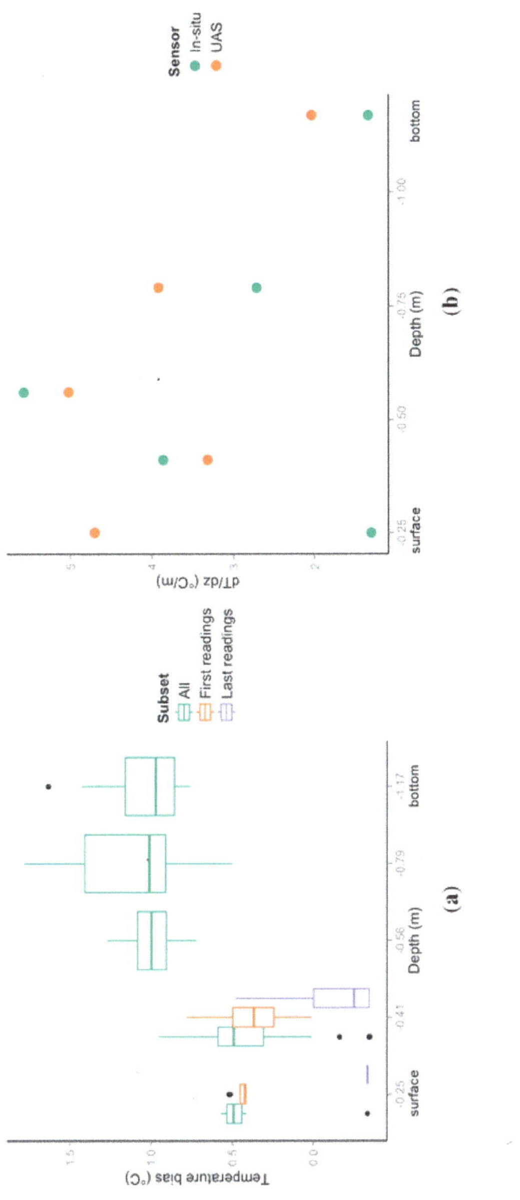

Figure 5: Comparison of UAS-borne and *in situ* temperature sensors, including logged data and detected thermal gradient. (**a**) Bias in temperature measured at each depth; (**b**) comparison of the derived temperature gradient at each depth.

As seen in Figure 5, the UAS sensor temperatures remain higher than the *in situ* sensor temperatures near the surface, but the bias is decreased in the first readings during a flight. The direction of the bias switches for the final shallow readings during a flight, with temperatures measured from the UAS being lower than the *in situ* array-recorded temperatures. We suggest that these observations are more consistent with the entrainment of cold water as the tube and sensor were moved up through the water column at the end of a flight, than with the effect of thermal lags. Thermal lags in the onboard UAS sensor would be expected to have led to large positive biases in temperature upon initial entry into the water column. Despite the temperature bias between the UAV and *in situ* sensors, the UAV sensor's temperature resolution of 0.01 °C is sufficient for capturing fine-scale thermal heterogeneities in the water column.

To remove systematic bias from the temperature sensors during the field trials and to consider only relative variations in temperature, we also compared estimates of the local thermal gradient in the water column as made from each of the sensing platforms, presenting the results in Figure 5. Below the water surface, both platforms produced comparable trends in the thermal gradient, again within 0.5–1 °C m^{-1} of each other. Both the peak in the thermal gradient associated with the thermocline, at a depth of approximately 0.6 m, and increased thermal stability with depth were identified by both sensor platforms. Significant disagreement between the measurements was only observed in the top 0.25 m of the water column, when the UAS measurements indicate a peak in thermal gradient comparable to that at the thermocline, while the *in situ*sensors indicate a fairly consistent temperature change with depth. The source of this high temperature gradient as sensed by the UAS is not completely resolved, but may be associated with a lack of elevation stability, causing the sensor to be suddenly pulled out of the water when at the surface and to undergo a rapid temperature change.

What Water Column Disturbance Is Induced by the UAS?

We hypothesized that water column disturbance induced by the UAS due to propeller wash and the motion of the tube would result in greater mixing and, thus, higher variance in temperatures measured below the UAS compared to a control. First, we confirmed that temperature variance measured at a given depth did not depend on the time window over which variance was calculated. We then computed temperature variances over 30-s periods for all depths and compared the profiles of the temperature variance at the test and control temperature arrays with a paired *t*-test. We did not find a significant difference in the temperature variances at any depth ($p > 0.05$), except for the shallowest *in situ* sensor depth: $t = -2.53$, $p = 0.01$. This suggests that any disturbance caused by the UAS in the water column is comparable in magnitude to natural sources of perturbation in the water profile. However, increased temperature variance at the surface of the water column provides evidence of UAS-induced disturbances due to propeller wash or mixing from the sensor occasionally being pulled out of the water.

Feasibility of Mapping 3D Temperature Structure

Figure 6 illustrates the interpolated temperature structure in a 10 × 10 × 2.5 m region of Big Lake. The UAS sensors allowed multiple sites and depths to be measured on a fine grid. The platform also resolved the thermal structure of Big Lake, including vertical trends and the more subtle lateral trends, specifically a general decrease in temperature away from shore. Obtaining these 30 data points required 26.5 min of flight time during two UAS flights, plus time for setup and deployment.

While still faster than some manual techniques, rapid development of comprehensive spatial maps over large areas with short time frames will require further advances. The obvious technical limitations include battery life, which, coupled with the relatively slow response time of the thermal sensors used, limited the number of locations that could be sampled per flight. Thermal sensors better suited to this application are available, and future deployments will focus on a thermocouple with response times of 5 s or less [39]. Using such a sensor, the UAS could measure temperatures every 0.5 m at depths from 0–4 m at twenty different sampling locations within the 18 min battery life. At this rate, approximately five flights would be needed to measure Big Lake on a 10 × 10 m grid. This would be achievable in under 2 h, and repeat flights would be sufficient to allow diurnal cycling to be resolved.

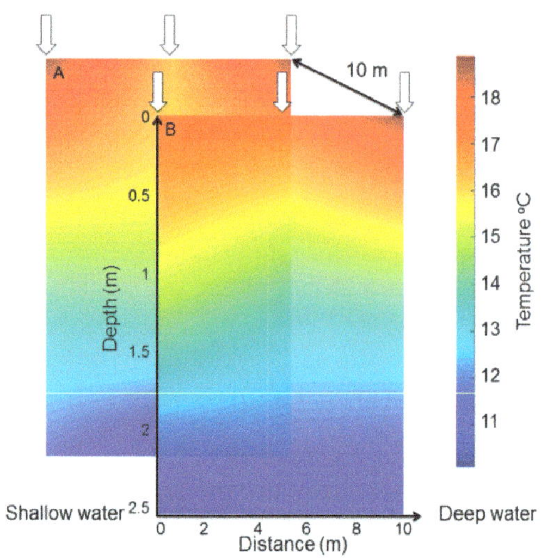

Figure 6: Thermal structure of a 10 × 10 × 2.5 m grid, interpolated and reconstructed from two UAS flights. White arrows indicate the UAS's sampling locations; depth measurements were made throughout the water column at these sites.

Use of the UAS platform, as with other mobile sensing methods, presents a tradeoff between the detailed information from the high spatial resolution of temperature measurements and the limitations of a discontinuous time frame in which measurements are taken. While *in situ* sensors provide a temporally-continuous record of temperature at a location, the UAS captures a snapshot of the thermal structure in time. To account for general trends in temperature changes over time and a large area, sampling on a less dense grid may be necessary to avoid spatial biases, because the time frame of measurement is significantly longer than the time frame of temperature adjustment. For high resolution sampling with a single UAS, a nested sampling grid approach may be appropriate: the UAS could be flown to points on a less dense grid to record an initial thermal structure, then flown to points on subsections of a denser grid during shorter, constrained flight times to avoid any significant warming or cooling during the flight. Data from a small set of *in situ* temperature loggers could then be incorporated to detrend warming during the flight time period; in practice, we expect to use both *in situ* and UAS sensors.

The quality of the resulting data and the ease of deployment would be enhanced by improved sensor calibration and improvements in stabilizing the UAS with respect to drift and altitude variation. Because this was an initial evaluation of the UAS platform's temperature probing capabilities, the sample size used in the experiment is not very large. To further assess the accuracy of this methodology, we plan to incorporate a new onboard thermocouple with a faster response time and to compare its measurements with *in situ* sensors to calibrate accordingly. Additionally, the latest commercially-available UASs are characterized by improved stability and altitude control, which would address these issues, and we expect that UAS and battery technology will continue to improve.

Operational Modes

The temperature readings were made using two different modes of flight: (1) flying down to within 1.2 m of the water surface, then unspooling the tube while hovering; and (2) flying with a completely unspooled tube and approaching the water from a higher altitude. We found no clear differences in the disturbances induced in the *in situ* lake temperatures between the flight modes, but made some operational observations. Specifically, flying with an unspooled tube enabled the UAS to remain high above the water surface. This offered a safer flight mode by increasing the buffer distance between the vehicle and lake and reducing the visual disturbance of the lake surface. However, altitude variation was more pronounced in this flight mode (typically above 1.8 m). At these altitudes, the UAS relies on a barometric altitude sensor that drifts due to wind and the continual natural variation in ambient air pressure. We often observed changes of ±0.2 m within less than a minute. However, when flying closer to the water surface (within 1.8 m), the UAS uses short distance ultrasonic rangefinders that enable steadier flight. In practice, we would recommend flying farther from the water to reduce operational stress on UAS pilots.

Pragmatic Limitations

Technological challenges aside, researchers must navigate a volatile regulatory landscape that varies by country [40,41]. Recently, some countries have proposed rules governing the use of UASs by academic researchers that are focused on limiting payload, altitude, visual range and proximity to densely-populated areas [42]. Researchers seeking official sanction should allow ample time to address local regulations or risk derailing a project's timeline for data collection. Clearer and simpler regulations for environmental sensing with UASs in low-risk, low-altitude and low-population areas would increase their practicality and usefulness.

CONCLUSIONS

This paper outlines a novel method using a UAS to characterize the thermal structure in shallow bodies of water. The preliminary experiments illustrate that physically lowering a sensor from air with a UAS can obtain reliable and robust measurements of ambient water temperature, with reasonable agreement with *in situ* measurements and without significantly disturbing the water column. The results suggest that operationally, the best results are obtained when a sensor is lowered into the water column from an unspooled tube. Furthermore, the most reliable measurements are made when the elevation change of the sensor is slow and the sensor is descending. Future research should focus on improving the UAS elevation stability and minimizing drift, along with incorporating faster temperature sensors to capture the thermal structure with high resolution even when moving quickly through the water. With improved temperature sensors, as many as twenty 4-m surface water profiles could be obtained in a single 18-min flight. This would be fast enough to allow multiple flights to capture key diurnal variations and sufficient to allow complete characterization of small water bodies (*circa* 1 ha) in less than two hours. Although the system tested here focused on a small lake, the aerial platform could be used to characterize the thermal structure and regimes in streams, rivers, wetlands and larger lake systems.

ACKNOWLEDGMENTS

We would like to thank Sebastian Elbaum for his insightful suggestions, Erik Viik of the Blue Oak Ranch Reserve for his assistance and hospitality and Rebecca Horzewski of the NIMBUSlab.

This work was partially supported by USDA-NIFA#2013-67021-20947 and NSF IIS-1116221. Michaella Chung is partially supported by the National Science Foundation Graduate Research Fellowship Program. Any opinions, findings, conclusions or recommendations expressed in this material are those of the authors and do not necessarily reflect the views of these agencies.

AUTHOR CONTRIBUTIONS

Carrick Detweiler, James Higgins and John-Paul Ore developed the UAS platform. Michaella Chung, Sally Thompson and Michael Hamilton conceived of and designed the field experiments. All performed the experiments, analyzed the data and contributed to writing the paper.

REFERENCES

1. Caissie, D. The thermal regime of rivers: A review. *Freshw. Biol.* 2006, *51*, 1389–1406.
2. Alabaster, J.S.; Lloyd, R.S. *Water Quality Criteria for Freshwater Fish*; Elsevier: London, UK, 2013.
3. Jankowski, T.; Livingstone, D.M.; Buhrer, H.; Forster, R.; Niederhauser, P. Consequences of the 2003 European heat wave for lake temperature profiles, thermal stability, and hypolimnetic oxygen depletion: Implications for a warmer world. *Limnol. Oceanogr.* 2006, *51*, 815–819.
4. Mooji, W.M.; Hulsmann, S.; de Senerpont Domis, L.N.; Nolet, B.A.; Bodelier, P.L.E.; Boers, P.C.M.; Pires, M.D.; Gons, H.J.; Ibelines, B.W.; Noordhuis, R.; *et al.* The impact of climate change on lakes in the Netherlands: A review. *Aquat. Ecol.* 2005, *39*, 381–400.
5. Bowler, D.E.; Mant, R.; Orr, H.; Hannah, D.M.; Pullin, A.S. What are the effects of wooded riparian zones on stream temperature? *Environ. Evid.* 2012, *1*, 1–9.
6. Vannote, R.L.; Sweeney, B.W. Geographic analysis of thermal equilibria: a conceptual model for evaluating the effect of natural and modified thermal regimes on aquatic insect communities. *Am. Nat.* 1980, *115*, 667–695.
7. Ward, J.V.; Stanford, J.A. Thermal responses in the evolutionary ecology of aquatic insects. *Annu. Rev. Entomol.* 1992, *27*, 97–117.
8. Wilby, R.L.; Orr, H.; Watts, G.; Battarbee, R.W.; Berry, P.M.; Chadd, R.; Dugdale, S.J.; Dunbar, M.J.; Elliott, J.A.; Extence, C.; *et al.* Evidence needed to manage freshwater ecosystems in a changing climate: Turning adaptation principles into practice. *Sci. Total Environ.* 2010, *408*, 4150–4164.
9. Tonolla, D.; Acuna, V.; Uehlinger, U.; Frank, T.; Tockner, K. Thermal heterogeneity in river floodplains. *Ecosystems*2010, *13*, 727–740.
10. Lewis, Q.W.; Rhoads, B.L. Rates and patterns of thermal mixing at a small stream confluence under variable incoming flow conditions. *Hydrol. Process.* 2015, *29*, 4442–4456.
11. Rice, S.P.; Kiffney, P.; Greene, C.; Pess, G.R. The ecological importance of tributaries and confluences. In *River Confluences, Tributaries and the Fluvial Network*; Rice, S., Roy, A., Rhoads, B., Eds.; John Wiley & Sons: Chichester, UK, 2008.
12. Ebersole, J.L.; Liss, W.J.; Frissell, C.A. Cold water patches in warm streams: Physicochemical characteristics and the influence of shading. *JAWRA J. Am. Water Resour. Assoc.* 2003, *39*, 355–368.
13. Holtby, L.B. Effects of logging on stream temperatures in Carnation Creek, British Columbia, and associated impacts on the coho salmon (Oncorhynchus kisutch). *Can. J. Fish. Aquat. Sci.* 1988, *45*, 502–515.
14. Torgersen, C.E.; Price, D.M.; Li, H.W.; McIntosh, B.A. Multiscale thermal refugia and stream habitat associations of chinook salmon in northeastern Oregon. *Ecol. Appl.* 1999, *9*, 301–319.
15. Ruesch, A.S.; Torgersen, C.E.; Lawler, J.J.; Olden, J.D.; Peterson, E.E.; Volk, C.J.; Lawrence, D.J. Projected climate-induced habitat loss for salmonids in the John Day River network, Oregon, USA. *Conserv. Biol.* 2012, *26*, 873–882.
16. Wüest, A.; Lorke, A. Small-scale hydrodynamics in lakes. *Annu. Rev. Fluid Mech.* 2003, *35*, 373–412.
17. Romero, J.R.; Kling, G.W. Spatial-temporal variability in surface layer deepening and lateral advection in an embayment of Lake Victoria, East Africa. *Limnol. Oceanogr.* 2002, *47*, 656–671.
18. Oldham, C.E.; Sturman, J.J. The effect of emergent vegetation on convective flushing in shallow wetlands: Scaling and experiments. *Limnol. Oceanogr.* 2001, *46*, 1486–1493.
19. Horsch, G.; Stefan, H. Convective circulation in littoral water due to surface cooling. *Limnol. Oceanogr.* 1988, *33*, 1068–1083.

20. Michael, C.; John, F. The radiatively driven natural convection beneath a floating plant layer. *Limnol. Oceanogr.* 1994, *39*, 1186–1194.

21. Rinke, K.; Huber, A.M.R.; Kempke, S.; Eder, M.; Wolf, T.; Probst, W.N.; Rothhaupta, K.O. Lake-wide distributions of temperature, phytoplankton, zooplankton, and fish in the pelagic zone of a large lake. *Limnol. Oceanogr.* 2009, *54*, 1306–1322.

22. Moore, R.D.; Spittlehouse, D.L.; Story, A. Riparian microclimate and stream temperature response to forest harvesting: A review. *J. Am. Water Resour. Assoc.* 2005, *41*, 813–834.

23. Selker, J.S.; Thevenaz, L.; Huwald, H.; Mallet, A.; Luxemburg, W.; van de Giesen, N.; Stejskal, M.; Zeman, J.; Westhoff, M.; Parlange, M.B. Distributed fiber-optic temperature sensing for hydrologic systems. *Water Resour. Res.* 2006, *42*, 1–8.

24. Torgerson, C.E.; Faux, R.N.; McIntosh, B.A.; Poage, N.J.; Norton, D.J. Airborne thermal remote sensing for water temperature assessment in rivers and streams. *Remote Sens. Environ.* 2001, *76*, 386–398.

25. Handcock, R.N.; Gillespie, A.R.; Cherkauer, K.A.; Kay, J.E.; Burges, S.J.; Kampf, S.K. Accuracy and uncertainty of thermal-infrared remote sensing of stream temperatures at multiple spatial scales. *Remote Sens. Environ.* 2006, *100*, 427–440.

26. Jensen, A.M.; Neilson, B.T.; McKee, M.; Chen, Y. Thermal remote sensing with an autonomous unmanned aerial remote sensing platform for surface stream temperatures. In Proceedings of the 2012 IEEE International Geoscience and Remote Sensing Symposium, Munich, Germany, 22–27 July 2012; pp. 5049–5052.

27. Schmugge, T.J.; Kustas, W.P.; Ritchie, J.C.; Jackson, T.J.; Rango, A. Remote sensing in hydrology. *Adv. Water Resour.*2002, *25*, 1367–1385.

28. Kay, J.E.; Kampf, S.K.; Handcock, R.N.; Cherkauer, K.A.; Gillespie, A.R.; Burges, S.J. Accuracy of lake and stream temperatures estimated from thermal infrared images. *J. Am. Water Resour. Assoc.* 2005, *41*, 1161–1175.

29. Dunbabin, M.; Grinham, A.; Udy, J. An autonomous surface vehicle for water quality monitoring. In Proceedings of the Australasian Conference on Robotics and Automation (ACRA), Sydney, Australia, 2–4 December 2009; pp. 2–4.

30. Laval, B.; Bird, J.S.; Helland, P.D. An autonomous underwater vehicle for the study of small lakes. *J. Atmos. Ocean. Technol.* 2000, *17*, 69–76.

31. Zhang, Y.; Godin, M.A.; Bellingham, J.G.; Ryan, J.P. Using an autonomous underwater vehicle to track a coastal upwelling front. *IEEE J. Ocean. Eng.* 2012, *37*, 338–347.

32. Marouchos, A.; Muir, B.; Babcock, R.; Dunbabin, M. A shallow water AUV for benthic and water column observations. In Proceedings of the OCEANS 2015—Genova, Genoa, Italy, 18–21 May 2015; pp. 1–7.

33. Zhang, F.; Wang, J.; Thon, J.; Thon, C.; Litchman, E.; Tan, X. Gliding robotic fish for mobile sampling of aquatic environments. In Proceedings of the IEEE 11th International Conference on Networking, Sensing and Control, Miami, FL, USA, 7–9 April 2014; pp. 167–172.

34. Ore, J.P.; Elbaum, S.; Burgin, A.; Detweiler, C. Autonomous Aerial Water Sampling. *J. Field Robot.* 2015.

35. Hardin, P.J.; Jensen, R.R. Small-scale unmanned aerial vehicles in environmental remote sensing: Challenges and opportunities. *GISci. Remote Sens.* 2011, *48*, 99–111.

36. Dunbabin, M.; Marques, L. Robots for environmental monitoring: Significant advancements and applications. *IEEE Robot. Autom. Mag.* 2012, *19*, 24–39.

37. Achtelik, M.C.; Doth, K.M.; Gurdan, D.; Stumpf, J. Design of a Multi Rotor MAV with regard to Efficiency, Dynamics and Redundancy. In Proceedings of the AIAA Guidance, Navigation, and Control Conference (AIAA), Minneapolis, MN, USA, 13–16 August 2012; pp. 1–17.

38. Tyler, S.W.; Selker, J.S.; Hausner, M.B.; Hatch, C.E.; Torgerson, T.; Thodal, C.E.; Schladow, S.G. Environmental temperature sensing using Raman spectra DTS fiber-optic methods. *Water Resour. Res.* 2009, *45*.

39. Garnier, B.; Lanzetta, F. *Tutorial 4: In situ Realization/Characterization of Temperature/Heat Flux Sensors*; Technical Report; Université de Nantes: Nantes, France, 2011.

40. Mulero-Pázmány, M.; Stolper, R.; van Essen, L.; Negro, J.J.; Sassen, T. Remotely piloted aircraft systems as a rhinoceros anti-poaching tool in Africa. *PLoS ONE* 2014, *9*.

41. Vincent, J.B.; Werden, L.K.; Ditmer, M.A. Barriers to adding UAVs to the ecologist's toolbox: Peer-reviewed letter. *Front. Ecol. Environ.* 2015, *13*, 74–75.

42. Whitehead, K.; Hugenholtz, C.H. Remote sensing of the environment with small unmanned aircraft systems (UASs), part 1: A review of progress and challenges. *J. Unmanned Veh. Syst.* 2014, *2*, 69–85.

INDEX

N

O

P

Q

R

S

T

U

V

W